NOTABLE WOMEN
IN THE
PHYSICAL SCIENCES

NOTABLE WOMEN
IN THE
PHYSICAL SCIENCES

A Biographical Dictionary

EDITED BY
Benjamin F. Shearer and Barbara S. Shearer

Greenwood Press
Westport, Connecticut • London

Library of Congress Cataloging-in-Publication Data

Notable women in the physical sciences : a biographical dictionary /
   edited by Benjamin F. Shearer and Barbara S. Shearer.
      p.   cm.
   Includes bibliographical references and index.
   ISBN 0–313–29303–1 (alk. paper)
   1. Women physical scientists—Biography.   I. Shearer, Benjamin F.
II. Shearer, Barbara Smith.
Q141.N734 1997
500.2'092'2—dc20
   [B]   96–9024

British Library Cataloguing in Publication Data is available.

Library of Congress Catalog Card Number: 96–9024
ISBN: 0–313–29303–1

First published in 1997

Greenwood Press, 88 Post Road West, Westport, CT 06881
An imprint of Greenwood Publishing Group, Inc.

Printed in the United States of America

The paper used in this book complies with the
Permanent Paper Standard issued by the National
Information Standards Organization (Z39.48–1984).

10 9 8 7 6 5 4 3 2 1

On the cover: (clockwise from top): Sarah Frances Whiting,
Gerty Cori, Mildred Cohn, and Paula Szkody.

Contents

Introduction

Notable Women in the Physical Sciences, a companion volume to *Notable Women in the Life Sciences*, provides biographical essays of 96 women (not all Americans) who have made significant contributions to the physical sciences from antiquity to the present. Forty-seven photographs accompany the essays. From long-recognized historical figures and women whose names were "starred" in the first editions of *American Men of Science*, to contemporary MacArthur Foundation "Genius" Award winners, National Medal of Science winners, Nobel Prize winners, and winners of other extraordinary awards such as the Federal Woman's Award, Annie J. Cannon prizes, and Garvan Medals, these prominent women have unique stories that can inspire talented young women to follow in their footsteps.

Taken as a whole, the biographical essays in this volume provide a broad historical sweep of the work of women in the physical sciences. The emphasis, however, is on twentieth-century women—most notably on those whose work continues. Disciplines covered in this volume include astronomy, astrophysics, biochemistry, chemistry, crystallography, physics, and related areas. Whenever extant sources allow, the essays go beyond the basic facts found in standard biographical dictionaries. This is most certainly true for those subjects who contributed their own words, through interviews or autobiographical contributions. The cooperation of these living scientists adds a vital and unique element to their biographical profiles.

Selection of subjects for inclusion in this volume began with a review of a number of excellent and widely known biographical sources, including Marilyn Bailey Ogilvie's *Women in Science: Antiquity through the Nineteenth Century*; H. J. Mozans's *Woman in Science*; Margaret Alic's *Hypatia's Heritage*; and Margaret Rossiter's pioneering *Women Scientists in America: Struggles and Strategies to 1940*. Caroline L. Herzenberg's *Women Scientists from Antiquity to the Present: An Index*, and *Women in the Scientific Search: An American Bio-Bibliography* by Patricia Joan Siegel and Kay Thomas Finley, were indispensable not only in building research sources but also in understanding the tremendous scope of women's achievements in science. The *World of Winners* was consulted for awards received, and the MacArthur Foundation kindly supplied its list of award-winning scientists. The annual "Women in Science" sections of the journal *Science* were helpful in identifying noted contemporary scientists, and the personal assistance of the staff of the Institute for Scientific Information was most appreciated in helping to identify "high impact" working scientists. From hundreds of possible subjects, a draft list appropriate to the size of this volume was created with the intent (1) to represent the disciplines of the physical sciences in a broad historical sweep but with an emphasis on the contemporary, (2) to present the stories of women scientists who have been recognized by their peers for the importance of their contributions, and (3) to ensure an international perspective. Decisions for inclusion necessarily had to take into account the existence of sufficient biographical sources for a chosen subject, a sad fact of working in the biography of women. The call for contributors brought an incredible response and the assistance of members of the Association for Women in Science, many of whom are contributors to this volume. Their perusal of the draft list of subjects resulted in further revisions and a nearly final list. Some living scientists who had been selected chose not to participate in the project or could not be contacted.

This volume presents the work of 70 contributors whose boundless enthusiasm for the project was born of the desire to recognize outstanding women in their own fields as well as an awareness of the paucity of biographical resources concerning women in science that has hampered scholarship in the past. Among the contributors—all active participants in their own fields—are astronomers, chemists, educators, physicians, biochemists, physicists, anthropologists, librarians, archivists, historians, and science writers.

The editors wish to thank the contributors for their dedication and hard work, without which this volume could not have come about. The editors also wish to thank the staff of the Frank A. Franco Library at Alvernia College, particularly John Kissinger, for kind and constant assistance in completing this project.

Special effort has been made to explain the work of each scientist clearly in terms familiar to general readers as well as to high school students, to whom this volume is aimed. The biographical essays, which profile the lives of the scientists, make reference to significant developmental influences, obstacles, and achievements whenever sources permit. Each is accompanied by a list of key dates in the scientist's life and a bibliography of sources for further information. Appendix I lists the subjects of the essays by discipline, and Appendix II lists them by awards they have received. The index listings assist in identifying the scientists by ethnic group and nationality. Cross-references to entries on other scientists appear in **bold** type.

NOTABLE WOMEN
IN THE
PHYSICAL SCIENCES

IDA BARNEY

(1886–1982)

Astronomer

Birth	November 6, 1886
1908	B.A., Smith College
1911	Ph.D., mathematics, Yale University
1911–12	Professor of Mathematics, Rollins College, Winter Park, FL
1912–17	Instructor in Mathematics, Smith College
1917–19	Professor, Lake Erie College, Plainsville, OH
1920–21	Assistant Professor, Smith College
1922–49	Research Assistant, Yale University Observatory
1949–55	Research Associate, Yale University Observatory
1952	Annie J. Cannon Prize
Death	March 7, 1982

Dr. Ida Barney spent 23 years producing 22 large volumes explaining the positions, magnitudes, and proper motions of about 150,000 stars. Her elemental and tedious work involved over half a million measurements of stars and remains a permanent contribution to our understanding of celestial bodies and their motion.

Ida Barney was born in 1886 in New Haven, Connecticut, to Samuel Eben Barney and Ida Bushnell Barney. She received a B.A. degree from Smith College in 1908 and a Ph.D. in mathematics from Yale University in 1911. She was an excellent student, having been initiated in the honor societies of Phi Beta Kappa and Sigma Xi. After completing her studies,

Ida Barney. Photo reprinted by permission of the Sophia Smith Collection, Smith College.

she spent ten years teaching mathematics. Between 1911 and 1921 she taught at Rollins College, Smith College, Lake Erie College, and again at Smith College. Her mathematical background set the stage for her contributions to astronomy.

In 1922, Barney joined the staff of the Yale Observatory. This was a time of explosive growth in astronomical measurement, as many universities were undertaking massive projects made possible by increased funding and the ability to mount special cameras on telescopes. Under the direction of Frank Schlesinger, Barney began work on an extensive program to measure stars on photographic plates and determine their exact positions. Her contributions included the measurement of the images and the supervision of the mathematical computations necessary to translate the positions to celestial coordinates. However, in keeping with

Schlesinger's preconceptions of what women in astronomy should do, she was not expected to introduce any innovations; at best, she was expected to make minor improvements based on practical (not theoretical) experience.[1] The measurements and calculations were time-consuming and tedious tasks.

In 1933, Schlesinger developed a new measuring engine to reduce both eyestrain and accidental errors owing to the observer's bumping the microscope. This innovation was a projection mechanism that allowed for the viewing and bisecting of images on a small screen rather than peering through a microscope. This increased both the speed and the accuracy of the measurements. Another innovation was the use of the emerging technology upon which computers were based. The catalogues that Schlesinger and Barney published between 1939 and 1943 were the first ones for which IBM punch-card machines were used for a large part of the necessary computations.

After Schlesinger retired in 1941, Barney took over full supervision of the project and was the sole author of subsequent volumes. Under her direction, the actual measurements of the photographic plates were carried out at the IBM Watson Scientific Laboratory with a new electronic device. The automatic image-centering device resulted in higher positional accuracy and less eye fatigue. In 1952, she was awarded the Annie J. Cannon Prize of the American Astronomical Society. This was given every three years to a woman of any nationality who had made outstanding contributions to astronomy.

Barney retired in 1955, although the zone catalogues that were in progress when she retired were published through 1959 and carry her name. She died on March 7, 1982, at the age of 95. Because of their accuracy, "the Yale positions will have . . . high weight in future determinations of the proper motions of these stars[;] hers is truly a lasting contribution to astronomy."[2]

Notes

1. Dorrit Hoffleit, "Appendix H: Women Astronomers at Yale through 1968," in *Astronomy at Yale 1701–1968* (New Haven: Connecticut Academy of Arts and Sciences, 1992), p. 210.

2. John S. Hall, "Ida Barney Remembered," *Sky and Telescope* 63, no. 6 (June 1982): 563.

Bibliography

American Women 1935–1940: A Composite Biographical Dictionary. Edited by Durwood Howes. Detroit, Mich.: Gale Research, 1981.

Barney, Ida. *Catalogue of the Positions and Proper Motions of . . . Stars* (Transactions of the Astronomical Observatory of Yale University, Vols. 16–27). New Haven, Conn.: The Observatory, 1945–1959.

————. "An Extension of Green's Theorum." *American Journal of Mathematics* 36, no. 2 (April 1914): 137–150.

Hall, John S. "Ida Barney Remembered." *Sky and Telescope* 63, no. 6 (June 1982): 563.

Hoffleit, Dorrit. "Appendix H: Women Astronomers at Yale through 1968," in *Astronomy at Yale 1701–1968*. New Haven: Connecticut Academy of Arts and Sciences, 1992.

————. "Positions, Proper Motions, Catalogues," in *Astronomy at Yale 1701–1968*. New Haven: Connecticut Academy of Arts and Sciences, 1992.

Schlesinger, Frank, and Ida Barney. *Catalogue of the Positions and Proper Motions of . . . Stars* (Transactions of the Astronomical Observatory of Yale University, Vols. 1–4, 7, 9–14). New Haven, Conn.: The Observatory, 1926–1943.

Warner, Deborah Jean. "Women Astronomers." *Natural History* 88, no. 5 (May 1979): 12–29.

Who's Who of American Women, 2nd ed. Chicago: Marquis Who's Who, 1961–1962.

NANCY SLIGHT-GIBNEY

LAURA BASSI

(1711–1778)

Physicist

Birth	October 29, 1711
1732	Received voting membership in Bologna's Academy of Sciences; first public debate, Palazzo Pubblico, on April 17; presented with a doctorate in philosophy from the University of Bologna on May 12; became a member of the University of Bologna faculty in the fall
1738	Married Giovanni Giuseppe Verati on February 7
1745	Appointed to full membership in the Benedettini Academics
1766	Joined the teaching faculty at the Collegio Montalto
1776	Professor of Experimental Physics, Institute of Sciences, Bologna
Death	February 20, 1778

Laura Bassi, an eighteenth-century Italian scientist, was one of the first women physicists in western history. Bassi's fame spread beyond Italy as she carved out a niche in a male-dominated academic world. Outside Italy, her achievements were particularly influential in French- and English-speaking scientific circles.

Laura Maria Caterina Bassi, daughter of Giovanni Bassi and Rosa Maria Cesari, was born in Bologna, Italy, in 1711. Because her father was a lawyer, Bassi had opportunities to meet some of Bologna's most learned citizens. The value her parents placed on education was evident as she grew up. At 5 years of age she began formal education under the auspices of her cousin, Lorenzo Stegani. Under Stegani, Bassi studied mathematics and mastered Latin. Her knowledge of Latin and her later proficiencies in French and Greek helped advance her career. At 13 years of age Bassi began studying natural philosophy and metaphysics under the direction of the family physician, Gaetano Tacconi.

By the time she reached age 20, Bassi was recognized in the Bologna academic community for her knowledge of philosophy and Latin. This led to a series of highly praised philosophical and scientific debates between Bassi and members of Bologna's learned society. In a time that placed great value on public deliberations, Bassi's successes in debate drew vital attention from the local community. This recognition exploded in 1732 with a series of events that established Bassi as a leading scientist. The first event occurred in March 1732, when she was presented with a membership in Bologna's Academy of Sciences, which was part of the Institute of Sciences. Several weeks later she was invited to debate philosophy publicly with professors from the University of Bologna; this occurred on April 17, 1732. The Palazzo Pubblico, which was traditionally used to present important university lectures to large audiences, was selected as the debate site. During the debate Bassi successfully defended forty-nine philosophical and scientific theses. Less than one month later, on May 12, 1732, she was recognized for her achievements by being presented with a doctorate in philosophy from the University of Bologna.

In recognition of her achievements and in an effort to bring attention to Bologna as a leading center for women in education, Bassi was offered a chair in philosophy at the University of Bologna in 1732. Even though she was named a university professor, the fact that she was a woman presented challenges to an academic community that, with few exceptions, had traditionally allowed only men to teach. In response, a compromise was reached under which Bassi was promised a faculty position in exchange for being allowed to give public lectures only by invitation.

Bassi's teaching career went beyond the halls of the University of Bologna. In 1766 she joined the teaching faculty at the Collegio Montalto, where she taught physics until her death. In 1776 she became a professor of experimental physics at Bologna's Institute of Sciences. She also offered many private lessons to students in her home. Private lessons were expected as part of her university teaching responsibilities, but they also served as a means of circumventing the ruling that restricted her public lectures. Over the years, in both European and North American scientific societies, Bassi became known as one of the finest instructors in the Italian scientific community. Having heard of her "celebrated" reputation, Dr. John Morgan attended a lecture she gave on "Light & Colours" while visiting Bologna in 1764. Following the lecture and a personal meeting with Bassi, Morgan reported that "she discoursed very learnedly on Electricity & other philosophical subjects."[1]

Much of Bassi's scientific career developed following her marriage to Giovanni Giuseppe Verati (1707–1793). Her marriage to Verati (sometimes spelled Veratti) took place on February 7, 1738. Bassi and Verati had at least eight children; some sources report they had twelve. In addition to the traditional reasons for marriage, Bassi recognized the benefits of marriage in a society that questioned the role of the single academic woman. As a scientist, she frequently met with her male counterparts to study, discuss, and conduct experiments. Because she was a young, unmarried woman, such meetings in public or private were scrutinized by eighteenth-century Italian society. And although Bassi had the support of many in the Vatican, the church leadership kept a watchful eye on her behavior and work as an unmarried woman scientist. Recognizing that her career would not achieve its full potential if she remained single, Bassi decided to marry Verati.

Bassi and Verati held similar intellectual interests that led to a variety of joint research projects. A medical doctor, anatomy instructor, and experimenter in the medical applications of electricity, Verati influenced Bassi to broaden her scientific interests. He also helped advance her teaching career through his connections in the scientific community. One of the most significant contributions Verati provided was the support he gave Bassi at the Institute of Sciences. In 1770, Verati became the teaching assistant to Paolo Balbi, an imminent professor of physics at the Institute. Through this connection Verati helped his wife acquire a teaching position in physics at the Institute. In 1772, Verati was given responsibility for the physics department because of Balbi's poor health. Following Balbi's death in 1776, Bassi was asked to co-chair the physics department with S. Canterzani. In support of his wife, Verati agreed to become Bassi's teaching assistant.

As one of the most progressive cities in Europe, Bologna offered a supportive environment for Bassi and her career as a woman scientist. Nonetheless, she met significant resistance from various circles as her career advanced. In part, her own determination and passion for the study of science overcame much of the opposition she encountered. Equally important to the success of her career was the personal support she received from members of the academic community, church leaders, and political figures. These included members of the Senate in Rome, such as Flaminio Scarselli, a Bassi family friend, and Filippo Aldrovandi, who played a key role in ensuring Bassi a university teaching position. Having won the support of key members of the Senate, which had authority over the University of Bologna, Bassi's position on the university faculty was protected. Although the Senate helped secure her academic career, the Vatican played a crucial role in guaranteeing her place in the scientific community. In this regard, one of the most influential supporters of Bassi and her work was Prospero Lorenzo Lambertini. A Bolognese noble and lawyer, Lambertini served as the archbishop of Bologna in the 1730s. In the same year that Lambertini became archbishop, he witnessed Bassi in a public debate. Her knowledge and debating skills impressed Lambertini to the extent that he became one of her most dedicated admirers. The role he played in the development of Bassi's career became particularly important after he was named Pope Benedict XIV in 1740. Using Vatican funds, Lambertini created the Benedettini Academics, which was designed to include the most prestigious scientists drawn from the Academy of Sciences. In 1745, to the dismay of some members of the scientific community and to the delight of others, Lambertini appointed Bassi to the newly formed Benedettini Academics. This appointment guaranteed her position as a respected scientist, making it more difficult for anyone to oppose her right to teach, conduct research, and utilize the resources of the Academy of Sciences.

As a researcher, Bassi made full use of the Academy's resources. Her commitment to scientific inquiry was partly motivated by her view that the study of science encouraged personal growth and served as a "guide to well-doing."[2] Her research resulted in the publication of five scientific papers. Her first publication, entitled *De acqua corpore naturali elemento aliorum corporum parte universi* (1732), resulted from one of her lectures on water. Four of her other publications appeared in *De Bononiensi Scientiarum et Artium Instituto atque Accademia Commentarii*. The four papers published in *Commentarii* were: "De aeris compressione" (1745), "De problemate quodam hydrometrico" (1757), "De problemate quodam mechanico" (1757), and "De immixto fluidis aere" (1791). The 1745 publication was a brief that studied air pressure. Of the two papers published in 1757, the first found a solution to a problem in hydraulics and the second demonstrated Bassi's use of mathematics to solve a trajectory problem. The 1791 paper, published following Bassi's death, was a summary of her research on bubbles formed from liquids in glass containers.

However, most of Bassi's research was never formally published. As a researcher, Bassi held an interest in a wide range of scientific studies, including fluid mechanics, Newtonian physics, and electricity. Much of her research on electricity, which partly supported the controversial Franklinian system, was a collaborative effort with her husband. Bassi's social position in Bologna also required her to write poetry for community events and to contribute to literary publications. Although Bassi was noted by her contemporaries as an exceptional poet, she saw herself primarily as a scientist.

Laura Bassi died unexpectedly on February 20, 1778. She remained an active scientist until the end of her life, as noted by her participation in an Academy of Sciences lecture only a few hours before her death. In recognition of her achievements, Bassi was buried at Bologna's Corpus Domini. Following her funeral, some of the most distinguished women of the city built a monument in her honor at the Institute of Sciences. Verati also honored his wife by assuming her teaching chair in experimental physics.

Notes

1. John Morgan, *The Journal of Dr. John Morgan of Philadelphia: From the City of Rome to the City of Londin, 1764, Together with a Fragment of a Journal Written at Rome, 1764, and a Biographical Sketch* (Philadelphia: J.B. Lippincott, 1907), pp. 98–99.

2. Edith E. Coulson James, *Bologna: Its History, Antiquities and Art* (London: Henry Frowde, 1909), p. 179.

Bibliography

Beard, Mary R. *On Understanding Women*. New York: Greenwood Press, 1968.

Elena, Alberto. " 'In Lode Della Filosofessa Di Bologna': An Introduction to Laura Bassi." *Isis* 82 (September 1991): 510–518.

Findlen, Paula. "Science as a Career in Enlightenment Italy: The Strategies of Laura Bassi." *Isis* 84 (September 1993): 441–469.

Heilbron, J.L. *Electricity in the 17th and 18th Centuries: A Study of Early Modern Physics*. Berkeley: University of California Press, 1979.

Hurd-Mead, Kate Campbell. *A History of Women in Medicine: From the Earliest Times to the Beginning of the Nineteenth Century*. Haddam, Conn.: Haddam Press, 1938.

James, Edith E. Coulson. *Bologna: Its History, Antiquities and Art*. London: Henry Frowde, 1909.

Logan, Gabriella Berti. "The Desire to Contribute: An Eighteenth-Century Italian Woman of Science." *American Historical Review* 99 (June 1994): 785–812.

Morgan, John. *The Journal of Dr. John Morgan of Philadelphia: From the City of Rome to the City of Londin, 1764, Together with a Fragment of a Journal Written at Rome, 1764, and a Biographical Sketch*. Philadelphia: J.B. Lippincott, 1907.

Schiebinger, Londa. *The Mind Has No Sex?: Women in the Origins of Modern Science*. Cambridge, Mass.: Harvard University Press, 1989.

Villari, Madame. "Learned Women of Bologna." *International Review* 5 (March/April 1878): 185–197.

Women of Mathematics: A Biobibliographic Sourcebook. Edited by Louise S. Grinstein and Paul J. Campbell. New York: Greenwood Press, 1987.

GARY L. CHEATHAM

JOCELYN BELL BURNELL

(1943–)

Astronomer

Birth	July 15, 1943
1965	B.Sc., University of Glasgow
1967	Observed first pulsar
1968	Ph.D., Cambridge University; married Martin Burnell
1968–73	Faculty, University of Southampton
1973	Awarded (jointly with Anthony Hewish) Albert A. Michelson Prize, Franklin Institute of Philadelphia
1974–82	Programmer and Associate Research Fellow, Mullard Space Science Laboratory of University College, London
1982–91	Senior Research Fellow, Royal Observatory, Edinburgh, Scotland
1987	Beatrice Tinsley Prize, American Astronomical Society
1989	Herschel Medal, Royal Astronomical Society; Burnells divorced
1986–90	Manager, James Clerk Maxwell Telescope
1991	Manager, EDISON (orbiting infrared observatory)
1991–	Professor of Physics, Open University, Milton Keynes, England

Jocelyn Bell Burnell is an astronomer whose career began with a discovery more momentous than most scientists make in a lifetime. As a 24–year-old graduate student she observed the first known pulsar, a star giving out an astonishingly rapid and regularly pulsed radio signal. Although Bell subsequently received other prestigious awards, the Nobel Prize awarded for the discovery went not to her but to her professor. Her career has involved many moves made to accommodate her husband's work; and since the discovery of pulsars, she has been distinguished for her administration of important instrument projects as well as for her teaching and research.

Born in Belfast, Northern Ireland, Susan Jocelyn Bell was attracted to astronomy as a child during frequent visits to the Armagh Observatory with her father, an architect. As an undergraduate, she was the only woman in her year to study physics at the University of Glasgow. In

1965 she began graduate work at Cambridge under the direction of Anthony Hewish, an astronomer who was designing a radio telescope to detect quasars. Quasars are compact sources; as their light passes through the irregularly ionized solar wind, they scintillate or twinkle more than extended sources do. Hewish's instrument was designed to detect this twinkling. Bell and other students spent two years sledgehammering and wiring to build the instrument, then she alone ran it and analyzed the data. This involved looking carefully at ink traces on literally miles of paper output, sorting out man-made interference and cataloging potential quasars.

In October 1967, Bell noticed a recurring bit of "scruff"—a signal that looked different from either quasar scintillation or interference. She noticed, moreover, that it showed up repeatedly at the same sidereal time, that is, when the telescope scanned the same location among the stars. "It only occupied a half an inch," she later said, "but unclassifiable things are too disturbing to be easily forgotten."[1]

Hewish and Bell agreed she should take a high-speed recording of the mysterious signal to see its structure better. It consisted of equally spaced pulses, one and one-third seconds apart. Known stars didn't vary nearly that regularly or that quickly. At first, Hewish was certain that the signal must be man-made. Ruling out that possibility, Bell and Hewish toyed with the idea that space aliens were broadcasting messages toward earth. She recalls being "really rather annoyed, thinking 'Here I am trying to get a Ph.D. out of a new technique and some silly lot of little green men have to choose *my* frequency and *my* aerial to try signaling us.' "[2] But if little green men ("LGM") were signaling from their planet, orbiting its own sun, telltale Doppler shifts should be evident in the signal received. They were not, making the "LGM" hypothesis seem unlikely. Subsequent observation of a second pulsar was conclusive: that two alien civilizations were signaling so similarly, from two such widely separated places, was too improbable to be true. The signal Bell had observed came from a star, but it was a star unlike any previously known.

The star lay far away from our solar system, but within our galaxy. To vary so rapidly, it had to be as small as a planet and at least as dense as a white dwarf. What was the mechanism of its accurate clock? When Hewish, Bell, and the rest of their team published their discovery in February 1968, Hewish suggested that the pulsation was caused by the vibration of the star.[3] The next years saw a flurry of interest among observers and theorists, and additional pulsars were found, one much faster than Bell's first. Astronomers came to agree with Cornell theorist Thomas Gold, who suggested that pulsars were spinning neutron stars— incredibly dense supernovae remnants whose existence had been predicted by theorists decades earlier, although none had previously been observed. Pulsars emit a radio beam from only part of their surface, and this beam flashes like the beam of a lighthouse. In 1974, Hewish shared the Nobel Prize in physics with his mentor, Martin Ryle, long-time head of the Cambridge radio astronomy group; Hewish was cited for his "decisive role" in the discovery of pulsars.[4]

That Bell did not share in the Nobel Prize has generated much debate.[5] In 1975, Fred Hoyle, the gadfly of British astronomy, precipitated public discussion when he wrote to the *Times* of London that the crucial elements of the discovery were clearly Bell's: the initial recognition of the signal as something distinctive and her realization that it was a sidereal, rather than terrestrial, phenomenon. The part of the process directed by Hewish, Hoyle insisted, would have had the same result if competently done by anyone:

> There has been a tendency to misunderstand the magnitude of Miss Bell's achievement, because it sounds so simple—just to search and search through a great mass of records. The achievement came from a willingness to contemplate as a serious possibility a phenomenon that all past experience suggested was impossible. I have to go back in my mind to the discovery of radioactivity by Henri Becquerel for a comparable example of a scientific bolt from the blue.[6]

Several things seem indisputable. The telescope conceived by Hewish and built by his students, Bell among them, was nearly ideal for the observation of pulsars. However, no one anticipated the discovery and all were, to varying degrees, skeptical even as observations accumulated. Hewish asked his students to compile data in such a way that if Bell had not noticed the pulsar signal, a subsequent student might well have done so. Yet Bell was the person who first singled out the pulsar signal for attention, and she was the readiest to believe that the signal came from a star.

Bell's response has been consistently gracious. While recognizing that boundary disputes between advisors and students are difficult to resolve, she believes that in setting up a lab and establishing a research program, an advisor assumes much of the risk of a project and so deserves much of the credit. From a personal point of view, however, a question more interesting than the award of a Nobel Prize may be: How does one pursue a career in science after having been (as one commentator has called Bell) "the most famous graduate student in the history of astronomy"?[7]

Immediately after finishing graduate school, Bell married Martin Burnell and took a job at the University of Southampton. This move required her to change the field of her research from radio to gamma-ray astronomy. As well as teaching physics, she worked on the development of a gamma-ray telescope and investigated the physics of the ionosphere with instruments sent aloft in immense balloons. "That move was the first of several that proved to be quite difficult. A lot of the moves I've made and the equally drastic changes of field within astronomy have been because I've moved as my husband moved around the country as a local government officer."[8]

In 1973 the Burnells had a son. Bell Burnell briefly quit science, then returned to work part-time at the Mullard Space Science Laboratory, located in Surrey. She returned to work as a programmer, later became an associate research fellow, and moved into yet another field, X-ray astronomy. She supervised analysis of the data returned by *Ariel V*, an X-ray astronomy satellite launched in 1974. Two years later, as data were flowing back from the satellite, she reflected:

> When I left radio astronomy and went into gamma-ray astronomy I told myself firmly that I had already had more than a lifetime's share of excitement and good luck and that I must settle down now and do some reliable and solid, undramatic science, though hopefully it would be interesting science. And certainly gamma-ray astronomy was suitably unspectacular (although I cannot help noticing how it has improved since I left the field three years ago). Then I went to MSSL and into X-ray astronomy, still telling myself that I had already had more than a lifetime's share. . . . I had not appreciated that X-ray astronomy was about to boom, and had not reckoned on the excitement of participating in a satellite project in those sorts of circumstances. . . .
>
> New, dramatic results have been rolling in thick and fast. X-ray transients have come to stay; many X-ray sources have been found to be highly variable; X-ray bursts have opened our eyes to yet another type of phenomenon; and X-ray emission from galaxies and clusters of galaxies is now well established. What will the Universe throw at us next? There is now a thirteenth commandment: "Thou shall not make predictions in X-ray astronomy, lest the Lord the God reveal the folly of thy ways unto all."[9]

In 1982 the Burnells moved again, Bell Burnell taking a job as senior research fellow at the Royal Observatory at Edinburgh, Scotland. In 1986 she became project manager for the James Clerk Maxwell Telescope, a multinational submillimeter telescope located on Mauna Kea in Hawaii, and owned by Great Britain, Canada, and the Netherlands. Managing the telescope's multimillion-dollar budget was an overwhelming administrative burden, requiring constant travel. In 1991 she resigned from the project, spending a year as manager of the orbiting infrared observatory, EDISON, before leaving the Royal Observatory for the Open University in Milton Keynes, near London. There she became one of only three female full professors of physical science in Great Britain. The move was hers alone, for the Burnells had divorced in 1989.

Founded in 1969, the Open University uses mail, television, and other media to offer home-study courses to adults who would otherwise be unable to get a university education. It serves an enormous number of students. Bell Burnell's astronomy course has enrolled more than a thousand students each year. She is committed to her students' cause: "When I meet the students I am tremendously encouraged by their ability—so stimulating, so keen. The number of people who have missed out on a conventional university education at the proper time—it's horrific!"[10] In addition to teaching, Bell Burnell continues research on topics including the physics of pulsars.

She is an active Quaker and has published her reflections on God and human suffering (inspired by the diagnosis of her son's diabetes) as *Broken for Life.*[11]

Notes

1. S. Jocelyn Bell Burnell, "Little Green Men, White Dwarfs, or What?" *Sky and Telescope* 55 (March 1978): 219. This gives Bell Burnell's account of the discovery; it is an expanded version of an after-dinner speech she gave at the 8th Texas Symposium on Relativistic Astrophysics, Boston, December 15, 1976, published as S. Jocelyn Bell Burnell, "Petit Four," *Annals of the New York Academy of Sciences* 302 (1977): 685–689.

2. Bell Burnell, "Little Green Men," p. 220.

3. A. Hewish, S.J. Bell, J.D.H. Pilkington, P.F. Scott, and R.A. Collins, "Observation of a Rapidly Pulsating Radio Source," *Nature* 217 (February 24, 1968): 709–713.

4. Hewish's Nobel lecture, giving his account of the pulsar discovery, has been published as A. Hewish, "Pulsars and High Density Physics," *Science* 188, no. 4193 (June 13, 1975): 1079–1083.

5. Nicholas Wade, "Discovery of Pulsars: A Graduate Student's Story," *Science* 189 (August 1975): 358–364; George Reed, "The Discovery of Pulsars: Was Credit Given Where It Was Due?" *Astronomy* 11 (December 1983): 24–28.

6. Fred Hoyle, in Wade, "Discovery of Pulsars," p. 358.

7. Reed, "The Discovery of Pulsars," p. 24.

8. S. Jocelyn Bell Burnell, quoted in Sharon Bertsch McGrayne, *Nobel Prize Women in Science: Their Lives, Struggles, and Momentous Discoveries* (Secaucus, N.J.: Carol Publishing, A Birch Lane Press Book, 1993), pp. 372–373.

9. Bell Burnell, "Petit Four," pp. 688–689.

10. Bell Burnell, quoted in Glyn Jones, "When Stardom Beckoned," *New Scientist* 135, no. 1830 (July 18, 1992): 39.

11. S. Jocelyn Burnell, *Broken for Life, 1989 Swarthmore Lecture to the London Yearly Meeting of the Society of Friends* (London: Quaker Home Service, 1989).

Bibliography

"Bell Burnell, Jocelyn." *Current Biography* 56, no. 5 (May 1995): 18–23.

Bell Burnell, S. Jocelyn. *Broken for Life: 1989 Swarthmore Lecture to the London Yearly Meeting of the Society of Friends.* London: Quaker Home Service, 1989.

————. "Little Green Men, White Dwarfs, or What?" *Sky and Telescope* 55 (March 1978): 218–221. [An expanded "Petit Four."]

————. "Petit Four." [After-dinner speech given at the 8th Texas Symposium on Relativistic Astrophysics, Boston, December 15, 1976.] *Annals of the New York Academy of Sciences* 302 (1977): 685–689.

Jones, Glyn. "When Stardom Beckoned." *New Scientist* 135, no. 1830 (July 18, 1992): 36–39.

McGrayne, Sharon Bertsch. *Nobel Prize Women in Science: Their Lives, Struggles, and Momentous Discoveries.* Secaucus, N.J.: Carol Publishing, A Birch Lane Press Book, 1993.

Reed, George. "The Discovery of Pulsars: Was Credit Given Where It Was Due?" *Astronomy* 11 (December 1983): 24–28.

Wade, Nicholas. "Discovery of Pulsars: A Graduate Student's Story." *Science* 189 (August 1975): 358–364.

JOANN EISBERG

RUTH MARY ROAN BENERITO

(1916–   )

Physical Chemist

Birth	January 12, 1916
1935	B.S., Sophie Newcomb College
1938	M.S., Tulane University
1940–43	Instructor, Randolph-Macon Woman's College, Lynchburg, VA
1943–47	Instructor (then Assistant Professor), Newcomb College, New Orleans, LA
1947–53	Assistant Professor, Tulane University
1948	Ph.D., University of Chicago
1950	Married Frank Henshaw Benerito on August 22
1953–58	Physical Chemist, Fat Emulsion Program, Southern Regional Lab, USDA, New Orleans
1958–60	Supervisory Physical Chemist, Head Organo-Phys. Investigations, Cotton Chemical Reactions Lab, Southern Regional Lab, USDA, New Orleans
1961–86	Supervisory Physical Chemist; Head Physical Chemistry Investigations, Natural Polymers Lab, Southern Regional Lab, USDA, New Orleans

1964	Distinguished Service Award, Department of Agriculture
1967	Distinguished Service Award, New Orleans Federal Executives Association
1968	Federal Woman's Award, U.S. Civil Service Commission; Southern Chemist Award, American Chemical Society
1970	Garvan Medal, American Chemical Society; Distinguished Service Award, Department of Agriculture
1971	Named as one of the 75 most outstanding women in the United States by *Ladies Home Journal*
1972	Southwest Regional Award, American Chemical Society
1981	D.Sc., Tulane University
1982	Outstanding Professional Award, Organization of Professional Employees, USDA
1984	Honored as a Woman of Achievement, 1984 World's Fair; Outstanding Federal Employee in Greater New Orleans

"Happiness comes only by contributing to the development and happiness of others; it abounds with selflessness and can be found without traveling to far-off places."

—Ruth Benerito[1]

Ruth Benerito's career in chemistry has impacted several different areas, including fat emulsions, epoxides, metallic salts, and diepoxy compounds, but her greatest contributions have been to the textile industry. She was born in New Orleans, Louisiana, in 1916, the third of six children. Her father, John Edward Roan, was a civil engineer and railroad official; her mother, Bernadette Elizardi, was an artist who was actively involved in civic and charitable projects.

Benerito graduated from high school in New Orleans at the age of 14 with a strong interest in science and mathematics. She began college at Sophie Newcomb College, the women's college of Tulane, at age 15, majoring in chemistry with minors in physics and mathematics. She completed her B.S. degree in 1935 and then spent one year at Bryn Mawr in graduate study.

Because she could not find a job when she returned to New Orleans during the middle of the Depression, she worked without pay as a laboratory technician at a hospital. She then went on to work as a social worker for the Works Progress Administration and to teach science and mathematics in the public school system of Jefferson Parish. Part of her job was to teach safety and driver's education, which required that she learn to drive herself.[2]

In the afternoons Benerito studied with Rose Mooney, an outstanding physicist at Newcomb; she subsequently was awarded an M.S. degree from Tulane in 1938. Benerito taught at Randolph-Macon Woman's College from 1940 to 1943 and then joined the faculty at Newcomb when she returned to New Orleans, teaching physical chemistry at Newcomb and Tulane until 1953. As was the case for many other brilliant women scientists of her day, women's colleges provided training and academic positions for Benerito that led to positions in traditional institutions. She taught physical chemistry at both graduate and undergraduate levels at Tulane until 1953. After entering government research in 1953, she became an adjunct professor in the graduate school of Tulane.

Benerito worked on her doctorate on a part-time basis, finally receiving her Ph.D. from the University of Chicago in 1948. She was promoted to assistant professor at Newcomb. In 1953, Benerito left full-time academic work to go to the Southern Regional Research Center of the USDA as a physical chemist in the Intravenous Fat Program of the Oilseed Laboratory. Southern Research Center was one of four USDA laboratories begun in 1940 to do research on regional products. The lab in New Orleans focused on cotton.

Benerito focused on two areas of investigation at the Southern Regional Research Center. While serving as project leader in the Intravenous Fat Program, she worked on the surface and interfacial properties of triglycerides. New methods of analysis of proteins and fats were determined. The results of her work in this area were used in determining the necessary caloric intake via intravenous feeding for long-term patients.[3] Benerito also studied cotton chemical reactions. In 1958 she became Research Leader of the Physical Chemistry Research Group of the Cotton Chemical Reactions Laboratory. During this time in her career, she developed methods that gave crease resistance to "drip-dry" cotton. This helped the cotton industry remain competitive with manufacturers of synthetic fabrics. The problem of chlorine scorch of cotton was also solved by her research.

Benerito also did research on the reactions of epoxides, leading to the production of a fabric that had dry crease resistance as well as wet crease resistance. This process also made a fabric that was more absorbent and, therefore, more comfortable to wear. Research results in this area have been used in wood preservation and not only by the cotton industry but also by the paper, film, and epoxy plastic industries. Other research resulted in a finishing agent that provided oil repellency to cottons. Benerito also studied flame-retardant chemicals for fabrics.[4]

During further work on cotton fabric, a glassy material was developed after treatment of cotton with sodium plumbite. After soaking the cotton in the solution, the lead content rose to 40 percent. The cloth was heated for about an hour at 1100°F in a porcelain container. The researchers expected to find a small pile of ash and a puddle of lead. Instead, they found a glasslike material stuck to the porcelain surface in the shape and weave pattern of the cotton.[5] When aluminum foil was used to cover the treated cotton, the glassy material became electroconductive.[6] When the lead content of the fabric was adjusted, the translucent glassy material adhered firmly to the surface of a glass object and retained both its shape and fiber pattern. This work was important to scientific glassblowers who make specialized laboratory equipment, as well as to the fields of electronics and solar energy.

In 1950, Benerito married Frank H. Benerito (who died in 1970 after only twenty years of marriage). She has served as adjunct professor at Tulane University since 1955 in the graduate school and the biochemistry department of the medical school, and also as a lecturer at the University of New Orleans since 1982. She retired from government service in 1986 and is now professor emerita at Tulane. During her research, Benerito was granted over 50 patents and authored 250 scientific publications.

Benerito received many honors during her career, including the Distinguished Service Award from the Department of Agriculture in 1964 and 1970 and the Distinguished Service Award from the New Orleans Federal Executives Association in 1967. She also received the Federal Woman's Award and the Southern Chemist Award in 1968. She was awarded the Garvan Medal in 1970 and the Southwest Regional Award in 1972 by the American Chemical Society. In 1982 she was given the Outstanding Professional Award by the USDA Organization of Professional Employees. In 1971 she was named as one of the 75 most outstanding women in the United States by *Ladies Home Journal,* and in 1984 she was among 102 women highlighted in a special pavilion at the 1984 World's Fair honoring the "labors, frustrations and achievements" of the women.[7] She was inducted as a Fellow into the American Institute of Chemists and an honorary member of Delta Kappa Gamma, an international teaching honorary society, and Iota Sigma Pi, the National Honor Society for Women Chemists.

Benerito's citation for the Southern Chemist Award in 1968 called her "an outstanding and inspiring teacher of chemistry, a brilliant research scientist, and an inspiring and untiring leader of research."[8] Because of her research, cotton fabrics are now more absorbent, wrinkle resistant, stain resistant, oil resistant, and flame retardant. Her contributions to the teaching and training of chemists, to the textile industry, as well as to other areas such as glassblowing and epoxy production, have had a great impact on modern life.

Notes

1. *Who's Who,* 1994, 48th ed. (Chicago: Marquis Who's Who, 1994), p. 250.

2. J.A. Miller, "Ruth Mary Roan Benerito (1916–   )," in *Women in Chemistry and Physics: A Biobibliographic Sourcebook* (Westport, Conn: Greenwood Press, 1993), p. 30.

3. Ibid., p. 32; "Garvan Medal to Ruth Benerito," *Chemical and Engineering News* 48, no. 3 (1970): 43.

4. Miller, "Ruth Mary Roan Benerito (1916–)," pp. 32–33.

5. " 'Magic' Turns Fabric Glassy," *Business Week* (December 14, 1978): 781.

6. "USDA Researchers Petrify Cotton," *Chemical Week: Technology Newsletter* (January 17, 1979): 27.

7. J. DeMers, "Women Honored at World's Fair Pavilion," UPI (April 24, 1984).

8. "Benerito Wins Southern Chemist Award," *Chemical and Engineering News* 46, no. 49 (1968): 72.

Bibliography

American Men and Women of Science, 1995–1996. 19th ed. New York: Bowker, 1995.

Benerito, R.R. "Modification of Cotton and Cotton-Polyester Blends with Epoxides." *Textile Research Journal* 44 (1974): 225–232.

Benerito, R.R., J.B. McKelvey, and R.J. Berni. "New Theory of Resiliency in Cotton." *Textile Research Journal* 36 (1966): 251–264.

"Benerito Wins Southern Chemist Award." *Chemical and Engineering News* 46 (1968): 72.

DeMers, J. "Women Honored at World's Fair Pavilion." UPI (April 24, 1984).

"Federal Woman's Award." *Chemistry and Engineering News* 46 (1968): 62.

"Garvan Medal to Ruth Benerito." *Chemical and Engineering News* 48 (1970): 43.

"Glass from Cotton Fabrics." *Fusion* 22 (1975): 7–9.

" 'Magic' Turns Fabric Glassy." *Business Week* (December 4, 1978): 781.

Miller, J.A. "Ruth Mary Roan Benerito (1916–)," in *Women in Chemistry and Physics: A Biobibliographic Sourcebook*. Edited by L.S. Grinstein, R.K. Rose, and M.H. Rafailovich. Westport, Conn: Greenwood Press, 1993.

"USDA Researchers Petrify Cotton." *Chemical Week: Technology Newsletter* (January 17, 1979): 27.

Who's Who. 48th ed. Chicago: Marquis Who's Who, 1994.

Who's Who in Government. 1st ed. Chicago: Marquis Who's Who, 1972.

JUDY F. BURNHAM

LEONORA NEUFFER BILGER

(1893–1975)

Chemist

Birth	February 3, 1893
1916	Ph.D., chemistry, University of Cincinnati

1916–18	Head, Dept. of Chemistry, Sweet Briar College
1919–29	Dept. of Chemistry, University of Cincinnati
1924	Sarah Berliner Fellow, Cambridge University
1929–54	Dept. of Chemistry, University of Hawaii (Dept. Head, 1943–54)
1953	Garvan Medal, American Chemical Society
1953–58	Senior Professor, University of Hawaii
1960–64	Professor Emerita, University of Hawaii
Death	February 19, 1975

Leonora Neuffer Bilger made a significant contribution not only to chemistry for her work with nitrogen compounds, but also to her institution by designing a chemistry laboratory for the University of Hawaii (UH). Dr. Bilger served on the faculty of the chemistry department at UH from 1929 until the early 1960s and was department head for eleven years.

Bilger received her Ph.D. at the University of Cincinnati in 1916, where she studied hydroxylamines and hydroxamic acids. Her first position upon receiving her doctorate was chair of the chemistry department at Sweet Briar College. She stayed there for two years. After that appointment, Bilger returned to the chemistry department at the University of Cincinnati and directed chemical research in the basic science research laboratory. That experience proved valuable in the challenge of designing a chemical laboratory facility later in her career.

The main research interest Bilger pursued was the study of asymmetric nitrogen compounds. In 1924 she received a Sarah Berliner Fellowship to continue research in that field at Cambridge University. After returning to the University of Cincinnati, Bilger took a leave of absence to spend two years as visiting professor of chemistry at the University of Hawaii. She married a fellow University of Cincinnati faculty member, Earl M. Bilger, and they moved to Hawaii in 1929 to take positions at the university. Bilger was appointed head of her department in 1943, a post she held for eleven years.

Leonora Bilger was awarded the Garvan Medal in 1953 in honor of her work on asymmetric nitrogen compounds and her service to the field of chemistry. Administered by the American Chemical Society, the Garvan Medal was established by Francis P. Garvan in 1936 with an endowment donated for the purpose of directing attention to women's contributions to chemistry. Honoring both research and distinguished service by women in the field, Garvan intended for medal candidates' work to be reviewed by a panel of unbiased men and women who were appointed by the Society.

In order to prepare herself to serve as the major consultant for designing the chemical laboratory facilities at the University of Hawaii, Bilger traveled with her husband to twenty-five chemical laboratories and interviewed people about the labs' features. Having responsibility for the project indicates that Bilger enjoyed great respect from her colleagues—they entrusted her with the design of their future surroundings for teaching and research. The facility was occupied in 1951, had 70,340 square feet, and cost $1.5 million to build. Bilger's goal for the structure, which she stated in an article in the *Journal of Chemical Education* in 1954, was "to provide an environment that arouses the enthusiasm of large numbers of students and research workers."[1] She was instrumental in setting high standards of training and worked to develop the chemistry program at the University of Hawaii so it received accreditation from the American Chemical Society. In 1959 the university named the lab facility after Bilger and her husband.

Note

1. Leonora Neuffer Bilger and Earl Matthias Bilger, "The New Chemical Laboratory of the University of Hawaii," *Journal of Chemical Education* 31 (1954): 300.

Bibliography

Bilger, Leonora Neuffer, and Earl Matthias Bilger. "The New Chemical Laboratory of the University of Hawaii." *Journal of Chemical Education* 31 (1954): 300.
Roscher, Nina Matheny. "Women Chemists." *Chemtech* 6 (December 1976): 738–743.

FLORA SHRODE

KATHARINE BURR BLODGETT
(1898–1979)
Physicist

Birth	January 10, 1898
1917	A.B., Bryn Mawr
1918	S.M., University of Chicago; joined General Electric
1926	Ph.D., physics, Cambridge University—first woman

Katharine Burr Blodgett. Photo reprinted by permission of Bryn Mawr College.

1933	Developed method to build up thin films
1938	Announced non-reflecting glass
1939	D.Sc., Elmira College
1942	D.Sc., Brown University and Western College
1944	D.Sc., Russell Sage College; "starred" in *American Men of Science*
1945	Achievement Award, American Association of University Women
1951	Garvan Medal; only scientist honored by Boston's First Assembly of American Women in Achievement; Katharine Blodgett Day in Schenectady, NY
1972	Progress Medal, Photographic Society of America
Death	October 12, 1979

Katharine Burr Blodgett had many firsts during her career. She was the first woman research scientist at General Electric (GE) and the first woman to attain a Ph.D. from Cambridge University. She invented non-reflecting glass, which is used in camera and optical equipment. She developed surface chemistry techniques with her mentor, Irving Langmuir. (In fact, the two were so instrumental that the field is called Langmuir-Blodgett films.) She was also credited with six U.S. patents and known for the clarity and precision of her writing.[1] Langmuir said she had a "rare combination of theoretical and practical ability."[2]

Blodgett was born in Schenectady, New York, in 1898. Her family had moved there from Boston (they had also lived in Maine) when her father, attorney George Bedington Blodgett, assumed the position of head of the GE patent department. He died before Blodgett was born. Her mother, Katharine Buchanen Burr, moved to New York City with her son and daughter and then three years later to France for four years so the children could become bilingual.[3] The family returned to New York City and then traveled to Germany.

In her teens, Blodgett returned to New York City and attended the Rayson School. The private school was run by three English sisters who taught Blodgett how to think and communicate clearly, skills she would be greatly commended for during her career.[4] Blodgett then won a scholarship to attend Bryn Mawr, where she loved the intellectual challenge and excelled, graduating second in her class.[5] Especially important to her there were Charlotte Angas Scott, who shared her love for math, and James Barnes, who guided her in physics.

Before graduating from college, Blodgett visited GE and met her future mentor, Irving Langmuir. The 1932 Nobel Prize winner in chemistry told her she must broaden her scientific education before working for him.[6] She next attended the University of Chicago, receiving her master's degree in 1918. Her choice of thesis topic was influenced by World War I and Germany's use of poisonous gas. She studied the adsorption (adhesion) of gases by coconut charcoal and thus worked on improving the chemical structure of gas masks.

With her new degree in hand, the 21–year-old began work as Langmuir's assistant. She was the first woman research scientist at GE. In her first years there, the pair worked on improving the electric light bulb. She overcame the obstacles against women scientists in part because there was a labor shortage resulting from World War I, because her father's former colleagues supported her, and most of all because she had great scientific skills. Her success in chemical research should not be underrated. "It was virtually impossible for women scientists to find professional-level jobs in corporations at that time."[7]

After six years Langmuir got Blodgett into England's Cavendish Laboratory to study physics under Sir Ernest Rutherford, a 1908 Nobel Prize winner. Upon completion of her studies, he communicated her doctoral research about electrons in ionized mercury vapor to the *Philosophical*

Magazine. In 1926, Blodgett became the first woman to earn a doctorate from Cambridge University. She also probably met Lucy Haynor, the other "of the best-trained women physicists of the time."[8] Blodgett encouraged Haynor to work at GE with Langmuir, which she did in 1925.

At GE, Blodgett began studying surface chemistry. Years earlier Langmuir had discovered that oil film on water was only one molecule thick (monomolecular) and behaved like a two-dimensional gas. This revolutionary concept earned him the Nobel Prize, but he had not developed any practical applications. Blodgett began investigating and, in 1933, developed a method of depositing a monomolecular fatty acid layer on a metal sheet by dipping it in water with the substance on the surface. The first person to build up layers of these thin films, she discovered that they reflected different colors depending on their thickness.

She invented a color gauge capable of measuring the thickness of a thin film to a millionth of an inch. Using the gauge, a person could compare the color of a film and read off the thickness, a great improvement over the expensive equipment needed previously for this measurement. Blodgett described this gauge in a radio program; her speech was published as a chapter entitled "A Gauge That Measures Millionths of an Inch."

In 1938 she gained much media attention and some honorary degrees upon announcing her invention of non-reflecting glass. She made it by placing liquid soap forty-four molecules thick on glass (four-millionths of an inch). The soap's thickness was the length of one quarter of a wave of light and prevented the refraction of the waves that caused reflection, thus permitting light to pass through the glass. *Life* magazine published several pictures showing the difference between normal and non-reflecting glass.[9] One of the first commercial applications of non-reflecting glass was in a projection lens contributing to better screen illumination for the film *Gone with the Wind*.[10]

At the Massachusetts Institute of Technology, another group announced the production of non-reflecting glass the day after Blodgett did. They went on to develop another method for making non-reflecting glass, incorporating Blodgett's ideas and their own. This method is still widely used. Of the happy invention, Blodgett said:

> You keep barking up so many wrong trees in research. It seems sometimes as if you're going to spend your whole life barking up wrong trees. And I think there is an element of luck if you happen to bark up the right one. This time I eventually happened to bark up one that held what I was looking for.[11]

As time passed, Blodgett worked less directly with Langmuir and began to mentor beginning scientists, mostly men. An especially interesting case is that of Vincent J. Schaefer, a lab assistant/equipment builder. Upon Blodgett's recommendation to Langmuir, he moved into performing experiments under Langmuir without any formal academic credentials.[12]

Women still found it hard to attain professional scientific positions at GE even though Blodgett had achieved such success.[13] It is also interesting to note the de-valuing of her work, as an article written in 1953 on the history of GE exemplifies. This article mentions all the male scientists who worked there but neglects to include Blodgett and her achievements.[14]

During World War II, Blodgett worked on plane wing deicing and also invented a smoke screen, which saved many lives in campaigns in North Africa and Italy. In 1947 she invented an instrument to measure humidity via weather balloons in the upper atmosphere. She developed the use of an "indicator" oil to show the edges of an invisible film.[15] She also built what was probably the first controlled-thickness X-ray grating.[16]

Appreciation for Blodgett's work continued. In 1945 she received the $2,500 Achievement Award from the American Association of University Women. In 1951 she won the Garvan Medal, given to honor an American woman for distinguished service in chemistry. She was the first industrial scientist to win the Garvan Medal. That year, Boston's First Assembly of American Women in Achievement honored her; and the town of Schenectady held a Katharine Blodgett Day, hailing her "as a tireless and disciplined worker, as a cheerful and witty colleague, as a leading citizen with a social conscience and a deep sense of civic responsibility."[17] In 1972 she won the Progress Medal from the Photographic Society of America. She was honored posthumously with the May 1, 1980, issue of *Thin Solid Films*, which was dedicated to her and included papers stemming from her work.

Standing about 5 feet tall, Blodgett had merry eyes and a deep, explosive laugh.[18] She was afraid of snakes but still loved to garden. She even performed experiments on the plants in her garden.[19] She kept a favorite battered old table and stool in her lab even when she moved into the new steel-and-glass GE building.[20] She believed women made good scientists, tending to be painstaking, thorough, enthusiastic, and conscientious.[21]

Though a private person, Blodgett participated in civic affairs in Schenectady and in the Travelers Aid Society. A Presbyterian, she was a member of a club called Zonta, even acting as its treasurer in 1936, and she presided over a GE employees' club.[22] She also acted in the Civic Players of Schenectady, where she is especially remembered for her comic and character roles.[23]

Katharine Blodgett bought and lived in a house with a view of the house where she was born. Although she was reputed to be an excellent cook, she tried to avoid cooking whenever possible. She had a one-room cabin retreat with an adjacent sleeping tent at Lake George, New York, "where she dart[ed] about the lake in a fast motor boat."[24] Langmuir, Schaefer, and other scientists also had cabins at Lake George. In fact, they socialized together. Blodgett's hobbies included astronomy, collecting antiques, and playing bridge, during which she is reputed to have remembered every card played.[25] She died on October 12, 1979, in her home. Co-worker Schaefer wrote in her obituary:

> Her associates remember her fondly and with sincere respect. The methods she developed have become classical tools of the science and technology of surfaces and thin films. She will be long—and rightly—hailed for their simplicity, elegance, and the definitive way in which she presented them to the world.[26]

Notes

1. K. Thomas Finley and Patricia J. Siegel, "Katharine Burr Blodgett," in *Women in Chemistry and Physics: A Biobibliographic Sourcebook* (Westport, Conn.: Greenwood Press, 1993), p. 69.

2. *Current Biography* (New York: H.W. Wilson, 1952), p. 55.

3. Shari Rudavsky, "Katharine Burr Blodgett," in *Notable Twentieth-Century Scientists* (Detroit: Gale Research, 1995), p. 197.

4. Edna Yost, *American Women of Science* (New York: Frederick A. Stokes, 1943), p. 198.

5. Alice C. Goff, "Katharine Burr Blodgett," in *Women Can Be Engineers* (Youngstown, Ohio: n.p., 1946), p. 172.

6. Rudavsky, "Katharine Burr Blodgett," p. 198.

7. Martha J. Bailey, *American Women in Science: A Biographical Dictionary* (Santa Barbara, Calif.: ABC-CLIO, 1994), p. 30.

8. Margaret W. Rossiter, *Women Scientists in America: Struggles and Strategies to 1940* (Baltimore: Johns Hopkins University Press, 1982), p. 257.

9. "Woman Makes Glass Invisible by Use of Film One Molecule Thick," *Life* 6 (January 23, 1939): 24–25.

10. Dortha Bailey Doolittle, "Women in Science," *Journal of Chemical Education* 22 (April 1945): 173.

11. Yost, *American Women of Science*, p. 207.

12. Albert Rosenfeld, "The Quintessence of Irving Langmuir," in *Langmuir, the Man, the Scientist*, Vol. 12: *The Collected Works of Irving Langmuir* (New York: Pergamon Press, 1962), p. 178.

13. *Women of Science: Righting the Record*, ed. G. Kass-Simon (Bloomington: Indiana University Press, 1990), p. 165.

14. Finley and Siegel, "Katharine Burr Blodgett," p. 69. See also C.G. Suits, "Seventy-Five Years of Research in General Electric," *Science* 118 (1953): 451–456.

15. Vincent J. Schaefer, "Obituary: Katharine Burr Blodgett 1898–1979," *Journal of Colloid and Interface Science* 76, no. 1 (July 1980): 270.

16. Ibid., p. 270.

17. *Making Contributions: An Historical Overview of Women's Role in Physics* (College Park: American Association of Physics Teachers, 1984), p. 16.

18. *Current Biography*, 1940, p. 90; Jo Chamberlin, "Mistress of the Thin Films," *Science Illustrated* 2 (December 1947): 9.

19. "Problem-Solver Katharine Blodgett Wins Garvan Medal," *Chemical and Engineering News* 29, no. 2 (April 9, 1951): 1408.

20. *Current Biography*, 1952, p. 57.

21. Ida L. Burleigh, "Sees Place for More Women in Research Laboratories," *Christian Science Monitor* 36 (October 24, 1944): 8.

22. *Current Biography*, 1952, p. 57.

23. Goff, "Katharine Burr Blodgett," p. 182.

24. "Problem-Solver Katharine Blodgett Wins Garvan Medal," p. 1408; Goff, "Katharine Burr Blodgett," p. 182.

25. Kathleen McLaughlin, "Creator of 'Invisible Glass' Woman of Many Interests," *New York Times* 89 (September 24, 1939): 4D.

26. Schaefer, "Obituary: Katharine Burr Blodgett 1898–1979," p. 271.

Bibliography

American Women, 1935–1940: A Composite Biographical Dictionary. Edited by Durward Howes. Detroit: Gale, 1981.

Bailey, Martha J. *American Women in Science: A Biographical Dictionary*. Santa Barbara, Calif.: ABC-CLIO, 1994.

Blodgett, Katharine. "Films Built by Depositing Successive Monomolecular Layers on a Solid Surface." *Journal of the American Chemical Society* 57 (1935): 1007–1022.

Blodgett, Katharine, and Irving Langmuir. "Built-Up Films of Barium Stearate and Their Optical Properties." *Physical Review* 51 (1937): 964–982.

Burleigh, Ida L. "Sees Place for More Women in Research Laboratories." *Christian Science Monitor* 36 (October 24, 1944): 8.

Chamberlin, Jo. "Mistress of the Thin Films." *Science Illustrated* 2 (December 1947): 8–11.

Current Biography. New York: H.W. Wilson, 1940, 1952.

Davis, Kathleen A. "Katharine Blodgett and Thin Films." *Journal of Chemical Education* 61, no. 5 (May 1984): 437–439.

Doolittle, Dortha Bailey. "Women in Science." *Journal of Chemical Education* 22 (April 1945): 171–174.

Finley, K. Thomas, and Patricia J. Siegel. "Katharine Burr Blodgett," in *Women in Chemistry and Physics: A Biobibliographic Sourcebook*. Westport, Conn.: Greenwood Press, 1993.

Goff, Alice C. "Katharine Burr Blodgett," in *Women Can Be Engineers*. Youngstown, Ohio: n.p., 1946.

Making Contributions: An Historical Overview of Women's Role in Physics. College Park, Md.: American Association of Physics Teachers, 1984.

McLaughlin, Kathleen. "Creator of 'Invisible Glass' Woman of Many Interests." *New York Times* 89 (September 24, 1939): 4D.

"Problem-Solver Katharine Blodgett Wins Garvan Medal." *Chemical and Engineering News* 29, no. 2 (April 9, 1951): 1408.

Roscher, Nina Matheny. "Women Scientists." *Chemtech* 6 (1976): 738–743.

Rosenfeld, Albert. "The Quintessence of Irving Langmuir," in *Langmuir, the Man, the Scientist*, Vol. 12: *The Collected Works of Irving Langmuir*. New York: Pergamon Press, 1962.

Rossiter, Margaret W. *Women Scientists in America: Struggles and Strategies to 1940.* Baltimore: Johns Hopkins University Press, 1982.

Rudavsky, Shari. "Katharine Burr Blodgett," in *Notable Twentieth-Century Scientists.* Detroit: Gale Research, 1995.

Schaefer, Vincent J. "Obituary: Katharine Burr Blodgett, 1898–1979." *Journal of Colloid and Interface Science* 76, no. 1 (July 1980): 269–271.

"Woman Makes Glass Invisible by Use of Film One Molecule Thick." *Life* 6 (January 23, 1939): 24–25.

Women of Science: Righting the Record. Edited by G. Kass-Simon. Bloomington: Indiana University Press, 1990.

The Women's Book of World Records and Achievements. Edited by Lois Decker O'Neill. Garden City, N.Y.: Anchor Press/Doubleday, 1979.

World Who's Who in Science: A Biographical Dictionary of Notable Scientists from Antiquity to the Present. Edited by Alan G. Debus. Chicago: Marquis Who's Who, 1968.

Yost, Edna. *American Women of Science.* New York: Frederick A. Stokes, 1943.

JILL HOLMAN

E. MARGARET BURBIDGE

(1919–   )

Astrophysicist

Birth	August 12, 1919
1939	B.Sc., University of London
1943	Ph.D., astrophysics, University of London
1946–51	Assistant Director and Acting Director, University of London Observatory
1954–57	Fellow, California Institute of Technology
1959	Associate Professor of Astronomy, University of Chicago; Helen B. Warner Prize
1964	Professor, University of California, San Diego (UCSD); Fellow, the Royal Society

1967	Co-author, *Quasi-Stellar Objects*
1968	Fellow, American Academy of Arts and Sciences
1972	Director, Royal Greenwich Observatory
1976–78	President, American Astronomical Society
1978	Fellow, National Academy of Sciences
1979–88	Director, Center for Astrophysics and Space Sciences, UCSD
1981	President, American Association for the Advancement of Science
1982	Catherine Wolfe Bruce Medal, Astronomical Society of the Pacific
1985	National Medal of Science
1988	Albert Einstein World Award of Science Medal
1990–	Professor Emerita, UCSD

Astrophysicist E. Margaret Burbidge enthusiastically describes her chosen field of astronomy as "a science of visualization as well as analysis and theory."[1] Although her visions include volcanic activity on a satellite of Jupiter and dry river channels on Mars, her theories emerge from "chemical, spectroscopic, and isotopic analysis" and embrace the evolution of stars, the rotation of galaxies, and the study of quasi-stellar objects. For this ground-breaking astrophysicist, each piece of data "whets the appetite for more detailed knowledge."[2]

Burbidge's interest in astronomy began in childhood. She was born Eleanor Margaret Peachey in Davenport, England, in 1919; her father was a lecturer in chemistry at the Manchester School of Technology, where her mother was a chemistry student. In the early 1920s the family relocated to London. Her father set up a laboratory there. Margaret's early interest in science, particularly astronomy, was encouraged by her parents. She examined the stars through her father's binoculars.

While at the University of London, Margaret took First Class Honors with her B.Sc. in 1939 followed by a Ph.D. in astrophysics in 1943. World War II interrupted her coursework, but she returned to school after the war. Margaret met and married Geoffrey Burbidge, a physics scholar and astronomer. They have one daughter.

From 1946 to 1951, Margaret Burbidge served as the assistant director and then acting director of the University of London Observatory. In 1951 the Burbidges traveled to the United States, where both held research fellowships. Margaret was a research fellow at the Yerkes Observatory (University of Chicago) and the Harvard College Observatory. After two years they returned to England and, working with astronomer Fred Hoyle and physicist William A. Fowler, commenced their study of the evolution of stars, or stellar formation, that culminated in the paper "Synthesis of the Elements in Stars."

Data for this theoretical study came from astronomical spectroscopy. A spectroscope converts light waves invisible to the human eye into colored bands of light and wavelength patterns that reveal the atomic composition, or elements, of the observed star. The surface composition of a star indicates the nuclear reactions occurring in the star's core. Spectroscopy reveals the presence of a variety of elements, both lighter and heavier, on star surfaces. This process of building heavier elements from lighter ones—nucleosynthesis—became the focal point of the research. Working with the premise that stars begin with hydrogen, and discovering how all the elements heavier than hydrogen could be produced from this lighter element, the foursome concluded "that the heavy element content of stars was a function of the age of the stars."[3] At first, a star consists largely of hydrogen. When stars become visible they are converting hydrogen into helium, and the starlight is released energy. Burbidge described the next stage:

> a star that is older than our sun or has gone through its life faster than our sun will start further reactions, like building carbon out of helium and so on, through heavier elements. And you can go on getting energy from nuclear reactions up until you build elements as heavy as iron.[4]

The stage of reactions not only indicated the life history of a star but was, in Burbidge's words, "very relevant to our attempts to decipher the formation and early history of the solar system and to the question of whether planetary systems about stars are common or rare."[5] For their work on the collaborative paper, Margaret and Geoffrey Burbidge received the Helen B. Warner Prize in 1959. Awarded by the American Astronomical Society, the Warner Prize is given for significant contributions by young astronomers.

Between 1954 and 1957, Margaret Burbidge returned to the United States with a full-time fellowship at the California Institute of Technology, while her husband was a Carnegie fellow at the Mount Wilson and Palomar Observatories. Since Margaret, despite her fellowship and expertise, was not officially allowed to use the telescopes on Mount Wilson, Geoffrey applied for telescope time in his name and Margaret accompanied him for the observing session. Margaret studied galaxies and determined how they rotate around an axis. Centrifugal force prevents their collapse. In a telephone interview with Joan Oleck on January 20, 1994, Burbidge explained: "For our galaxy, the force of gravity holds the material in place and the rotational speed counterbalances this. So a galaxy takes up a structure in which gravitational pull is equated to centrifugal force."[6]

In a collaborative paper written in 1961, the Burbidges noted that although most galaxies are of spiral, spherical, or ellipsoidal configurations, some are irregular. Such irregularity may indicate that a galaxy is in an early stage of evolution. Margaret produced some of the first accurate estimates of the weight and mass of other galaxies.

In 1957, Margaret returned to the Yerkes Observatory as a research fellow. Two years later she was appointed associate professor of astronomy at the University of Chicago. In 1962 she was hired by the chemistry department at University of California, San Diego. When nepotism rules that prevented her employment in the same department as her husband were abolished two years later, Burbidge was transferred to physics as a full professor.

Margaret directed the University's Center for Astrophysics and Space Sciences from 1979 to 1988 and became professor emerita in 1990. During these years she co-authored a book, returned to England, assumed a host of professional responsibilities, and received many awards including honorary degrees. The book, *Quasi-Stellar Objects* (1967), was another successful collaborative effort with her astronomer husband. Begun in the spring of 1966, this survey of quasars concentrated on quasi-stellar objects that "vary in light and in radio flux over periods sometimes as short as days." Although they recognized the "complexities and uncertainties" of these objects, the Burbidges wanted to "collect together in a coherent fashion as much of the observational material as we could and see how it could be fitted together at this early stage."[7]

The contents of this 235–page book include identification of quasi-stellar objects, descriptions of variations in their radio flux emissions, their proper motions, distribution, and the nature of redshifts, or spectral lines denoting ordinary elements that have shifted to the red side of the spectrum. Redshifts indicate that quasars, among the most distant objects known, are receding from our galaxy at tremendous speeds.

Margaret returned to England in 1972 and remained until 1973 to serve as the director of the Royal Greenwich Observatory. Although she was the first woman director of this observatory, she was nonetheless denied the honorary title of Astronomer Royal, traditionally granted to its director. The title was given to another astronomer. Following her return to the United States, she became quite active in professional associations. She served as president of the American Astronomical Society—the first woman to do so—from 1976 to 1978, during which period she became a U.S. citizen. In 1981, Margaret was selected president of the American Association for the Advancement of Science. She also holds membership in the Royal Astronomical Society, the American Philosophical Society, the Société Royale des Sciences de Liège, the Astronomical Society of the Pacific, and the New York Academy of Sciences.

Margaret has been awarded twelve honorary degrees and a number of other honors. In addition to the previously mentioned Helen B. Warner Prize, she was elected a fellow of the Royal Society in 1964, the American Academy of Arts and Sciences in 1968, and the National Academy of Sciences in 1978. Margaret was the first woman recipient of the Catherine Wolfe Bruce Medal from the Astronomical Society of the Pacific in 1982, honoring a lifetime of achievement and distinguished service to astronomy. On February 27, 1985, she was one of 19 prominent scientists awarded the National Medal of Science; in 1988 she received the Albert Einstein World Award of Science Medal. In 1971, Margaret had declined to accept the Annie J. Cannon Prize given by the American Astronomical Society because the award was restricted to women. She explained, "It is high time that discrimination in favor of, as well as against, women in professional life be removed."[8]

Appointed to the National Aeronautics and Space Administration (NASA) team, Burbidge worked to perfect a "faint-object spectrograph" for the 1984 shuttle.[9] She has also served on scientific committees that planned and equipped the Hubble space telescope.

Notes

1. E. Margaret Burbidge, "Adventure into Space," *Science* (July 29, 1983): 421.
2. Ibid., p. 422.
3. Joan Oleck, "E. Margaret Burbidge, English-Born American Astrophysicist," in *Notable Twentieth-Century Scientists,* Vol. 1, ed. Emily J. McMurray (Detroit, Mich.: Gale Research, 1995), p. 279.
4. Ibid.
5. Burbidge, "Adventure into Space," p. 424.
6. Oleck, "E. Margaret Burbidge," p. 279.
7. Geoffrey R. Burbidge and E. Margaret Burbidge, *Quasi-Stellar Objects* (San Francisco: W.H. Freeman, 1967), p. vi.
8. Sally Stephens, *Women in Astronomy* (San Francisco: Astronomical Society of the Pacific, 1992), p. 28.
9. Jennifer S. Uglow, comp. and ed., *The International Dictionary of Women's Biography* (New York: Continuum, 1982), p. 86.

Bibliography

The Biographical Dictionary of Scientists. Roy Porter, consultant editor. New York: Oxford University Press, 1994.
Burbidge, E. Margaret. "Adventure into Space." *Science* (July 29, 1983): 421–426.
Burbidge, Geoffrey, and E. Margaret Burbidge. "Peculiar Galaxies." (February 1961; downloaded from America Online, December 7, 1995).
———. *Quasi-Stellar Objects.* San Francisco: W.H. Freeman, 1967.

Daintith, John, Sarah Mitchell, Elizabeth Tootill, and Derek Gjertsen. *Biographical Encyclopedia of Scientists*, Vol. 1. 2nd ed. Bristol and Philadelphia: Institute of Physics Publishing, 1994.

A Dictionary of Twentieth Century World Biography. Asa Briggs, consultant editor. Oxford and New York: Oxford University Press, 1992.

The International Dictionary of Women's Biography. Edited and compiled by Jennifer S. Uglow. New York: Continuum, 1982.

The International Who's Who 1995–96. 59th ed. London: Europa Publications Limited, 1995.

Larousse Dictionary of Scientists. Edited by Hazel Muir. New York: Larousse, 1994.

Norman, Colin. "White House Awards Science, Technology Medals." *Science* (March 8, 1985): 1183.

Oleck, Joan. "E. Margaret Burbidge, English-Born American Astrophysicist," in *Notable Twentieth-Century Scientists*, Vol. 1. Edited by Emily J. McMurray. Detroit, Mich.: Gale Research, 1995.

Stephens, Sally. *Women in Astronomy.* San Francisco: Astronomical Society of the Pacific, 1992.

Vernoff, Edward, and Rima Shore. *The International Dictionary of 20th Century Biography.* New York and Scarborough, Ontario: New American Library, 1987.

REBECCA LOWE WARREN

MARY LETITIA CALDWELL

(1890–1972)

Nutritional Chemist

Birth	December 18, 1890
1913	A.B., Western College for Women, Oxford, OH
1914	Instructor, Western College for Women
1917	Assistant Professor, Western College for Women
1919	Master's degree, Columbia University
1921	Ph.D., Columbia University; stayed on to teach
1948–59	Professor, Columbia University
1959	Retired from teaching
1960	Garvan Medal, American Chemical Society
1961	D.Sc., Columbia University
Death	July 1, 1972

Mary Letitia Caldwell, a celebrated role model for women pursuing careers in chemistry, was born in Bogotá, Colombia, in 1890. Her parents, Milton and Susannah (Adams) Caldwell, were citizens of the United States, but the family was living in Colombia while Milton, a Presbyterian minister, was doing missionary work there. The family returned to the United States in time for Mary to attend high school.

There were few role models for young women interested in chemistry in the early part of the twentieth century. **Marie Curie** had already won two Nobel Prizes by the time Mary entered high school, and Caldwell probably knew of her, but a large part of her fame outside the scientific community was due to the oddity of her being a woman and a scientist. Yet Mary had two extra sources of encouragement that many women of her day did not. Her family obviously valued education and achievement: all five of the Caldwell children went on to college and became teachers or scholars in their chosen fields. And she attended an all-women's college, Western College for Women in Oxford, Ohio.

Little is known about Caldwell's college days. What is indisputable is that before World War I (and long after) many of the country's finest colleges and universities did not allow women as students or as faculty members. Women's colleges may not have offered world-class faculties or equipment, but they provided a place for women to continue their education and, more important, to be taken seriously as scholars.

Caldwell graduated in 1913 and stayed on at Western College for four years as an instructor in chemistry. Here she began what would be an important part of her professional life for the next six decades: serving as a role model and mentor for other women.

In 1918, Caldwell entered the graduate program in chemistry at Columbia University in New York City, where the faculty thought enough of her potential to offer her a fellowship. Under the guidance of her teacher, Dr. Henry C. Sherman, she began studying nutrition and biological chemistry. Nutritional chemistry was a new field when she took it on, and many of her discoveries over the next decades provided the foundation for the work of those who came after. She concentrated on amylases, dedicating some sixty years to carefully and thoroughly isolating them and discovering how they acted with different chemical groups. An amylase is an enzyme that breaks down carbohydrates in plants or animals to make sugar. In animals, the highest concentration of amylases is in the saliva and the pancreas. Caldwell focused on amylase from the pig pancreas, an especially rich source. Today, purified amylases are used widely in industry for fermentation, for preparing textiles, for paper sizing, and for preparing various foods.

One of Caldwell's early frustrations was that the commercially available materials used to follow the action of the amylases were not pure enough. It was impossible to determine accurately what factors influenced an amylase's stability and activity, because any observed activity might be caused by an impurity in the enzyme. She set out to find a way to produce her own, purer, materials. The methods she developed and perfected became the standard methods in American and European laboratories. Her research was eventually supported by grants from large industrial firms that increased their own profits by using her methods. Her work also attracted the interest and the financial support of the National Institutes of Health.

Caldwell had completed her master's degree in 1919 and her Ph.D. in 1921. She joined the Columbia faculty and served there for thirty-eight years as a teacher, researcher, advisor, and departmental administrator. In 1948 she was promoted to full professor, the only woman professor or associate professor in Columbia's large chemistry department at the time. Her specialty course was "Chemistry of Food and Nutrition," which she took over from her mentor, Dr. Sherman, when he retired. Nearly all of her own students, assistants, and collaborators were women. Although little is known of those who encouraged Caldwell to pursue her interest in chemistry, her own role in supporting aspiring women chemists has been gratefully acknowledged by her students.

Of the graduate students she directed, eighteen completed the Ph.D. at Columbia. Her students also managed to publish some fifty papers in chemical journals. One former student, Marie M. Daly, reflected after Caldwell's death that she

> inspired her students with respect for technical excellence as well as fine scholarship. Her manners were rather formal; she rarely addressed students by first names and scrupulously changed the "Miss" or "Mr." to "Dr." immediately following a successful thesis defense. Despite her formal manner, she conveyed a sense of concern for a student's personal welfare. She could summon a bright word of encouragement when the work was not progressing fast enough, often ending her comments with a philosophical "Well, child, that's research!"[1]

Although she was known for being diligent and thorough in her research and dedicated to her students, Caldwell had a full life, not confined only to her laboratory. She never married but stayed close to her siblings and her nieces and nephews. She was an avid hiker until a muscular disability required her to use a cane and then a wheelchair. She also enjoyed gardening.

In 1959, Caldwell retired from Columbia University. She left her home on Park Avenue in New York City and moved to the country in Fishkill, New York. Here she had more room for gardening and bird watching, but her chief interest remained chemistry.

Throughout her career Caldwell received many honors and awards from her peers. Two of her most meaningful moments of recognition came after her retirement. In 1960 she was awarded the Garvan Medal from the American Chemical Society, established to honor each year "an American woman for distinguished service in chemistry." The following year, Columbia University presented her with an honorary Doctor of Science degree.

Caldwell continued to prepare papers for publication until her death at home on July 1, 1972.

Note

1. Marie M. Daly, "Mary Letitia Caldwell," in *American Chemists and Chemical Engineers*, ed. Wyndham D. Miles (Washington, D.C.: American Chemical Society, 1976), p. 62.

Bibliography

Caldwell, M.L. "An Experimental Study of Certain Amino Acids." Ph.D. dissertation, Columbia University, 1921.

———. "A Study of the Influence of the New Sulfur-Containing Amino Acids upon the Stability and Enzymic Activity of Pancreatic Amylase." *Journal of the American Chemical Society* 49 (1922): 2957–2966.

Caldwell, M.L., and J.T. Kung. "Influence of Factors on the Stability and Activity of Pancreatic Amylase." *Journal of the American Chemical Society* 75 (1953): 3132–3135.

Caldwell, M.L., and J.E. Little. "Action of Pancreatic Amylase." *Journal of Biological Chemistry* 142 (1942): 585–595.

Caldwell, M.L., H.C. Sherman, and J.E. Dale. "Enzyme Purification—Further Experiments with Pancreatic Amylase." *Journal of Biological Chemistry* 88 (1930): 295–304.

Daly, Marie M. "Mary Letitia Caldwell," in *American Chemists and Chemical Engineers.* Edited by Wyndham D. Miles, pp. 62–63. Washington, D.C.: American Chemical Society, 1976.

"Garvan Medal, Dr. Mary L. Caldwell." *Chemical and Engineering News* 38 (April 18, 1960): 86.

"Mary L. Caldwell." *Chemical and Engineering News* 50 (1972): 64.

Svoronos, Soraya. "Mary Letitia Caldwell (1890–1972)," in *Women in Chemistry and Physics.* Edited by Louise S. Grinstein et al., pp. 72–76. Westport, Conn.: Greenwood Press, 1993.

CYNTHIA A. BILY

ANNIE JUMP CANNON

(1863–1941)

Astronomer

Birth	December 11, 1863
1884	B.S., Wellesley College
1895–97	Special Student, Radcliffe College
1896–1911	Assistant, Harvard College Observatory
1911–38	Curator of Astronomical Photographs, Harvard College Observatory
1914	Honorary Member, Royal Astronomical Society
1918	D.Sc., University of Delaware
1921	D.Sc., University of Groningen
1922	Nova Medal, American Association of Variable Star Observers
1925	D.Sc., Oxford University; LL.D., Wellesley College
1931	Draper Gold Medal, National Academy of Sciences
1932	Ellen Richards Research Prize, Association to Aid Scientific Research by Women
1935	D.Sc., Oglethorpe University
1937	D.Sc., Mt. Holyoke College
1938–40	William Cranch Bond Astronomer, Harvard College Observatory
Death	April 13, 1941

Annie Jump Cannon was recognized even during her lifetime as the world's expert in identifying and classifying stars, tasks she performed with incredible accuracy and speed.

Annie Cannon was born on December 11, 1863, in Dover, Delaware, and was the eldest child of prosperous ship-builder, merchant, and state senator Wilson Lee Cannon and his second wife, Mary Elizabeth. Annie had two younger brothers and a number of older half-siblings. She learned her love of astronomy through her mother, who spent time on the house roof showing Annie the constellations.[1] She was educated in the Dover public schools and at Wilmington Conference Academy and later studied physics at Wellesley College as a student of famed physics and astronomy professor **Sarah Whiting**.

After graduating in 1884, Cannon returned to Dover and spent the next ten years studying piano and traveling, resulting in a book of photographs and text entitled *In the Footsteps of Columbus*.[2] Her reported deafness was attributed to the scarlet fever that she apparently contracted during this time.[3] After her mother's death in 1894, she renewed her interest in science and returned to Wellesley as a lab assistant for Whiting. In 1895 she decided to study astronomy seriously and became a special student at Radcliffe for two years through the assistance of Edward C. Pickering, the Harvard College Observatory (HCO) director who was famous for hiring women as computers. (Computers ascertained the identity of a star on a photographic plate and calculated its position.)[4] She spent her free time in the observatory studying photographic plates and making observations of variable stars. She officially joined the HCO staff in 1896 and remained there until her death.

Cannon's original assigned tasks were to catalogue variable stars and classify the spectra of southern stars (to complement **Antonia Maury's** work on northern stars). Cannon agreed with Antonia Maury's rearrangement of the original Fleming classes, but she dropped Maury's line width scheme and instead added numbers signifying subclasses within a spectral class. The resulting classes were ordered (from high surface temperature to low surface temperature) OBAFGKM.[5] Her "Spectra of Bright Southern Stars," containing classifications of 1,122 stars, was published in 1901 as Volume 28, Part II, of the *Annals of the Harvard College Observatory*. Colleague **Cecilia Payne-Gaposchkin** characterizes this catalogue as "a treasure-house of information for the student of stellar spectra, but its greatest interest probably lies in the fact that the system of spectral classification that later came into general currency was here crystallized for the first time."[6] This system was so "user-friendly" that it was universally adopted by the International Astronomical Union in 1910 and, with minor modification, is still used today as the Harvard Spectral Classification. In 1903, Cannon published her "Provisional Catalogue" of 1,227 variable stars and their discoverers. The updated version in 1907 contained 1,957 stars and was the most complete catalogue of its kind.

Upon Fleming's death in 1911, Pickering asked the Harvard Corporation to grant Cannon an official appointment (as they had Fleming). The Corporation considered Fleming a "special case," and Pickering bestowed on Cannon the somewhat unofficial appointment of Curator of Astronomical Photographs. She began Pickering's next project of revising the *Henry Draper Catalogue* for stars down to 8th magnitude, which would encompass all stars previously classified by herself, **Williamina Fleming**, and Antonia Maury as well as fainter stars. This became her major work, *The Henry Draper Catalogue*, published in pieces between 1918 and 1924 as Volumes 91–99 of the *Annals of the Harvard College*

Observatory and including classifications of 225,300 stars. Later *Extensions* to the catalogue by Cannon brought the number of classified spectra to 350,000. The immensity and importance of this project was recognized in Cannon's lifetime: "there is probably no other instance in the history of science where so great a mass of important data has been obtained by a single observer on a homogeneous and uniform system."[7] HCO's later director, Harlow Shapley, said that her work made "a structure that probably will never be duplicated in kind or extent by a single individual."[8]

Cannon's lack of a Corporation position at the HCO was noticed by the Observatory Visiting Committee that observed her work during the production of the *Draper Catalogue*:

> At the present time she is the one person in the world who can do this work quickly and accurately. Through familiarity with it she has acquired such a perfect mental picture not only of the general types, but of their minute subdivisions, that she is able to classify the stars from a spectrum plate instantly upon inspection without any comparison photographs of the typical stars.
>
> This gives her great speed in classification, amounting to no less than 300 stars an hour. . . . At the same time her great speed in no way limits the accuracy of her estimates. From an investigation of her probable error upon parts of the sky where she has made independent duplicate estimates, it is found that her average deviation amounts to only $\frac{1}{10}$ of a unit. . . .
>
> It is an anomaly that though she is recognized the world over as the greatest living expert in this area of investigation, and her services to the Observatory as so important, yet she holds no official position in the University. . . . It is the unanimous opinion of the Visiting Committee that the University would be honoring itself and doing a simple act of justice to confer upon her an official position which would be a recognition of her scientific attainments.[9]

She continued her spectroscopy work until shortly before her death. Other discoveries include almost three hundred variable stars and five novae, all found by peculiarities in their spectra.

Cannon was well known and respected by the astronomical community and received numerous awards, including six honorary degrees, one of which was the first honorary Doctor of Science degree from Oxford University awarded to a woman (1925); a lunar crater named in her honor; and an honorary membership in the Royal Astronomical Society. She was awarded the Nova Medal of the American Association of Variable Star Observers in 1922; she was the first woman to be awarded the Draper Gold Medal of the National Academy of Sciences in 1931 (although she was never made a member); and in 1932 she shared the Ellen Richards Research Prize of the Association to Aid Scientific Research by Women. Cannon used the money to establish an award given triennially through the American Astronomical Society that would honor contributions made by women astronomers. Despite her achievements she was not formally recognized by Harvard during most of her career, although she finally received an official appointment as William Cranch Bond Astronomer in 1938.

Cannon was a friend and inspiration to a generation of women astronomers, "a human being of the first order."[10] She was a member of the National Women's Party and supported women's suffrage. Cecilia Payne-Gaposchkin described her life as "reflected in the furtherance of the education of women. Miss Cannon took pleasure in this result, and several younger generations of women scientists have owed much to her kindness, help and encouragement."[11] Cannon died at age 77 of heart failure and arteriosclerosis. She summed up her research in this way: "We do our best to increase the sum of human knowledge as pertains to the story of starlight."[12]

Notes

1. R.L. Waterfield, "Dr. Annie J. Cannon," *Nature* 147 (1941): 738.
2. The book was published in Boston in 1893.
3. Bessie Zaban Jones and Lyle Gifford Boyd, *The Harvard College Observatory: The First Four Directorships, 1839–1919* (Cambridge, Mass.: Harvard University Press, 1971).
4. For more information on the hiring of women at the HCO, see Dorrit Hoffleit, *The Education of American Women Astronomers before 1960* (Cambridge, Mass.: AAVSO, 1994); Jones and Boyd, *The Harvard College Observatory*; Pamela E. Mack, "Straying from Their Orbits: Women in Astronomy in America," in *Women of Science, Righting the Record*, eds. G. Kass-Simon and Patricia Farnes (Bloomington: Indiana University Press, 1990); Margaret W. Rossiter, *Women Scientists in America: Struggles and Strategies to 1940* (Baltimore: Johns Hopkins University Press, 1982).
5. For more information on Fleming and Maury's classification schemes, see entries elsewhere in this volume. For details on Cannon's classification scheme, see Jones and Boyd, *The Harvard College Observatory*; Solon I. Bailey, *The History and Work of Harvard Observatory, 1839 to 1927* (New York: McGraw-Hill, 1931); Dorrit Hoffleit, "The Evolution of the Henry Draper Memorial," *Vistas in Astronomy* 34 (1991): 107–162.
6. Cecilia Payne-Gaposchkin, "Annie Jump Cannon," *Science* 93 (1941): 443.
7. "Award of Gold Medals to Dr. Annie J. Cannon and Professor Henry B. Bigelow," *Science* 74 (1931): 645.
8. Edna Yost, *American Women of Science* (Philadelphia: Lippincott, 1955), p. 42.
9. Jones and Boyd, *The Harvard College Observatory*, p. 408.

10. Yost, *American Women of Science*, p. 43.
11. Payne-Gaposchkin, "Annie Jump Cannon," p. 444.
12. Annie J. Cannon, "The Story of Starlight," *Telescope* 8 (1941): 61.

Bibliography

"Award of Gold Medals to Dr. Annie J. Cannon and Professor Henry B. Bigelow." *Science* 74 (1931): 644–647.

Bailey, Solon I. *The History and Work of Harvard Observatory, 1839 to 1927*. New York: McGraw-Hill, 1931.

Campbell, Leon. "Annie Jump Cannon." *Popular Astronomy* 49 (1941): 345–347.

Cannon, Annie Jump. "The Henry Draper Catalog." *Annals of the Harvard College Observatory* 91–99 (1918–1924).

———. "The Henry Draper Extension." *Annals of the Harvard College Observatory* 100 (1925).

———. *In the Footsteps of Columbus*. Boston: L. Barta, 1893.

———. "Second Catalogue of Variable Stars." *Annals of the Harvard College Observatory*, 55, pt. 1 (1907).

———. "Spectra of Bright Southern Stars." *Annals of the Harvard College Observatory*, 28, pt. 2 (1901).

———. "The Story of Starlight." *Telescope* 8 (1941): 56–61.

Jones, Bessie Zaban, and Lyle Gifford Boyd. *The Harvard College Observatory: The First Four Directorships, 1839–1919*. Cambridge, Mass.: Harvard University Press, 1971.

Payne-Gaposchkin, Cecilia. "Annie Jump Cannon." *Science* 93 (1941): 443–444.

———. "Miss Cannon and Stellar Spectroscopy." *Telescope* 8 (1941): 62–63.

Waterfield, R.L. "Dr. Annie J. Cannon." *Nature* 147 (1941): 738.

Yost, Edna. *American Women of Science*. Philadelphia: Lippincott, 1955.

KRISTINE M. LARSEN

EMMA PERRY CARR

(1880–1972)

Chemist

Birth	July 23, 1880
1898–99	Student, Ohio State University
1899–1904	Student (then Assistant) in Dept. of Chemistry, Mt. Holyoke College
1904–05	B.S., University of Chicago

Emma Perry Carr. Photo reprinted by permission of The Mount Holyoke College Archives and Special Collections.

1905–08	Instructor in Chemistry, Mt. Holyoke College
1908–10	Ph.D., University of Chicago
1910–13	Associate Professor of Chemistry, Mt. Holyoke College
1913–46	Professor of Chemistry; Chair, Dept. of Chemistry, Mt. Holyoke College
1924	Cooperating Expert, International Critical Tables
1937	Awarded first Garvan Medal
1946	Retired from Mt. Holyoke
Death	January 7, 1972

Emma Carr was an expert in ultraviolet spectroscopy who helped demonstrate the relationship between the ultraviolet spectra of organic compounds and their molecular structure. Her work helped to solidify Mt. Holyoke College's reputation for educating women in science on a very high level.

Emma Perry Carr was born in 1880 in Holmesville, Ohio, into a family with a tradition of sons becoming country doctors. Her brother James followed this path after his father and grandfather. Her other brother,

Edmund, became a businessman. The three Carr daughters took various paths available to women at the time. Lida married and made a career as a minister's wife. Grace remained single and stayed at home to care for her parents. Emma, however, went to college and became a professional.

Emma attended high school in Coshocton, Ohio, a town of 10,000 to which her family had moved when she was an infant. She was known in the town as "Emmy the smart one." She started college at nearby Ohio State University, where she studied chemistry under William McPherson, who was professor of chemistry there from 1897 to 1937. As far as anyone in her family can remember, this was the beginning of her interest in chemistry. None of her living relatives knows why Emma decided after one year to "go east" to Mt. Holyoke College in Massachusetts, since no one in the family had done so before, but it may have been at McPherson's suggestion.[1]

Carr studied chemistry at Mt. Holyoke from 1899 to 1901, but she did not graduate with what she would always consider her class, the class of 1902. Instead, she worked as an assistant in the chemistry department for three years, then transferred to the University of Chicago, where she received a bachelor's degree in 1905. In 1905 she went back to Mt. Holyoke as an instructor, returning after three years to Chicago to study for her Ph.D. in chemistry, which she obtained in 1910. At the University of Chicago she worked under the mentorship of Julius Stieglitz, who remained a lifelong friend. In 1910 she returned again to Mt. Holyoke as an associate professor; in 1913 she became a full professor and head of the chemistry department, a post she held until her retirement in 1946. One of her graduate students, remembering meeting Carr for the first time in 1914, said of her, "Suddenly there entered a tall dark slim young woman with a disarming smile. She looked so young—indeed she was so young—that it seemed incredible that at the age of 33 she had already become a professor and head of the Department of Chemistry, probably the youngest person ever to receive such an appointment."[2]

Throughout most of her career Carr returned to Coshocton each summer, where she would become again maiden aunt, Emma. Known as an elegant dresser, Aunt Emma's opinion on matters of fashion and domestic decoration were valued highly by her sisters and nieces. No one in her family really understood what her work involved, but she was viewed by her family as a woman of accomplishment who associated with women like Eleanor Roosevelt and Frances Perkins (also from the Mt. Holyoke class of 1902 and a chemist who became a member of Franklin Roosevelt's cabinet).[3]

At Mt. Holyoke, Emma Carr was primarily thought of as a gifted teacher and the leader of a research team that was one of the most productive at any liberal arts college in the United States. With her colleagues, she established at Mt. Holyoke a chemistry curriculum as rigorous as those of nearby men's colleges. Using textbooks she had used at the University of Chicago, she established the introductory course in chemistry as one in which every faculty member in the department would take part. Carr's philosophy, which became the philosophy of the department, was that the most experienced teachers should staff the introductory course and that every member of the department should be responsible for at least one section of the course. Thus the first course in chemistry became a magnet for students, and Carr's lectures enticed many young women into a major in chemistry.[4]

Carr was extremely popular with students, even those who were not enrolled in her classes. She lived in a dormitory for many years and her table in the dining room was a popular one. Because she was known as a talented storyteller, "Miss Carr's table was always hilarious."[5] She was "a prime mover in dinner conversations, which ranged in subject from international politics to the respective advantages of being a monkey or a cow. And it seemed to us that no one could possibly make a funny story funnier than could Miss Carr, especially one on herself."[6]

Carr took the futures of her students very seriously, often keeping up with their careers for years. Those who became chemists were constantly in touch about job opportunities, and others continued to correspond about husbands and children. One student recalled that a Mt. Holyoke graduate "did not think it inappropriate at all to ask Miss Carr's advice when she had to decide whether or not to spend her last fifty dollars on a new suit in order to make a good impression on the president of the college to whom she was applying for a position."[7] While Carr was department chair, 43 of her undergraduate students and 25 of her master's students went on to the Ph.D.

At Mt. Holyoke, Carr found a special friend in **Mary Sherrill**, who came to Mt. Holyoke in 1921 after getting her Ph.D. from the University of Chicago, also under the guidance of Julius Stieglitz. Carr and Sherrill did research together and shared a house from 1935 to 1961. Carr retired from teaching in 1946 but continued to live in the house. Sherrill retired in 1954 and the two traveled together frequently after that.

Through her research publications Emma Carr became well known as the American expert in ultraviolet spectroscopy, a field that was well developed in Europe but virtually unknown in the United States. This was not the field that she had studied for her doctorate at Chicago, but Carr recognized it as a field that was wide open as well as a subject that would allow contributions from physical chemists, organic chemists, and physicists working together. Carr said of this decision:

I knew nothing about the technique except what I had read in the foreign journals but we went ahead and ordered our first Hilger spectrograph and began work in 1913. I had hoped to go abroad for study but with the outbreak of war in 1914 this was not possible until 1919 when I worked for a semester with A.W. Stewart at Queen's University, Belfast. Our first publication in this field was in 1918.[8]

For many years, Mt. Holyoke was the center of this work. Chemistry departments from all over the United States sent samples for analysis. Through this work Carr helped demonstrate the relationship between ultraviolet spectra of organic compounds and their molecular structure.

Having already attained national status for her work, Carr gained international status when she was asked to work as one of the three spectroscopy experts for the International Critical Tables in the 1920s, the other two being Jean Becquerel of the Collège de France in Paris and Victor Henri of the University of Zurich. Carr traveled to Zurich to work with Henri and, while in Europe, became the first woman to represent the United States at a meeting of the International Union of Pure and Applied Chemists. She served in this capacity again in 1927 (in Washington, D.C.) and in 1936 (in Lucerne, Switzerland).

In 1935, Francis Garvan of the Chemical Foundation, an organization that held patents and trademarks from which it had an annual income of $9 million, wondered why there were no women chemists. When he was informed that although women chemists were rare, they certainly did exist, Garvan offered to endow a medal to recognize an American woman who had made a distinguished contribution to chemistry. A committee was formed to examine possible candidates, and Carr reluctantly agreed to be a member. She did not believe that women in chemistry should be separated out and judged differently from the men. While she was in Europe in 1936, Carr received word that the committee had unanimously selected her as the first recipient of the Garvan Medal.[9]

Carr continued to teach and lead her research team until 1946. After her retirement, she lived near Mt. Holyoke in South Hadley for eighteen years and remained active in research and civic affairs, publishing her last paper in 1957. In 1964 she moved to Evanston, Illinois, where she died in 1972.

Robert Mulliken, winner of the 1966 Nobel Prize in chemistry, refers to Carr's work in his autobiography.[10] On her death, he credited Carr with being one of the pioneers in the field of ultraviolet spectra, on which he was able to build the molecular orbital theory. Earlier, one of her students had said of her, "She has shown those of us who have tried to follow her not just what was the structure of organic substances but what is the structure of the substance of a teacher."[11]

Notes

1. Anna Mary Wells (niece of Emma Carr), interview with author, July 11, 1991; a tape of this interview is in the Mount Holyoke College Library Archives.

2. C. Pauline Burt, "Emma Perry Carr," *Nucleus* (June 1957): 214.

3. Helen Wells Smits, M.D. (greatniece of Emma Carr), interview with author, July 18, 1991.

4. Emma Perry Carr, "Chemical Education in American Institutions: Mount Holyoke College," *Journal of Chemical Education* 25 (1948): 11–15.

5. Margaret Chapin, letter dated December 9, 1923, Chapin files, Mount Holyoke College Library Archives.

6. "Emma Perry Carr," *Mount Holyoke Alumnae Quarterly* 30 (August 1946): 54.

7. Ibid., p. 55.

8. Emma Perry Carr, "Scientific Research in a Liberal Arts College," *Nucleus* (June 1957): 216–220.

9. Awarded three times between 1937 and 1946 and then annually by the American Chemical Society, the Garvan Medal went in 1947 to Mary Sherrill and in 1957 to Lucy Pickett, another member of the research group at Mount Holyoke.

10. Robert S. Mulliken, *Life of a Scientist* (Berlin and New York: Springer Verlag 1989), p. 106.

11. Burt, "Emma Perry Carr," p. 216.

Bibliography

Burt, C. Pauline. "Emma Perry Carr." *Nucleus* (June 1957): 214–216.

Carr, Emma Perry. "Chemical Education in American Institutions: Mount Holyoke College." *Journal of Chemical Education* 25 (1948): 11–15.

———. "One Hundred Years of Science at Mount Holyoke College." *Mount Holyoke Alumnae Quarterly* 20 (1936): 135–138.

———. "Scientific Research in a Liberal Arts College." *Nucleus* (June 1957): 216–220.

Jennings, Bojan. "The Professional Life of Emma Perry Carr." *Journal of Chemical Education* 63 (1986): 923–927.

Rossiter, Margaret W. *Women Scientists in America: Struggles and Strategies to 1940.* Baltimore: Johns Hopkins University Press, 1982.

Shmurak, Carole B. "Emma Perry Carr: The Spectrum of a Life." *Ambix* 41 (July 1994): 75–86.

CAROLE B. SHMURAK

MARJORIE CASERIO

(1929–)

Chemist

Birth	February 26, 1929
1950	B.S., Chelsea College, University of London
1951	M.A., Bryn Mawr
1952–53	Associate Chemist, Fulmer Research Institute, England
1956	Ph.D., Bryn Mawr
1956–64	Research Fellow, California Institute of Technology
1957	Married Frederick Caserio
1965–90	Assistant Professor (then Professor) of Chemistry, University of California, Irvine
1975	Garvan Medal, American Chemical Society
1975–76	John S. Guggenheim Foundation Fellow
1990–95	Vice Chancellor of Academic Affairs, University of California, San Diego
1995–96	Interim Chancellor, University of California, San Diego

Marjorie Caserio has had a distinguished career as a researcher, educator, author, and academic administrator. Born in London, England, in 1929, Caserio was 15 years old when she entered Chelsea Polytechnic (now Chelsea College) of the University of London, where originally she planned to study podiatry. "It was not a glorious beginning as a scientist," says Caserio, "but a pragmatic approach to receive training as quickly as possible to begin generating an income."[1] These were war years, after all, and hard times deprived young people of the luxury of growing up slowly. Caserio's last year of high school was spent in a bomb shelter, a bleak experience from which she "learned nothing." She left school knowing only that she did not want to be a secretary and enter "a holding pattern until marriage," but instead wanted the career options a university degree could provide.

The choice to pursue her education was not made easy for Caserio. She received a number of rejections from colleges because she was female and because spaces were reserved for former servicemen returning from the war effort. "This treatment was not prejudicial," explains Caserio, "but simply the mores of the time." Her spirit of determination is evident

Marjorie Caserio. Photo courtesy of Marjorie Caserio.

in the way she continued to apply for admission until she finally found a college that would accept her, even if it meant occupying a unique position as the only woman in her class. The difficulties that arose from this status were not academic in nature, but social. "If you want to achieve anything on your own," says Caserio, "you have to be careful not to submerge your own personality into the society in which you are a part."

At Chelsea College, Caserio began to re-evaluate her interest in podiatry, ultimately deciding that "cutting people's feet isn't particularly exciting." However, the degree requirements for podiatry did include study in the sciences at a level much more serious than Caserio had previously realized. She found that she liked science, was good at it, and enjoyed tutoring other students, so she transferred to the chemistry department and completed her B.S. degree in 1950.

At the time she was graduating, Caserio became aware of and applied for a fellowship that was to be offered to three British women to continue graduate study in the sciences in the United States. Caserio's grandparents had lived for a while in Kingston, New York, and her father had been born in the United States. Her grandfather had always encouraged her to live in America. Also, from a previous trip to the United States, Caserio had met "some enlightened people who put the seed in my mind to come back as a graduate student." Competition for the Sir John Dill Fellowship of the English Speaking Union was intense. Caserio credits her award to the letters of reference written by her professors at Chelsea College. "I have long since realized the value of those letters," says Caserio, "and I would emphasize to young people the impact a good recommendation can have on your career." In addition to her fellowship, she received a Fulbright travel grant.

Caserio pursued her graduate degrees in organic chemistry at Bryn Mawr, where she earned an M.A. in 1951 and a Ph.D. in 1956. Between degrees, she returned briefly to England for a year (1952–53) to work as an associate chemist at the Fulmer Research Institute in Stoke Poges. Caserio's doctoral work was under the guidance of Ernst Berliner, chair of the chemistry department, whom she describes as "an exceptionally fine man." Caserio also admired the work of his wife, Frances Berliner, whom she calls "extremely inspiring and an absolute perfectionist in the lab." She has stayed in contact with the couple throughout her career.

Another mentor in Caserio's life has been John "Jack" Roberts, whose research group at the California Institute of Technology she joined upon completion of her doctorate. She enjoyed her collaboration with Roberts because of his outstanding scientific abilities and acumen and because he encouraged her to plan actively for a career in chemistry. "Jack was extraordinarily helpful in terms of my confidence as a scientist," says Caserio. "He advised me not to put myself down, which is so easy for women to do." Caserio found her research to be very rewarding. "When you discover something significant, it is enormously exciting," she says. "It brings you one microstep further to discovering the mysteries of the world." With Roberts, Caserio co-authored *Basic Principles of Organic Chemistry*, which became, in the 1960s and 1970s, one of the most popular chemistry textbooks in the country.[2] It has been translated into the Russian language and has been praised as an influential text that changed the emphasis of undergraduate studies in organic chemistry, in part by demonstrating the application of emerging technologies such as chromatography and nuclear magnetic resonance spectroscopy to understanding chemical structures and mechanisms.[3]

Caserio spent eight years at the California Institute of Technology, two as a postdoctoral research fellow and six as a senior research fellow. Then she began to realize she would have to leave the school if she wanted a faculty position. Caserio acknowledges that even though it is much easier in today's environment for women to get an education in the sciences, faculty positions are as competitive as ever. In 1965, though, the University of California system was expanding and three new campuses opened, offering greater opportunities for teaching. Caserio believes the favorable reception to her textbook not only gave her an edge in obtaining a position as assistant professor of chemistry at the newly constructed Irvine campus, but also provided her with an enduring link to future generations of chemists. She notes that

> nothing we ever do is lost. It comes back to profit us in unexpected ways. . . . I must have taught thousands of young people over the years. It has always been a most gratifying experience to meet the occasional student years later, usually under circumstances far removed from a classroom setting, who volunteers how much s/he enjoyed the course or the book, and that it contributed positively to their own success. Many times I have met people at conferences and seminars who were not former students of mine but who studied from Roberts and Caserio and mention how much it helped them. This is what makes a career as an educator so worthwhile.

Caserio remained at the University of California, Irvine, for twenty-five years, rising through the academic ranks from assistant to full professor. Her work as a chemist and scholar has been nationally recognized. She was awarded a John S. Guggenheim fellowship from 1975 to 1976. That same year, Caserio was presented with the American Chemical Society's prestigious Garvan Medal for her threefold contribution to research, education, and professional organizations in chemistry. Caserio's research in physical organic chemistry has focused on reaction mechanisms, particularly in the fields of allene and organosulfur chemistry. Her work seeks to elucidate the precise events that occur when organic material undergoes a transformation. Her studies in sulfur-containing organic materials have also had implications for biological chemistry. Caserio has co-authored four textbooks and has published over seventy articles.

Caserio served as chair of the Irvine Division of the Academic Senate from 1982 to 1984; chair of the statewide, nine-campus University of California (UC) Academic Senate from 1984 to 1986; faculty representative to the UC Board of Regents from 1984 to 1986; and as chair of UC Irvine's department of chemistry from 1987 to 1990. She was a keynote speaker for the Women in Bioscience Conference in San Diego in 1993. In addition, Caserio has served as chair of the American Chemical Society's Committee on Professional Training, with responsibility for setting guidelines for the teaching of chemistry at 580 colleges and universities in the United States. She holds memberships in the American Association for the Advancement of Science, the American Chemical Society, and Sigma Xi.

Caserio's work with the Academic Senate brought her high visibility within the UC system. During her tenure as head of the Irvine campus Academic Senate, the Nixon Archives Foundation made the decision not to build the Richard M. Nixon library at the school because of opposition from the faculty.[4] In 1990, Caserio was appointed vice chancellor of academic affairs at UC San Diego. Unwilling to give up her interaction with students in the classroom, Caserio continued to teach while being an administrator. Although faced with controversies (such as issues of academic freedom during the Gulf War) as well as financial challenges (such as downsizing, budget cuts, and the loss of key faculty to early retirement incentive programs), Caserio found her new position to be a "very gratifying, rewarding job." During her tenure as vice chancellor, UC San Diego ranked sixth among U.S. universities in the amount of federal funding spent on research and development and first among the nine campuses in the UC system.[5] Caserio retired from her position as vice chancellor in July 1995, with the impression that UC San Diego is "stronger than ever."

Marjorie (nee Beckett) married Frederick Caserio in 1957. They have two sons, Alan and Brian. She describes her husband as an equal partner who has honored and supported her career. They both enjoy outdoor activities, including skiing, hiking, and backpacking.

Notes

1. Marjorie Caserio and Nancy Allee, telephone conversation, August 19, 1995. Unless otherwise noted, quotations in this essay are taken from this source.

2. See "Marjorie Caserio, Professor of Chemistry, Appointed as Vice Chancellor for Academic Affairs at UCSD," *UCSD Chemistry Graduates' Association* [newsletter], no. 6 (Winter 1991).

3. See *McGraw-Hill Modern Scientists and Engineers* (New York: McGraw-Hill Book Co., 1980).

4. "Names and Faces," *Boston Globe* (March 17, 1983).

5. Richard Acello, "UCSD Was a Billion-Dollar Business in '94," *San Diego Daily Transcript* (December 28, 1994): 1.

Bibliography

Acello, Richard. "UCSD Was a Billion-Dollar Business in '94." *San Diego Daily Transcript* (December 28, 1994): 1.

Association for Women in Science and Connect (The UCSD Program in Technology and Entrepreneurship). "Women in Bioscience." University of California at San Diego, October 16, 1993.

Baringa, Marcia. "Early Retirement Program Cuts Deep into UC Faculties." *Science* 264, no. 5162 (1994): 1074–1076.

Caserio, Marjorie. *Experimental Organic Chemistry*. New York: W.A. Benjamin, 1967.

———. "The Intersegmental Committee of the Academic Senate," in "Issues in California Community Colleges." Edited by Karen Sue Grosz. *Forum* 4 (Summer 1987): 17–22. ED 287536.

"Chemistry Chair at UC Irvine Named UCSD Vice Chancellor." *Los Angeles Times* [San Diego County edition], March 8, 1990.

Kennedy, Michael J. "UC Officials Call Faculty Underpaid." *Los Angeles Times*, (October 21, 1994): A3.

McGraw-Hill Modern Scientists and Engineers. New York: McGraw-Hill, 1980.

"Names and Faces." *Boston Globe* (March 17, 1983).

"1975 Winners of ACS National Awards Named." *Chemistry and Engineering News* 9 (September 1974): 18–19.

Richardson, James. "UC Is Outbid for Professors, Regents Are Told." *Sacramento Bee* (October 21, 1994): B1.

Roberts, John D., and Marjorie C. Caserio. *Basic Principles of Organic Chemistry*. 3 vols. Pasadena: California Institute of Technology, 1961–1963.

———. *Modern Organic Chemistry*. New York: W.A. Benjamin, 1967.

———. *Organic Chemistry: Methane to Macromolecules*. Menlo Park, Calif.: W.A. Benjamin, 1971.

Wallace, Amy. "Battle Erupts at UCSD over Gulf War Memo." *Los Angeles Times* [San Diego County edition], April 29, 1991.

The Women's Book of World Records and Achievements. Edited by Lois Decker O'Neill. Garden City, N.Y.: Anchor Press/Doubleday, 1979.

NANCY ALLEE

MILDRED COHN

(1913–   )

Biochemist, Biophysicist

Birth	July 12, 1913
1931	B.A., Hunter College
1938	Ph.D., physical chemistry, Columbia University
1938–46	Research Associate, biochemistry, Cornell University Medical School
1946–58	Research Associate, Washington University Medical School
1950–51	Research Associate, Harvard Medical School

Mildred Cohn. Photo courtesy of The University of Pennsylvania Archives.

1958–60	Associate Professor, Washington University Medical School
1960–78	Professor, University of Pennsylvania Medical School
1963	Garvan Medal, American Chemical Society
1964–78	Career Investigator, American Heart Association
1975	Cresson Medal, Franklin Institute
1978–82	Benjamin Bush Professor, University of Pennsylvania Medical School
1979	Biochemists Award, International Organization of Women
1982	National Medal of Science
1982–	Professor Emerita, University of Pennsylvania Medical School

1982–85	Senior Member, Institute for Cancer Research, Fox Chase
1987	Distinguished Award, College of Physicians
1988	Remsen Award, American Chemical Society

Mildred Cohn has had a rich and rewarding career in the fields of biochemistry and biophysics. She has made important contributions in both fields, using nuclear magnetic resonance (NMR) to study enzyme-catalyzed reactions and energy transduction within cellular reactions. Her work has been recognized by her election to the National Academy of Sciences, the American Academy of Arts and Sciences, and the American Philosophical Society. In 1982 she was presented the National Medal of Science by President Ronald Reagan.

Cohn was born on July 12, 1913, in New York City to Russian Jewish emigres Isidore and Bertha Klein. She was the second of two children. At an early age, she was encouraged by her father to do anything that she wanted. Her father was a man of learning with a strong respect for scholarship. Her mother's goals for her were more modest and conventional.

In public school, Cohn found that she enjoyed chemistry and performed well in mathematics. At the age of 15 she attended Hunter College, the free college for women in New York. She decided to major in chemistry and to minor in physics in her sophomore year, having once stated that "I could study the humanities and social sciences on my own, but for the physical sciences I needed formal instruction."[1] Cohn emerged from Hunter with a B.A. degree and with her interest in chemistry intact, despite what she considered to have been an inferior science education. The chairman of Hunter's chemistry department did not consider it "ladylike" for women to study chemistry except to become science teachers.[2] In Cohn's senior year, her persistence was rewarded by more challenging courses in physical chemistry and an introduction to modern physics.[3] These courses inspired her interest in chemistry and physics, and she decided to pursue graduate work in both disciplines. Her father encouraged her in this endeavor, although her mother wanted her to become a schoolteacher.

Cohn faced many difficulties on the road to graduate school. Her father's business failed during the Great Depression, so he was unable to help her financially. Her applications for financial assistance or scholarships, despite her good grades, were turned down by twenty graduate schools. She decided to attend Columbia University and paid the $300 tuition out of her savings from summer jobs. She applied for a teaching assistantship but was turned down because she was a woman. She supported herself with baby-sitting money.

In 1932, Cohn received her master's degree in chemistry after one year of coursework. Academically this was a very important year for her. She was introduced to thermodynamics, classical mechanics, and molecular spectroscopy. Her only disappointments were that first-year students were limited to coursework and were not allowed to perform research and that her undergraduate work had not provided a more comprehensive portrayal of chemistry. She also met Harold Urey, her professor for both thermodynamics and molecular spectroscopy courses, whom she selected for her mentor as she pursued a Ph.D. in chemistry.

Unfortunately, however, Cohn was forced to leave Columbia in 1932 because she did not have sufficient funds to continue her studies. She reluctantly accepted tedious computational work at the National Advisory Committee for Aeronautics at Langley Field in Virginia, but she soon transferred to a research position working on airplane engines.[4] She gained a healthy respect for applied research while working in this area. Applied research was well ahead of theoretical research in the area of engine combustion. During the two-year period Cohn was at Langley, she published two reports, for one of which she was the primary author. She was the only woman among seventy men, and after two years she reached the top salary for her rank. She was informed that she would not be promoted above her present rank because she was a woman.[5] At this point Cohn decided to use the savings she had accumulated to pursue a Ph.D. in chemical engineering.

Cohn returned to Columbia but was told by the chairman of the chemical engineering department that no women had ever been admitted to the program and that the policy was not going to change.[6] She then decided to return to physical chemistry and asked Urey to accept her as his graduate student. Urey was reluctant, telling her that "You don't want to be my graduate student; I don't pay attention to my graduate students."[7] Cohn was persistent, however, and felt confident enough to perform her research independently. Urey and Cohn worked on separating the stable isotopes 12C and 13C.[8] Though ultimately unsuccessful owing to equipment problems with a mass spectrometer, "the theoretical and experimental approaches" she learned under Urey "were to guide her throughout her career."[9] Cohn passed her qualifying exams after a single semester of coursework. In 1934 she became an official candidate for a Ph.D. under Urey's supervision.

Cohn completed her thesis entitled "Oxygen Exchange Reactions of Organic Compounds and Water" in 1937, but she was not able to find a job after graduation.[10] The economy was still poor, and the industrial jobs that many of her male colleagues secured were not available to a Jewish woman. Urey was also not able to find her a job, but he offered her a postdoctoral position to continue her research in the area of her Ph.D. Fortunately an offer emerged to perform postdoctoral work at George Washington University Medical School under the biochemist

Vincent du Vigneaud. After consulting with Urey, Cohn accepted the position and began her work in applying "isotopic tracers to biological problems."[11] In du Vigneaud's lab, Cohn worked as part of a research team with each member contributing in individual areas of expertise. Cohn reported no regrets about changing her field from straight chemical physics to applying chemical physics to biology.[12]

Cohn's work in du Vigneaud's lab involved using isotopic tracers to study the sulfur–amino acid metabolism. Isotopic tracers are forms of chemical elements with differences in their nuclear structure. They allow mass or radioactivity to be traced as they progress through metabolic processes. By tracing the isotope, it is possible to understand the chemical reactions occurring in animals' metabolism.[13]

Cohn married physicist Henry Primakoff on May 31, 1938. Shortly after her marriage, she moved with du Vigneaud to Cornell Medical College in New York. Her husband was offered a position in New York at about the same time. During World War II, Cohn continued her basic research with the draft-exempt men from du Vigneaud's lab. The rest of the lab focused on military-based research.[14]

By 1946, Cohn felt that she had been initiated into the bioorganic approach, but also that she had worked in this area of biochemistry for long enough.[15] It was time for a change. Primakoff was offered a position at Washington University that she urged him to take. Du Vigneaud helped Cohn to obtain a position at Washington University Medical School in St. Louis in the biochemistry department, chaired by Carl Cori, a Nobel Prize–winning scientist. Cori shared the Nobel with his co-researcher and wife, **Gerty Cori**, an expert in enzymology. Cohn had two objectives for her future research: to work independently rather than as a member of a team, and to use isotopes for insight into the mechanisms of enzyme-catalyzed reactions in isolated systems.[16] Gerty Cori took Cohn under her wing and trained her in enzymology. Cohn used nuclear magnetic resonance (NMR) and an isotope of oxygen to study the enzyme-catalyzed reactions of phosphates.[17]

In 1960, Primakoff moved to the University of Pennsylvania to become Donner professor of physics. Cohn followed her husband and joined the biophysics department there. She continued her work with NMR, researching energy transduction within cell and cellular reactions in which adenosinetriphosphate (ATP) is utilized.[18] After one year at the University of Pennsylvania, she was promoted to full professor. For the previous 21 years she had been a research associate. In 1964 she was named a career investigator for the American Heart Association, an organization that supports people rather than just research projects. She held this position until 1978. From 1978 to 1979 she was president of the American Society of Biological Chemistry. Then in 1982 she was named Benjamin Bush professor in biochemistry and biophysics at the University of Pennsylvania. She is currently a professor emerita of biochemistry and biophysics there as well as visiting professor of biochemistry at the Johns Hopkins Medical School. She continues to conduct research and to publish her results.

Cohn and Primakoff have three children. Her extracurricular activities include theater, hiking, writing, and reading. Cohn considers the most important aspect of her career to be the overall sense of enjoyment she has experienced—the joy of predicted results developing, the even more rewarding experience of serendipitously discovering an entirely unexpected phenomenon, and the special gratification of having the results applied to medical problems.[19] She considers herself fortunate to have worked in stimulating milieux and to have interacted with first-class minds, from her mentors to her colleagues to her postdoctoral fellows and students.[20] She considers herself especially lucky to have married Primakoff, who treats her as an intellectual equal and who always assumed that she would pursue her scientific career.[21]

Notes

1. Mildred Cohn, "Atomic and Nuclear Probes of Enzyme Systems," *Annual Review of Biophysics and Biomolecular Structure* 21 (1992): 3.

2. Ibid., p. 3.

3. Ibid.

4. Ibid., p. 4.

5. Ibid.

6. Ibid., p. 5.

7. Ibid.

8. Ibid., p. 6.

9. Ibid.

10. John Henry Dreyfuss, "Mildred Cohn," in *Notable Twentieth Century Scientists*, ed. Emily J. McMurray (New York: Gale Research, 1995), p. 381.

11. Cohn, "Atomic and Nuclear Probes," p. 7.

12. Ibid.

13. Dreyfuss, "Mildred Cohn," p. 382.

14. Cohn, "Atomic and Nuclear Probes," p. 9.

15. Ibid.

16. Ibid., p. 10.

17. Dreyfuss, "Mildred Cohn," p. 382.

18. Ibid.

19. Cohn, "Atomic and Nuclear Probes," p. 22.

20. Ibid.

21. Ibid.

Bibliography

American Men and Women of Science. New York: R.R. Bowker, 1995/1996.

Cohn, Mildred. "Atomic and Nuclear Probes of Enzyme Systems." *Annual Review of Biophysics and Biomolecular Structure* 21 (1992): 1–24.

Dreyfuss, John Henry. "Mildred Cohn," in *Notable Twentieth Century Scientists.*
Edited by Emily J. McMurray. New York: Gale Research, 1995.
"Garvan Medal: Dr. Mildred Cohn." *Chemical and Engineering News* 41 (February
4, 1963): 92.

TIMOTHY WILLIAM KLASSEN

GERTY CORI

(1896–1957)

Biochemist

Birth	August 1896
1920	M.D., Carl Ferdinand University, Prague
1946	Midwest Award, American Chemical Society
1947	Nobel Prize; Squibb Award in Endocrinology
1948	Garvan Medal, American Chemical Society; St. Louis Award; D.Sc., Boston University
1949	D.Sc., Smith College
1950	Sugar Research Prize
1951	Borden Award, Association of Medical Colleges; D.Sc., Yale University
1952	Appointed to the science board of the National Science Foundation by President Truman
1954	D.Sc., Columbia University
1955	D.Sc., Rochester University
Death	October 26, 1957

In 1947, Dr. Gerty Theresa Radnitz Cori became the first American woman to win a Nobel Prize in science. One of her co-laureates was Dr. Carl Ferdinand Cori, her husband and long-time colleague. They were awarded the Nobel Prize in Physiology or Medicine "for their discovery of the course of the catalytic conversion of glycogen"—that is, for explaining the physiological mechanism by which the body metabolizes sugar. This marked the Coris as the third husband and wife team to win a Nobel Prize. As Carl explained at the Nobel banquet, "Our collaboration began thirty years ago when we were still medical students at the University of Prague and has continued ever since. Our efforts have been

Gerty Cori. Photo courtesy of Archives, Washington University
School of Medicine.

largely complementary, and one without the other would not have gone
as far as in combination."[1]

The eldest of three daughters of Otto and Martha (Neustadt) Radnitz,
Gerty Theresa Radnitz was born in Prague in 1896 and schooled at home
until age 10. Her Uncle Robert, a professor of pediatrics at Carl Ferdinand University, nurtured in her an interest in mathematics and sciences
that might not have developed during her time at a girls' lyceum, an
environment emphasizing social graces over academic knowledge. Encouraged by her uncle, Gerty considered the possibility of attending university and specializing in medicine. To do so, however, would require
more math, physics, and chemistry as well as solid knowledge of Latin,
which she had never studied. Determined to gain a place at university
to study medicine, Gerty enrolled in the Realgymnasium at Tetschen and

mastered these subjects in one year, graduating in 1914. At age 18 she passed what she called "the hardest examinations I was ever called upon to take" and enrolled in the German branch of the medical school at Prague's Carl Ferdinand University.[2]

It was in 1914, during her first-semester anatomy class at university, that Gerty met Carl Cori. The two shared a seriousness of purpose about their studies; a desire to pursue medical research, not medical practice; and a determination to gain medical certification (a six-year course) before marriage. Their first research collaboration was an immunological study of complement in human serum, which they published jointly. In 1920 both received their medical degrees and were married. Gerty and Carl were drawn to one another not only because of their shared academic interests but also because they shared a love of sporting and the outdoors. First as medical students and then as husband and wife, they climbed glaciers and mountains (both the Austrian Alps and the American Rockies), skated, swam, and played tennis. They even gardened together, Cori tending the flowers and Carl the vegetables.

With their medical degrees in hand, the Coris moved to Vienna. Gerty worked at the Karolinen Children's Hospital from 1920 until 1922, doing biochemical research. She published several papers on the thyroid and spleen. In 1922, Carl received an offer to work in the United States, a far more appealing prospect to him than working in war-torn Austria. In a decision demonstrating great independence, Gerty did not immediately resign her job. Rather, she remained in Vienna until she found employment as assistant pathologist in the New York State Institute for the Study of Malignant Diseases in Buffalo. Once again she worked alongside her husband, a biochemist in the Institute. Gerty particularly enjoyed the position because the Institute offered "good equipment and complete freedom" to pursue the research that intrigued her.[3] In 1925 she was promoted to assistant biochemist.

Gerty and Carl collaborated on much of their work. At first they focused on the metabolism of tumors. But then Gerty's interest in sugars (her father managed several sugar refineries), and Banting and Macleod's recent discovery of insulin, suggested a new direction for their research. They set out to explore the chemical process of carbohydrate metabolism; that is, how the body uses sugars for energy and life. Specifically, Gerty and Carl aimed to describe the chemical processes of the sugar cycle in depth, which they did in such detail that their findings became known as the Cori cycle. And Gerty found a new substance in muscle tissue, glucose-1-phosphate, now known as Cori ester.

Despite the successes of their collaborative research, resistance to Gerty's presence in science became evident at a few key moments in her career. For example, the director of the Institute threatened to fire Gerty if she did not end collaborative work with her husband. A few years later, when another university offered Carl a position, one of the stipulations was that he end collaborative work with his wife. The reasoning? It was un-American for a man to work with his wife—Gerty was standing in the way of her husband's career advancement.[4] Yet the two persisted in their work together.

In 1928 the Coris became naturalized American citizens. In 1931 they moved to St. Louis to Washington University, where Gerty and Carl were appointed research associate in pharmacology and professor of pharmacology, respectively. In this powerhouse laboratory the Coris worked alongside Severo Ochoa, Arthur Kornberg, Earl Sutherland Jr., Christian DeDuve, and Luis Leloir—all future Nobel laureates. During her years in St. Louis, Gerty gave birth to a son, Thomas, who eventually followed in his parents' footsteps to become a research chemist. From 1936, Gerty managed the challenging combination of being a mother, a wife, and a scientist. As Bernardo Houssay, one of her co-laureates, described her, these roles were the "triple crown that adorned her life."[5]

Gerty's work as a scientist was well known, yet historians and biographers have written that were she a man, she might have held even higher level positions. If, as the New York Post reported, "it is hard to tell where the work of one leaves off and that of the other begins," it would seem unclear why Carl began at Washington University as professor of pharmacology and biochemistry whereas Gerty began as research associate in pharmacology.[6] Gerty did, however, receive two promotions during her time in St. Louis. She became research associate professor in biochemistry in 1943, the year she and Carl achieved the synthesis of glycogen in a test tube, confirming their understanding of the biosynthesis of glycogen. Resistance to allowing women to hold senior faculty appointments—in combination with anti-nepotism rules— kept Gerty from being promoted to full professor of biochemistry until 1947, shortly before she was awarded the Nobel Prize. In that year, Carl became head of the department.

The Coris' discovery was not simply a breakthrough to dazzle the scientific community. Their work had practical implications for the treatment of disease. For example, whereas the effects of insulin in alleviating diabetes had already been discussed widely, the physiological mechanisms of sugar metabolism were not fully understood. The Coris' discovery that insulin decreases the amount of sugar stored as glycogen in the liver and increases the amount oxidized in muscle was crucial for doctors treating diabetes.

In 1947, Gerty was one of twelve women honored at Hobart and William Smith College at a celebration marking the centennial of the awarding of the first medical degree to a woman. Gerty's life is often described in conjunction with her husband's, since in many cases their research was indistinguishable. However, Gerty pursued research and published many articles on her own as well as in cooperation with other scientists. In 1951, for example, she worked with J. Larner to identify a new enzyme, amylo-1, 6-glucosidase—or more commonly, the debrancher. This helped her to define chemically the structure of glycogen, and it also helped with applications of her research. In 1955 she furthered her long-held interest in hereditary glycogen storage diseases in children. The first of its kind, her work showed how enzyme defects or deficiencies can lead to congenital metabolic diseases.

Gerty Cori died on October 26, 1957, after a decade-long battle with myelosclerosis, a bone-marrow disease. In her final years, she reduced her sporting and climbing activities but did not avoid the lab and her students, particularly foreign students with whom she felt a special kinship. Born an Austrian, Gerty Cori died an American. In her own words, "I believe that the benefits of two civilizations . . . have been essential to whatever contributions I have been able to make to science."[7]

Notes

1. Olga S. Opfell, *The Lady Laureates: Women Who Have Won the Nobel Prize* (London: Scarecrow Press, 1986), p. 215.

2. Ibid., p. 216.

3. Ibid., p. 218.

4. See Lois N. Magner, "Gerty Theresa Radnitz Cori," in Daniel Fox, Marcia Meldrum, and Ira Rezack, eds., *Biographical Dictionary of Nobel Prize Laureates in Medicine or Physiology* (New York: Garland Publishing, 1990).

5. Opfell, *Lady Laureates*, p. 220.

6. Ibid., p. 219.

7. Ibid., p. 222.

Bibliography

Cori, C.F., and G.T. Cori. "Glucose 6–Phosphate of the Liver in Glycogen Storage Disease." *Journal of Biological Chemistry* 199 (1952): 661–667.

Cori, G.T., C.F. Cori., and G. Schmidt. "The Role of Glucose-1-Phosphate in the Formation of Blood Sugar and Synthesis of Glycogen in the Liver." *Journal of Biological Chemistry* 129 (1939): 629–639.

Current Biography Yearbook, 1947. New York: H.W. Wilson, 1948.

"Enzymes and Metabolism: A Collection of Papers Dedicated to Carl F. and Gerty T. Cori on the Occasion of the 60th Birthday." *Biochimica Biophysica Acta* 20 (1956).

Fruton, Joseph S. "Cori, Gerty Theresa Radnitz," in *Dictionary of Scientific Biography*, Vol. 3, p. 415–416. New York: Charles Scribner's, 1971.

Houssay, B.A. "Carl F. and Gerty T. Cori." *Biochimica Biophysica Acta* 20 (1956): 16.

Larner, J., and G.T. Cori. "Action of Amylo-1,6-Glycosidase and Phosphorylase on Glycogen and Amylospectin." *Journal of Biological Chemistry* 188 (1951): 17–29.

Magner, Lois N. "Gerty Theresa Radnitz Cori," in *Biographical Dictionary of Nobel Prize Laureates in Medicine or Physiology*. Edited by Daniel Fox, Marcia Meldrum, and Ira Rezack. New York: Garland Publishing, 1990.

Ochoa, Severo, and H.M. Kalckar. "Gerty T. Cori, Biochemist." *Science* 128 (1958): 16–17.

Opfell, Olga S. *The Lady Laureates: Women Who Have Won the Nobel Prize*. London: Scarecrow Press, 1986.

Parascandola, John. "Cori, Gerty Theresa Radnitz," in *Notable American Women: The Modern Period*. Edited by Barbara Sicherman and Carol Hurd Green. Cambridge, Mass.: Belknap Press, 1980.

Yost, Edna. *Women of Modern Science*. New York: Dodd Mead, 1959.

JENNIFER LIGHT

MARIE SKLODOWSKA CURIE
(1867–1934)
Physicist, Chemist

Birth	November 7, 1867
1893	*Licenciée ès Sciences Physiques*, University of Paris
1894	*Licenciée ès Sciences Mathématiques*, University of Paris
1895	Married Pierre Curie on July 26
1898	Discovered polonium (July) and radium (December); Prix Prize, French Academy of Sciences
1900	Prix Prize, French Academy of Sciences
1900–06	Professor, Normal Superior School for Girls at Sèvres
1902	Isolated pure radium; Prix Prize, French Academy of Sciences; Berthelot Medal, French Academy of Sciences
1903	Ph.D., physical science, University of Paris; Nobel Prize in Physics; Davy Medal, Royal Society of London
1906	Pierre Curie killed in an accident; Assistant Professor, the Sorbonne
1907	Actonian Prize, Royal Institution of Great Britain
1908–34	Professor, University of Paris
1909	Elliott Gresson Gold Medal, Franklin Institute
1911	Nobel Prize in Chemistry
1912	Radium Institute founded in Paris (opened in 1914)

1922	Elected to the French Academy of Medicine
1931	Cameron Prize, University of Edinburgh
Death	July 4, 1934

The French physicist Marie Curie is known for having discovered radium with her husband, Pierre Curie. This discovery not only led to the use of radium in medicine but, more important, set the foundation for our modern understanding of the structure of the atom, which can be split to release enormous energy.

Born in Warsaw, Poland, in 1867, Marya Sklodowska (known to the world later as Marie Curie or Madame Curie) was the youngest of Wladyslaw Sklodowski and Bronislawa Sklodowska's five children. Marya's father taught physics and mathematics in secondary schools. Her mother was the director of one of the best schools for young girls in Warsaw. Partly because of her poor health, her mother resigned shortly after Marya was born to devote herself to raising the family. Two years younger than her classmates, Marya ranked first in history, literature, German, and French in her class in a private school for girls. When she was 9 years old her eldest sister died of typhus, which spread to the Sklodowskis from students who were boarding with the family. Within two years Marya's mother died of tuberculosis.

In 1883, Marya graduated first in her class from high school with a gold medal. At the time Marya graduated from high school, women were not admitted to university in Russian-ruled Poland. Marya and her elder sister Bronya dreamed of going to Paris to study, but their father could not afford to send them to universities abroad. They attended the floating university that was secretly organized in response to young Polish women's yearnings for higher education. From 1885 to 1889, Marya worked as a governess in the country and sent her savings to Bronya so that she could study medicine in Paris. During these years Marya read books to further her own study, tried gradually to discover her real preferences, and "finally turned toward mathematics and physics."[1] But only when she returned to Warsaw after finishing her governess job could she conduct her first experiment in chemistry.

In 1891, Marya went to Paris to study at the Sorbonne, part of the University of Paris. She registered as Marie Sklodowska for a degree of *Licenciée ès Sciences*. She had little money and suffered from hunger and cold. She frequented the light, warm library to avoid the cold of her simple apartment. After two years of hard work, in July 1893 she received her *Licenciée ès Sciences Physiques*. She was the first woman to receive such a degree at the Sorbonne and graduated number one in the class of thirty. In July 1894 she received her second degree, *Licenciée ès Sciences Mathématiques*, and placed second in her class.

In the spring of 1894, Marie needed laboratory space to study the magnetic properties of different types of steel. A friend introduced her to Pierre Curie, a most promising young French physicist who was the director of the laboratory in the School of Industrial Physics and Chemistry of the City of Paris. Their mutual interests in physics and their similar family backgrounds immediately drew them together. Their friendship blossomed and they married on July 26, 1895. In September 1897 their first daughter, Irène (**Irène Joliot-Curie**, Nobel Prize–winner in Chemistry in 1935) was born.

The German physicist Wilhelm Conrad Roentgen had discovered X-rays in 1895. In 1896, Henri Becquerel, a French physicist, found that uranium, a rare chemical element, emitted rays, the glow of fluorescence. Pierre and Marie were very interested in this mysterious radiation. As Pierre was occupied with research on crystals, he suggested that Marie investigate radiation; she resolved to make it the theme of her doctoral thesis. At the end of 1897, Marie began testing the chemical elements in pitchblende to identify the substance causing the glow. In July 1898, Marie and Pierre announced the discovery of a new radioactive element. Marie named it polonium in honor of her motherland, Poland. She also unexpectedly discovered that pitchblende, a mineral ore, emitted as much as four times the radiation as pure uranium. In December 1898, the Curies announced the discovery of a second, more radioactive element: radium. The next step was to isolate the pure radium and polonium to prove their existence.

However, the Curies were very poorly equipped. They "had no money, no suitable laboratory and no personal help" for the difficult task they were undertaking.[2] The Curies spent four more years studying these elements. Using an old, leaky shed as their laboratory, Marie devoted herself to the purification of radium and Pierre concentrated on the study of the physical properties of the rays emitted by these new radioactive substances. Finally, in 1902, after treating one ton of pitchblende residues, Marie produced one-tenth of a gram of pure radium. Polonium, which was a product of radioactive decay, was also isolated. She later recalled this difficult period as the happiest in her life.

While doing this research, Marie continued to teach physics at Normal Superior School for Girls at Sèvres. On June 25, 1903, she was awarded her Doctor of Physical Science degree on the topic of "Researches on Radioactive Substances" from the University of Paris. In November 1903 the Nobel Prize in physics was awarded to Henri Becquerel and to Pierre and Marie Curie for their joint research on the radiation phenomena. Marie Curie became the first woman to receive a Nobel Prize and the only woman Nobel laureate in science for many years, until her daughter Irène Joliot-Curie won the prize in 1935. A chair of physics was created for Pierre at the Sorbonne and he was given a laboratory with three assistants. Marie was given the job of chief of laboratory work, the first paid job for her research on radioactive elements. In October 1904 their second daughter, Eve, was born, who later wrote her mother's biography.

On April 19, 1906, Pierre was killed by a horse-drawn wagon while crossing the street. Marie suffered great grief and pain at the loss of her husband, closest companion, and best friend. Only by continuing the research they had begun together did Marie finally overcome this severe blow. She was offered Pierre's position to teach at the Sorbonne as an assistant professor and director of the laboratory. Even though she was a celebrated scientist and Nobel Prize–winner, Marie was not given Pierre's rank of full professor; but she was the first woman to teach at the Sorbonne. She was finally appointed full professor in 1908. In December 1910, Marie was nominated for membership in the nation's most prestigious scientific society, the Académie des Sciences, one of the five academies of the Institut de France. Although many of her colleagues supported her, she was rejected because, as one member insisted, "women cannot be part of the Institut de France."[3] However, Marie's contribution to science was recognized by the world. In December 1911 the Royal Swedish Academy of Sciences awarded her the Nobel Prize in Chemistry for the discovery and isolation of polonium and radium. No other laureate had ever received two Nobel prizes. In 1912 the University of Paris and the Pasteur Institute founded the Radium Institute, which opened in 1914. The founding of the Institute enabled Madame Curie to host outstanding young scientists from all over the world to do research in the Curie Laboratory at the Institute.

During World War I the use of X-rays in medicine was not well known. Marie Curie dedicated herself to this application. She served as the director of the Red Cross's Radiological Service and installed X-ray apparatus in many hospitals in Paris and on the western fronts. She also directed the transformation of a fleet of ordinary cars into a mobile X-ray service. Those Radiology Cars, called "little Curies," sped the process of locating the metal fragments and broken bones of the wounded and saved many lives.

The Curies believed that radium was an element and, thus, that it belonged to all people. They filed no patent and "published, since the beginning, without any reserve, the process of preparing radium."[4] While some industries made millions of dollars on radium, Marie Curie was still working in an inadequate lab with less than one gram of radium. In 1921, Missy Meloney, an American journalist, collected a "Marie Curie Radium Fund" in the United States and offered Madame Curie the gift of one gram of radium, worth $100,000, for scientific research. Marie Curie went to the United States to receive the gift and visited a number of universities and colleges, which bestowed honorary degrees on her.

In 1922, Marie Curie was elected to the Académie de Médecine for her contributions to radiological medicine. She became the first woman member in the 227-year history of the Institut de France. She spent the rest of her life doing research on the chemistry of radioactive materials and their medical applications. She also participated in many international activities in science in her later years. She prepared the first international standard of radium for the International Bureau of Weights and Measures and served on the International Committee on Intellectual Cooperation until her death. On July 4, 1934, Marie Curie died of leukemia as the consequence of life-long exposure to radium.

Notes

1. Marie Curie, *Pierre Curie with Autobiographical Notes*, trans. Charlotte and Vernon Kellogg (New York: Macmillan, 1923), p. 82.

2. Ibid., p. 91.

3. Eve Curie, *Madame Curie*, trans. Vincent Sheean (New York: Doubleday, Doran, 1937), p. 291.

4. Curie, *Pierre Curie with Autobiographical Notes*, p. 110.

Bibliography

Curie, Eve. *Madame Curie*. Translated by Vincent Sheean. New York: Doubleday, Doran, 1937.

Curie, Marie. *Pierre Curie with Autobiographical Notes*. Translated by Charlotte Kellogg and Vernon Kellogg. New York: Macmillan, 1923.

"Marie Curie," in *Nobel Prize Winners*. Edited by T. Wasson. New York: H.W. Wilson, 1987.

Pflaum, Rosalynd. *Grand Obsession: Madame Curie and Her World*. New York: Doubleday, 1989.

Quinn, Susan. *Marie Curie: A Life*. New York: Simon & Schuster, 1995.

Reid, Robert. *Marie Curie*. New York: Saturday Review Press, 1974.

JIE LI

INGRID DAUBECHIES
(1954–)
Physicist

Birth	August 17, 1954
1975	B.S., physics, Vrige University Brussels
1975–84	Research Assistant, Dept. for Theoretical Physics, Free University Brussels
1980	Ph.D., physics, Vrige University Brussels
1984	Louis Empain Prize for physics
1984–87	Assistant Professor, Dept. for Theoretical Physics, Free University Brussels
1987–94	Professor of Mathematics, Rutgers University (on leave after July 1993, returned full-time to AT&T Bell Laboratories for academic year 1993–94)
1992	MacArthur Foundation Fellowship
1993–	Member, American Academy of Arts and Sciences; Professor of Mathematics, Princeton University
1994	Steele Prize, American Mathematical Society

Ingrid Daubechies has been recognized internationally for her research in the area of wavelets. In 1992 the MacArthur Foundation named her a Fellow for her contributions to the study of wavelets. Her work has focused on the mathematical properties of wavelet transforms, for which wavelets form the basis. Wavelet transforms assist in the analysis of functions with transient high frequency phenomena, such as the sound of a scratch on a musical recording.[1]

"Wavelets can be thought of as building blocks," Daubechies explained in an interview. "The small blocks have fine features; bigger blocks can be described as having coarse features. The family of building blocks are the wavelets. The wavelets help bring complicated objects into simple terms by viewing it as something that's composed of many building blocks."[2] A prominent feature of Daubechies's work in the area of wavelets is its cross-disciplinary applications. "It can," she stated, "be a bridge between different communities. For example, both a mathematician and an engineer can understand it, appreciate the significance of it, and apply it to their respective fields."

Daubechies was born in Houthalen, Belgium, on August 17, 1954. As a young girl she had a strong interest in math and science. She attended all-girls schools through high school and was strongly motivated and determined to pursue a career. She was encouraged by her mother to be independent, and her father helped nurture her love of science.

Before entering college Daubechies intended to pursue a career in engineering, but at the last minute she decided she wanted to become a scientist. The areas of research and academics greatly interested her. Ever since childhood she had been fascinated by the idea that one can discover new things by thinking and experimenting—things no one else would know.

From the Free University Brussels, Daubechies earned a bachelor's degree in 1975 and a Ph.D. in 1980. Both degrees were in physics. From 1975 to 1984 she served as a research assistant in the department for theoretical physics at the Free University Brussels. In 1984 she won the Louis Empain prize for physics, an award presented every five years to a young Belgian scientist on the basis of work done before the age of 29. Daubechies received it for her work in quantum mechanics and quantization. In July 1984 she received tenure. She held this position until 1987, although she was on leave frequently.

At the age of 26, Daubechies came to the United States to do postdoctoral work. During this period in her life she faced many difficulties and challenges and came close to returning to Belgium. Characteristic of her determination, however, she viewed her difficulties as an incentive to try harder. In 1987 she became a member of the technical staff at the Mathematics Research Center, AT&T Bell Laboratories. Her work there involved mathematical research related to signal analysis. She worked at the Research Center until 1994.

During this time Daubechies gradually found herself switching from theoretical physics to applied mathematics, even though this was not the result of a conscious decision. Her physics work had led to a thorough study of certain mathematical techniques, which she found to be ideally suited for investigating problems in signal analysis. She found that she particularly enjoyed this type of crossover.

From 1991 to 1994 Daubechies served as a full professor in the mathematics department at Rutgers University. In 1993 she was appointed professor in the mathematics department at Princeton University. She was on academic leave from both university positions during the 1993–1994 academic year.

Daubechies is an excellent teacher who creates a genuine rapport with her students. She finds the challenge of developing mathematics courses for non-science majors very rewarding. In her teaching she emphasizes the importance of asking why and seeking answers for oneself. Daubechies also lectures and publishes extensively. She has served as an editor for several journals. She was elected a member of the American Academy of Arts and Sciences in 1993 and is listed in *American Men & Women of Science 1995–96*. She was awarded the American Mathematical Society's Steele Prize for Exposition in 1994. This award recognized her book, *Ten Lectures on Wavelets*. Daubechies is an outstanding scientist who has made lasting contributions to the fields of mathematics and physics. She is married to A. Robert Calderbank and has two children, Michael and Carolyn.

Notes

1. "Daubechies and Schneider among 1992 MacArthur Fellows," *Physics Today* 46, no. 1 (January 1995): 83 (2).
2. Ingrid Daubechies, interview with Anne Marie Perrault, 1995.

Bibliography

American Men and Women of Science, 1995–96. New York: R.R. Bowker, 1995.
Daubechies, Ingrid. *Ten Lectures on Wavelets*. Philadelphia: Society for Industrial and Applied Mathematics, 1992.
———. "The Wavelet Transform, Time-Frequency Localization and Signal Analysis." *IEEE Transactions on Information Theory* 36 (1990): 961–1005.
———. "Wavelets and Signal Analysis." Paper presented at the 1992 symposium on "Frontiers of Science" organized by the National Academy of Science.
"Daubechies and Schneider among 1992 MacArthur Fellows." *Physics Today* 46, no. 1 (January 1995): 83(2).

ANNE MARIE PERRAULT

ALLIE VIBERT DOUGLAS

(1894–1988)

Astrophysicist, Astronomer

Birth	1894
1912	Graduated in first place, Westmount High School, Montreal; awarded a McGill University entrance scholarship
1915–18	Chief of Women Clerks, Statistical Branch, Dept. of Recruiting, War Office, London, England

Allie Vibert Douglas. Photograph by William Notman & Son, Ltd.
Courtesy of Queen's University Archives, Allie Vibert Douglas
Fonds.

1918	Member of the Order of the British Empire (MBE)
1918–19	Registrar, Khaki University of Canada (England)
1920	B.A., first class honors, mathematics and physics, McGill University; Anne Molson Gold Medal
1921	M.Sc., physics, McGill University; Imperial Order Daughters of the Empire (IODE) War Memorial Overseas Scholarship
1921–22	Cavendish Laboratory at Cambridge (Newnham College)
1925	Research Assistant, Yerkes Observatory, WI

1926	Ph.D, physics, McGill University
1927–39	Lecturer, physics and astrophysics, McGill University
1939–59	Dean of Women, Queen's University
1939–60	Professor of Astronomy and Physics, Queen's University
1943–45	President, Royal Astronomical Society
1947–50	President, International Federation of University Women (IFUW)
1954	Delegate to the UNESCO Conference in Uruguay
1965	D.Sc., University of Queensland, Australia; LL.D., McGill and Queen's
1967	Officer of the Order of Canada; named "Woman of the Century" by the National Council of Jewish Women
1984	Distinguished Service Award, University Council of Queen's University
Death	July 2, 1988

Allie Vibert Douglas, a distinguished Canadian scientist, was dedicated to the advancement of scholarly research and to the improvement of women's status throughout the world. She was also the first woman in Canada to become an astrophysicist.

Born in Montreal in 1894, Allie first lived in London, England, with her Irish grandmother and older brother, George Vibert Douglas (both parents had died the same year she was born). Growing up in the custody of caring relatives, the two children returned to Montreal in 1904, where Allie completed her schooling at Westmount Academy, graduated in first place, and earned a McGill University entrance scholarship. Her education was thoughtfully supplemented by her two aunts, Mary and Mina, who frequented the public lectures held at the Natural History Society, McGill Physics Building, Theological College, and the Montreal Fine Arts Gallery.

Interested in science at a young age, Allie felt disadvantaged by being female. In a small high school club that was partially organized by her brother, she was refused admission because of her gender, a pattern that would be repeated in other places. George, however, secretly allowed her to listen to discussions of science and literature from the hallway by purposefully leaving the door ajar. This experience taught her early on that a woman must be innovative and determined if she wished to pursue science. Her brother, a sensitive and supportive person, understood but did not accept the social reality of women's inequality and became her main supporter during the difficult years ahead.

In 1912, Allie enrolled in honors mathematics and physics at McGill, but toward the end of her third year her studies were interrupted by the outbreak of war. George enlisted as an officer, having anticipated that his regiment would be stationed near London during that winter. He suggested that Allie, Mary, and Mina join him. This eventually became a four-year adventure as Allie was invited to participate in the war effort by (Sir) Auckland Geddes, a family acquaintance. She then served as chief of women clerks at the Statistical Branch, Department of Recruiting, War Office, London. In these dangerous and unsettled times, bombs fell precariously close to her lodgings and workplace, but she persevered, becoming the highest paid temporary woman civil servant in the National Service.[1] Her efforts were rewarded with a silver cross as Member of the Order of the British Empire (MBE).

After the Civil Service appointment came to a close, Allie was appointed registrar of the Khaki University of Canada in 1918, under Henry Marshall Tory.[2] This correspondence school offered courses in literacy, matriculations, and first- and second-year university for soldiers in Great Britain and France during demobilization. Allie also devised her own program, which consisted of auditing lectures at the Royal College in South Kensington, made possible by her McGill professors. In her spare time she visited wounded soldiers at Endell Street Hospital, attended cultural events in London, and explored historical sites with her aunts, even though Mary was seriously ill. That summer, Allie and Mina accompanied George on a trip to France and Flanders to view the battlefields, many of which he had known intimately. The experience of seeing "ravaged, desolate villages and bomb-pocked fields" affected Allie greatly and influenced her personal mission to promote world friendship and understanding.[3]

Although it was difficult to regain lost study habits, Allie finished her B.A. in 1920 with first class honors, earning the Anne Molson Gold Medal.[4] Feeling slightly out of touch with her younger classmates, she pursued many individual sports and organized and competed in the university's first swim meet for women. Her other extracurricular activities included public speaking and debating. The next year she enrolled as a graduate student in physics and worked as a demonstrator in the department under professors J. A. Gray and A. S. Eve. She studied the range of B-rays from an isotope of radium (E), for which she received an M.Sc. degree in 1921 alongside George, who had specialized in mining and geology.

Having been awarded a postgraduate fellowship sponsored by the Imperial Order Daughters of the Empire (IODE), Allie returned to England with Mary and Mina, whom she was reluctant to leave in Montreal. Having almost total responsibility for their welfare, Allie made suitable arrangements for their housing and daily care. Then she reported to the Cavendish Labs at Cambridge to begin work under Ernest Rutherford, whom she described as "gracious, affable and friendly at my first interview."[5] Interested in radioactivity, she was assigned a project that dealt with heat emission from a particular radium disintegration product, which brought her into contact with leaders in that particular field of research.

Although she followed all the prerequisites for a Ph.D., Allie became increasingly discouraged with her studies. Women at that time could obtain all the academic and resident requirements but still be denied a Cambridge degree. "I became more aware of my own limitations and more doubtful that I would ever make myself into a research physicist," she noted.[6] Unfair conditions often made concentration difficult. In a freezing cold lecture room at Trinity College, the only available heat came from a small stove beside the teacher's desk (the first two rows were reserved for male students; women were relegated to the back). On the advice of advisor Sir Geoffry Butler, Allie diversified her program and took courses in English, fine arts, and archaeology. An affiliation with Newnham College (for women) as a nonresident enabled her to associate with some "remarkably fine, scholarly women who were tutors and dons, as well as a few undergraduates."[7]

It was at the Observatory where Professors Arthur S. Eddington and W. M. Smart were conducting a course in practical astronomy and the use of telescopes that Allie's natural curiosity and talents prevailed. She happily resigned from Rutherford's lab and embarked on a statistical study of the relationship between stellar velocity and the absolute magnitude of a large group of stars. Providing material proof for a question that Eddington had already answered theoretically, Allie's path toward the Ph.D. now seemed possible. Her work, "which brought both proper motion and radial velocity conclusions into harmony (as predicted by Eddington)," was submitted to the spring meeting of the Royal Astronomical Society.[8] Concurrently, she analyzed the sedimentation rates of pelagic samples brought back from the Antarctic voyage *Quest*, on which George had signed as geologist. Rutherford had given her an empty room on the top floor of his building in which to work; between this and stellar spectra, she was occupied all winter. The results that Allie and George generated were highly regarded, and Rutherford presented their findings at the Royal Society of Edinburgh.[9]

Returning to McGill in 1923, Allie realized that "no one at McGill was really interested in my chosen field."[10] She continued as a lecturer and demonstrator and began a smaller study of cloud chamber patterns. She was much sought after as a speaker for scientific societies; in fact, one of her talks, entitled "The Chemistry of the Star," was so good that A. S. Eve remarked: "If you were a man I would say that was a damn fine lecture, you took them by storm."[11] Later, in 1925, Ludwig Silberstein of Eastman Kodak persuaded Allie to go to Yerkes Observatory in Wisconsin as a volunteer research assistant under E. B. Frost. There she developed an interest in spectroscopic absolute magnitudes of Class A stars, gaining access to the observatory's entire collection of I-prism spectrograms. This work made possible the determination of absolute magnitudes and parallaxes of approximately 200 Class A stars and led to her McGill doctorate in 1926.

Allie Douglas's career brought her into contact with astronomers **Helen Hogg** at the Dominion Observatory in Victoria, British Columbia, and **Annie J. Cannon** at Harvard. In 1954 at Princeton she met Albert Einstein, with whom she had a major disagreement. Whereas Einstein believed that "a scientist should not try to popularize his theories . . . [and] it is the duty of the scientist to remain obscure," Allie argued that "a scientist has a duty to try to educate the public at least to an appreciation of what the scientist is attempting to do."[12] This commitment to public education translated into one of her main activities. In addition to writing 12 technical papers, Allie published 24 articles in journals of the Royal Astronomical Society and authored over 30 popular ones, many for the *Atlantic Monthly*. Her best-known book, a biography of mentor and professor Arthur S. Eddington, was praised for its scientific accuracy and literary excellence.

Allie lectured in astrophysics at McGill for many years but was never offered a proper academic position. In 1939 she left for Queen's University to become dean of women and a full professor in astronomy and astrophysics. This was a difficult challenge, as another world war seemed inevitable and Kingston housed an important military barracks.

Sympathetic to the plight of displaced scholars in Europe after both wars, Allie believed that much good could be accomplished by women with university training. Having met the president of the IFUW in 1919, she joined the Canadian branch as a life member and subsequently was the only Canadian to serve as president of the organization. Her tenure in this position was from 1947 to 1950, a challenging postwar period.[13] As dean of women she helped make the path easier for younger women entering university and the sciences, particularly engineering.[14] Officially recognized as an outstanding scholar in 1965, she received an honorary D.Sc. from the University of Queensland, Australia, and LL.D.'s from McGill and Queen's. In 1967 the National Council of Jewish Women named her "Woman of the Century," and she was celebrated as an Officer of the Order of Canada.[15]

When a colleague offered to assist Allie with her belongings at a meeting in Kyoto in 1974, she firmly replied: "No thank you. When I am too old to carry my own suitcase, I shall be too old to attend conferences."[16] She was 79 at the time. Throughout her long and productive career, Allie was an avid and enthusiastic conference participant. She attended more general assemblies of the International Astronomical Union than any other Canadian astronomer; and with thirty-four meetings of the IFUW, numerous UNESCO tours, and various astronomy conferences, she traveled in all but six countries.[17] Whether it was urging a driver on through impossible roads in Ghana's interior or crossing the Khyber Pass three times in one day by bus, taxi, and lorry, nothing seemed to dampen her adventurous spirit. Indeed, whenever Allie Douglas's name is mentioned, the word "intrepid" is not far behind.[18]

She tied her hair in a bun, wore comfortable tweeds and sturdy walking shoes, and had a strong handshake. Her graceful manner led her many friends, colleagues, and students to believe her quite ageless. She enjoyed the company of her nieces and nephews and was a frequent visitor at their summer cottage, which had formerly belonged to her parents. In one of her last interviews in 1984, she insisted: "I can't think of anything more dismal than to have to earn your living doing something that was distasteful or didn't really call forth your enthusiasm or joy in doing it. It makes all the difference if you love what you're doing. I have been very lucky."[19] In 1988, astronomers named the recently discovered planet Vibert Douglas in her memory.

Notes

1. Allie Vibert Douglas, "Memoirs," [typescript] in Allie Vibert Douglas Papers, Archives, Queen's University Archives, Kingston, Ontario: Chapter 2, p. 14.

2. Ibid., p. 15.

3. Ibid., p. 22.

4. "Prize Winners at McGill," *Montreal Daily Star* (May 12, 1920): 5.

5. Douglas, "Memoirs," Chapter 3, p. 4.

6. Ibid., "Cambridge 1921–23," p. 1.

7. Ibid., Chapter 3, p. 5.

8. Ibid., "Cambridge 1921–23," p. 17.

9. Allie Vibert Douglas, "The Sizes of Particles in Certain Pelagic Deposits," *Proceedings of the Royal Society of Edinburgh* 43 (1922–1923): 16; Allie Vibert Douglas, *Deep Sea Deposits and Dredgings* (London: Trustees of the British Museum, 1930).

10. Douglas, "Memoirs," Chapter 4, p. 8.

11. Ibid., p. 5.

12. Lynn Messerschmidt, "Old Time Astronomer," *Whig Standard* [Kingston, Ontario] (December 15, 1984): 10–11.

13. Eileen Clark, "Dr. A. Vibert Douglas," *Canadian Federation of University Women Journal* 24, no. 9 (1988): 11.

14. Shirley Brooks, "Dr. A. Vibert Douglas, A Personal Memory," *Canadian Federation of University Women Journal* 24, no. 12 (1988): 3.

15. "Eleven Are Named Women of the Century," *Globe and Mail* [Ottawa, Ontario] (June 7, 1967): 11; "Dr. Vibert Douglas Named as 'Woman of the Century,' " *Whig Standard* [Kingston, Ontario], no. 158 (June 7, 1967): 23; Robert M. Stamp, *Canadian Obituary Record* (Toronto: Dundurn Press, 1988), pp. 46–47; A.H. Batten, "Douglas, Allie Vibert," in *Canadian Encyclopedia*, 2nd ed., Vol. 1 (Edmonton:

Hurtig Publishers, 1988), p. 614; "Queen's Astrophysicist Popular Dean of Women," *Globe and Mail* [Ottowa, Ontario] (July 5, 1988): A19.

16. Eileen Clark, "Dr. A. Vibert Douglas Remembered," *Canadian Federation of University Women Journal* 24, no. 12 (1988): 3.

17. Brooks, "Dr. A. Vibert Douglas, A Personal Memory," p. 3.

18. Peter M. Millman, "Allie Vibert Douglas, 1894–1988," *Journal of the Royal Astronomical Society of Canada* 82, no. 6 (December 1988): 309–311.

19. Messerschmidt, "Old Time Astronomer."

Bibliography

Douglas, Allie Vibert. "Between the Stars." *Atlantic Monthly* 147 (1931): 75–79.
——. "From Atoms to Stars." *Atlantic Monthly* 144 (1929): 158–165.
——. *The Life of Arthur Stanley Eddington*. London: Thomas Nelson & Sons, 1956.
——. *Meteors*. Toronto: University of Toronto Press, 1932.
——. *The 1932 Total Solar Eclipse*. Toronto: University of Toronto Press, 1932.
——. "Other Little Ships." *Atlantic Monthly* 136 (1925): 169–174.
——. "A Tool, Not a Creed." *Atlantic Monthly* 170 (1942): 73–74.
——. "Two Sciences." *Atlantic Monthly* 140 (1927): 488–491.
Eve, A.S. *The Universe as a Whole; What We Know about the Stars*. Toronto: University of Toronto Press, 1932. [A.V.D. was a contributor.]
Foster, John S. *Stack Effect in B Stars*. Edinburgh: Neill, 1939. [A.V.D. was a contributor.]

E. TINA CROSSFIELD

HELEN M. DYER

(1895–   )

Biochemist

Birth	May 26, 1895
1917	B.A., biology, Goucher College
1919–20	Instructor in Physiology, Mt. Holyoke College
1920–28	Research Assistant, Pharmacologist Hygienic Lab., U.S. Public Health Service
1929	M.S., biochemistry, George Washington University
1929–35	Teaching Fellow, George Washington University
1935	Ph.D., biochemistry, George Washington University

1935–42	Assistant Professor of Biochemistry, George Washington University
1942–65	Chemist, National Cancer Institute
1954	Achievement and Service Award for Teaching and Research, Goucher College
1958	Alumni Achievement Award for Biochemical Research in the Field of Cancer, George Washington University
1962	Garvan Medal, American Chemical Society
1965–67	Research Biochemist, Life Sciences Research Office, Federation of American Societies for Experimental Biology

Helen Dyer was a pioneer in cancer research. She is best known for her research on the mechanisms of carcinogenesis, although she also made major discoveries in the fields of metabolism and nutrition.[1] For example, she made a major discovery in 1938 when she synthesized the ethyl analog of methionine as a possible substitute for methionine, an amino acid used both in medicine and as a food supplement. She discovered later, however, that this analog was toxic and that it failed to replace methionine. This finding had an impact in the field of medicine, especially on sulfa drugs.[2]

Helen Dyer was born in 1895 in Washington, D.C., to Joseph E. and Florence (Robertson) Dyer. She received a B.A. in biology from Goucher College in 1917. In 1919–1920, Dyer went to Mt. Holyoke College to teach physiology. While teaching there, she took extra chemistry courses. In 1920–1928, she was employed by the Pharmacologist Hygienic Lab, U.S. Public Health Service, in Washington as a research assistant.[3] Dyer earned her master's degree in biochemistry from George Washington University in 1929. From 1930 to 1935, she held a teaching fellowship at George Washington medical school, where she worked with Dr. Vincent du Vigneaud, a future Nobel laureate (1955), on sulfur compounds while pursuing her doctoral degree. She received her doctorate in biochemistry from George Washington University (GWU) in 1935. Dyer taught biochemistry and the chemistry of nutrition for a total of twelve years at GWU before joining the National Cancer Institute in 1942. Her students admired her thorough knowledge of practically every branch in biochemistry.[4]

In 1942, Dyer left her teaching position to work for Carl Voegtlin, then head of the National Cancer Institute. There she was to continue the experimentation on dogs to study gastric cancer. She also studied vitamin B_6 and its antimetabolite, which itself causes cancer in certain animals.[5] Dyer was the first woman chemist to discover that animals ingesting N-2-fluorenylacetamide, a potent liver carcinogen, required increased amounts of vitamin B_6 to prevent increased excretion of xanthurenic acid and other abnormal metabolites of tryptophan, a crystalline amino acid. Dyer's findings stimulated extensive studies of the role of vitamin B_6 in preventing N-2-fluorenylacetamide carcinogenesis.[6] In 1949, Dyer compiled the first comprehensive index of tumor chemotherapy, which served as one of the most essential aids to the National Cancer Institute in the development of its cancer chemotherapy program. She has been author or co-author of more than sixty published technical articles.[7]

During her tenure at the National Cancer Institute, Dyer collaborated with many scientists and did numerous studies on tumors and enzymes of liver cancers. Through these studies she has made significant contributions to the field of cancer research.[8] Dyer has received numerous awards and recognitions for her work. In 1962 she was awarded the Garvan Medal of the American Chemical Society. (The Garvan Medal, established in 1937, is awarded for outstanding achievement to one American woman chemist annually.) In 1961, Goucher College awarded her an honorary Doctor of Science degree for her distinguished service to the field of chemistry. Among her other awards were the Goucher College Achievement and Service Award for Teaching and Research in 1954, and the Alumni Achievement Award for Biochemical Research in the Field of Cancer from George Washington University in 1958.[9]

It should be noted that although women scientists who were employed in government, industry, and academia from the 1920s to the 1940s encountered many obstacles and faced discrimination in their places of employment, they continued to make major discoveries and contributions in their fields. Helen Dyer was no exception—she made key discoveries in biochemistry, metabolism, and nutrition throughout her career of four decades. Like most other women scientists, she received less prominence in the press and experienced discrimination that prevented her from being promoted to higher ranks. Women scientists who worked for governmental agencies received lower salaries than their male counterparts. This was especially true for single women who worked in the National Institutes of Health.[10] Similarly, in academia most women scientists (including Dyer) did not have the opportunity to move up in academic ranks. Despite her active research and significant publications, Helen Dyer was still an assistant professor at age 46.[11]

Dyer continued to work actively in the field of cancer causation after her retirement in 1965. She has been a fellow of the American Association for the Advancement of Sciences (AAAS) and a member of scientific societies such as the American Chemical Society, Sigma Xi, Sigma Delta Epsilon, Iota Sigma Pi, the Society of Experimental Biology and Medicine, the American Association for Cancer Research, and the American Association of Biological Chemistry.[12]

Notes

1. *The Women's Book of World Records and Achievements*, ed. Lois Decker O'Neill (Garden City, N.Y.: Anchor Press/Doubleday, 1979), p. 220.

2. "Garvan Medal," [portrait] *Chemical and Engineering News* 40 (March 19, 1962): 102.

3. *Who's Who of American Women 1970–71*, 6th ed. (Chicago: Marquis Who's Who, 1970–1971), p. 350.

4. Nina M. Roscher and Chinh K. Nguyen, "Helen M. Dyer, A Pioneer in Cancer Research," *Journal of Chemical Education* 63 (1986): 253–255.

5. Roscher and Nguyen, "Helen M. Dyer, A Pioneer in Cancer Research," p. 254.

6. "Garvan Medal," p. 102.

7. Roscher and Nguyen, "Helen M. Dyer, A Pioneer in Cancer Research," pp. 253–254.

8. "Garvan Medal," p. 102.

9. Roscher and Nguyen, "Helen M. Dyer, A Pioneer in Cancer Research," pp. 254–255.

10. Margaret W. Rossiter, *Women Scientists in America: Struggles and Strategies to 1940* (Baltimore: Johns Hopkins University Press, 1982), p. 234.

11. Molly Gleiser, "The Garvan Women," *Journal of Chemical Education* 62 (1985): 1065–1068.

12. Roscher and Nguyen, "Helen M. Dyer, A Pioneer in Cancer Research," p. 255.

Bibliography

Dyer, Helen M. *An Index of Tumor Chemotherapy: A Tabulated Compilation of Data from the Literature on Clinical and Experimental Investigation.* Washington, D.C.: Federal Security Agency, Public Health Service, 1949.

"Garvan Medal." *Chemical and Engineering News* 40 (March 19, 1962): 102.

Gleiser, Molly. "The Garvan Women." *Journal of Chemical Education* 62 (1985): 1065–1068.

Roscher, Nina M., and Chinh K. Nguyen. "Helen M. Dyer, A Pioneer in Cancer Research." *Journal of Chemical Education* 63 (1986): 253–255.

Rossiter, Margaret W. *Women Scientists in America: Struggles and Strategies to 1940.* Baltimore: Johns Hopkins University Press, 1982.

Who's Who of American Women 1970–71. 6th ed. Chicago: Marquis Who's Who, 1970–1971.

The Women's Book of World Records and Achievements. Edited by Lois Decker O'Neill. Garden City, N.Y.: Anchor Press/Doubleday, 1979.

MAY MONTASER JAFARI

HELEN T. EDWARDS

(1936–)

Physicist

Birth	May 27, 1936
1957	B.A., Cornell University
1966	Ph.D., Cornell University
1966–70	Research Associate, Cornell's Laboratory for Nuclear Studies, Ithaca, NY
1970–71	Head, Booster Group, Fermi National Accelerator Laboratory, Batavia, IL
1975–79	Head, Switchyard Extraction Group and Leader, Tevatron Design Group, Fermi National Accelerator Laboratory
1980–88	Deputy Head, SAVER Division, Fermi National Accelerator Laboratory
1986	Ernest O. Lawrence Award, U.S. Department of Energy
1988	MacArthur Fellowship
1989–92	Head and Associate Director, Superconducting Division, Superconducting Supercollider Laboratory, Dallas, TX
1992–	Guest Scientist, Fermi National Accelerator Laboratory

"It has never been a one woman show," said Dr. Helen T. Edwards, who works as a high energy physicist for Fermi National Accelerator Laboratory (Fermilab) in Batavia, Illinois. "To accomplish the kinds of projects in which I have been involved requires the supervision and participation of many talented individuals."[1] It is her study of physics, perhaps, that gives Edwards a perspective grounded in realism. She is not one to romanticize science, but she gravitated toward this area of study at a very young age as she grappled with her weaknesses and attempted to discover where her strengths and talents lie. This pursuit led her to a career in physics and the opportunity to coordinate the construction of the Tevetron particle accelerator, the most powerful proton accelerator in the world.

Helen Edwards was born in 1936 in Detroit, Michigan. "In school, grade school, and junior high, I was lousy at spelling and lousy at reading," she said. "It took me forever to read, but I found I enjoyed mechanical tasks, figuring out how things were put together and how they

Helen T. Edwards. Photo courtesy of Donald Sena, Public Affairs, Fermilab.

worked; I also enjoyed nature, biology, and the natural sciences." Edwards spent her childhood days in the comfort of Detroit suburbia and then attended Madeira High School, an all-girls school outside of Washington, D.C. The youngest of a family not at all involved in science and research, she staked out a career path that was independent of the interests of her parents and siblings. Her father worked in the investment and business world. Her one brother pursued interests in literature and newspaper work. Out of a family of five children, only one of her sisters shared similar technical interests and earned a degree in architecture.

Edwards recalls having applied to the top engineering schools in the country by the time her high school graduation took place—MIT, Duke, and Cornell, among others. Up to this point she was fortunate not to have encountered much gender-based bias toward her pursuit of the sciences, for she had attended a high school in which all students in the science classes were female and her two greatest mentors and role models, her mathematics and biology teachers, were also female. But in the early 1950s the American universities presented a very different environment. "My parents really didn't want me to go to MIT because of the isolation they feared I would experience. MIT engineering and physics programs were practically 100 percent male. They wanted me to go to Cornell, which had a few more female students," Edwards said. Of the fifteen physics students in her class at Cornell, Edwards recalls there being about two other female students besides herself. And although she remembers experiencing at times the unspoken but pervasive attitude that "educating women in the sciences was a waste of time because they would choose marriage over a research career," Edwards expresses nothing but praise for her professors' willingness to help each student succeed: "Those guys were immensely helpful," she recalled. Dr. B. D. McDaniels, whom Edwards met while she was an undergraduate, would later become her thesis advisor as she worked toward her doctorate; to this day, she considers him a close colleague and long-time friend in the profession.

Falling in love with the beauty of Ithaca, New York, the town in which Cornell University is situated, and realizing the tremendous opportunity that lay before her, Edwards decided to stay at Cornell for graduate studies after completing her bachelor's degree in 1957. It was an exciting time for physicists. The work at Los Alamos during World War II had infused the field with new possibilities and new research in the areas of quantum and particle physics. Cornell had already acquired an international reputation for its pioneering work in the construction of particle accelerators. Edwards received her Ph.D. in 1966 and then served as a research associate at Cornell's Laboratory for Nuclear Studies for the next four years. In 1967 she was primarily responsible for commissioning Cornell's 12–GeV electron synchrotron. The laboratory director at Cornell, Dr. Robert Wilson, became another important colleague in her life, for he later headed the staff at Fermilab and invited Edwards to her next professional position. It was at Fermilab that Edwards advanced her career to the leading edge of accelerator technology, having held various technical and administrative positions within the laboratory.

As scientists and researchers began to realize the theoretical implications of being able to understand the nature of subatomic particles, the demand greatly increased for new and better data about the elusive behavior of these fundamental constituencies of the atom. The significance of Edwards's work resides in the fact that it has provided the instrumentation for collecting these valuable data, which are derived from the analysis of subatomic particles after they have been accelerated to great speeds and then collided with one another or with stationary obstacles.

These collisions can release great amounts of energy and cause all types of changes to occur to the structure of the particles. At Fermilab, Edwards was instrumental in commissioning the 400–GeV main accelerator and also led the switchyard/extraction commissioning effort for the 400–GeV accelerator. (The GeV is a unit of measurement for the energy level of accelerated particles equivalent to a billion electron volts.) In 1987, Edwards was one of the supervisors to oversee the completion of the world's highest energy superconducting particle accelerator, referred to as the Tevatron. This accelerator can produce an energy level of 1 TeV (equivalent to 1,000 GeV) as it collides protons and antiprotons moving in opposite directions. The field of particle acceleration has come a long way, Edwards points out: "The first accelerators developed were no more than a few feet in length, and now they span distances measured in kilometers."

The Director of Fermilab, Dr. Leon Lederman, had this to say in a press release about Edwards's extensive contribution toward the completion of the Tevatron:

> It is difficult to select any single subset within the entire Tevatron construction project that Dr. Edwards was not intimately involved with in a creative and technical manner. For illustration we can divide the effort into the following: superconducting magnet production, magnet testing, cryogenics (in turn divided into Central Helium Liquefier and 24 satellite refrigerators at 600W each), radio frequency, vacuum systems, beam diagnostics, controls, lattice design and orbit theory, magnet parameter acceptance criteria, magnetic correction elements, power supplies, reporting, documentation and cost control, quench protection, extraction system, injection systems and beam transport, beam-abort systems, radiation protection and installation.[2]

This lengthy and technical list of responsibilities demonstrates just how complicated and complex is the construction of a particle accelerator, and Lederman acknowledges the importance of Edwards's role as "group leader, (one of the) chief designer(s) as well as overall project coordinator."[3] For her work on the Tevetron, Edwards received the U.S. Department of Energy's Ernest O. Lawrence Award in 1986; and in 1988, she was awarded a John D. and Catherine T. MacArthur Fellowship. In 1989, Edwards was a co-recipient of the President's National Medal of Technology. In addition, she is a fellow of the American Physical Society and a member of the National Academy of Engineering.

Although Edwards works in a very technical capacity, if you ask her what she believes to be her greatest contribution, she answers by talking about the importance of people and cooperation. "The most important part of what I do is getting people to work together and to coordinate their efforts with one another." Only through the power of teamwork can the miracle of modern science express itself, and Edwards often notices the indirect results of her kind of research surfacing in many modern products and applications—especially in the medical field with the new CAT scan and magnetic resonance imaging technologies. When asked if she would choose the same career path over again, she hesitates only slightly, then answers, "Yes. Probably, but I think it's getting harder and harder today to conduct new research with so many bureaucratic obstacles and high costs in the way." Helen Edwards is, however, one to persevere. She has already overcome a tremendous number of obstacles by rising to her position in the field of physics. Her contributions are extraordinary and well acknowledged.

Notes

1. Unless otherwise noted, all quotations in this essay are taken from a telephone interview with Helen Edwards by Darin Savage on August 25, 1995.

2. Leon Lederman, "Fermilab press release on Helen Edwards," June 1, 1987, n.p.

3. Ibid.

Bibliography

Edwards, H. "The Tevatron Energy Saver." *Physics Today* 38 (January 1985): S36–S37.

Elliot, M. "SSC's Culture Clash." *Science* 251 (March 29, 1991): 1551.

"Five Honored for Particle-Accelerator Advances." *Physics Today* 39 (December 1985): 92.

"Five Physicists Receive Lawrence Awards (James J. Duderstadt, Helen T. Edwards, Joe W. Gray, G. Bradley Moore, and James L. Smith)." *Physics Today* 39 (October 1986): 137–138.

DARIN C. SAVAGE

GERTRUDE BELLE ELION

(1918–　)

Biochemist

Birth	January 23, 1918
1937	A.B., chemistry, Hunter College

1941	M.S., Chemistry, New York University
1942	Food Chemistry Analyst, Quaker Maid Co.
1943	Research Chemist, Organic Chemistry, Johnson & Johnson
1944–83	Burroughs Wellcome Co.: Research Chemist, 1944–63; Assistant to Director, Chemotherapy Division, 1963–67; Head, Experimental Therapy, 1967–83
1968	Garvan Medal, American Chemical Society
1983	Emerita Scientist, Burroughs Wellcome Co.; Judd Award, Sloan-Kettering Institute
1984	Cain Award, American Association of Cancer Research
1985	Distinguished North Carolina Chemist Award, American Chemical Society
1988	Nobel Prize in Physiology/Medicine
1990	Medal of Honor, American Cancer Society
1991	National Medal of Science; National Inventors Hall of Fame; National Women's Hall of Fame
1992	Engineering and Science Hall of Fame
1995	Foreign Member, The Royal Society

The drug research carried out by Gertrude Elion has saved the lives of thousands of people. Despite discrimination against her in academia and the workplace because of her gender, Elion's hard work and determination enabled her to discover new drugs that saved lives. In 1988, Elion won the Nobel Prize for her contributions to medical research.

Gertrude Elion was born in New York, New York, in 1918. Her father, Robert Elion, immigrated from Lithuania when he was 12 years old. By working as a drugstore assistant, he saved enough money to attend dental school at New York University. After graduating in 1914, he ran a successful dentistry practice and invested in real estate in the city. Her mother, Bertha Cohen, came alone to New York from an area of Russia (now Poland) at age 14. She attended night school to learn English and worked as a seamstress during the day. She married Robert when she was 19. Both of Gertrude's parents came from scholarly Jewish families that included generations of rabbis. She had a close relationship with her maternal grandfather, who arrived from Russia when she was 3 years old. Gertrude and her brother Herbert, who was born in 1923, were gifted scholastically. Herbert eventually became the owner of a bioengineering and communications engineering company.[1]

Robert Elion lost most of his investments in the stock market crash of 1929. As a result of his bankruptcy, Gertrude's opportunities for higher education were limited. Although it was extremely competitive, the City College of New York offered free tuition. She applied and was accepted into Hunter College (for women). Elion's decision to study chemistry was influenced by the death of her grandfather from stomach cancer. While visiting him in the hospital, she became convinced that she wanted to work toward a cure for cancer.[2]

Elion graduated from Hunter College with highest honors in chemistry in 1937. Despite her high academic achievement, however, she was turned down for graduate assistantships at fifteen graduate schools. Discouraged from seeking a position in a laboratory because she was told that as a woman she might distract the other chemists, Elion took secretarial classes for a short time. In 1938 she volunteered in a chemistry lab, eventually earning a small salary for her work there. She saved enough money to attend graduate school for one year.[3]

As a way to pay for a master's degree at New York University, Elion worked as a substitute teacher in New York City public high schools for two years. She then found a position with the quality control laboratories of A&P analyzing fruit, vanilla beans, and pickles.[4] Another position with Johnson & Johnson ended after six months because the lab she was working in closed. In 1944, Elion called the Burroughs Wellcome Company to inquire if they had a research laboratory. This call resulted in an interview, which landed her a job with that company at the age of 26.[5]

George Hitchings hired Gertrude Elion and collaborated with her for the next thirty-two years. Hitchings had earned his Ph.D. in biochemistry at Harvard University[6] and was working to develop a scientific approach to the discovery of new drugs that could replace the trial and error approach, which dominated drug research at the time.[7] Elion was one of the first researchers to work on the antimetabolites of purines, the building blocks of nucleic acid.[8] Her research led to the publication of more than 225 papers over the course of her career.[9] The research by Hitchings and Elion fulfilled the philosophy of the founders of Burroughs Wellcome: that the company focus on developing drugs for diseases for which there was as yet no cure. The scientists studied leukemia, malaria, bacterial infections, herpes, gout, and the rejection of transplanted organs.[10]

In 1950, Elion synthesized the purine compound that hindered the multiplication of leukemia cells. Her discovery of the drug known as 6-MP dramatically changed cancer therapy and research. Elion used each discovery as a stepping stone. A derivative of 6-MP, called Imuran, allowed successful organ transplantation and is still used to prevent rejection of kidney transplants. Her development of Allopurinol prevented kidney blockage in cancer patients treated with radiation and chemotherapy, as well as in gout patients; this has saved thousands of lives.[11]

Elion became head of the Department of Experimental Therapy at Burroughs Wellcome in 1967. When forced to choose between a job or a full-time Ph.D. program in 1950, Elion chose to continue working at Burroughs Wellcome.[12] In 1968 she won the Garvan Medal from the American Chemical Society for her research in drugs for chemotherapy. Continued research in related purine compounds led to the discovery of the drug Acyclovir in 1974. This drug is used successfully to treat herpes virus infections.[13]

One year after Elion retired in 1983 her research group introduced Azidothymidine, or AZT. AZT is used to treat the AIDS virus. In 1984, Elion received the Cain Award from the American Association of Cancer Research. The following year, she received the Distinguished North Carolina Chemist award from the American Chemical Society. In 1988, Elion received the Nobel Prize in Physiology or Medicine. She shared the award with George Hitchings and Sir James Black of the University of London. All three were recognized for contributing to a new method of drug development for a variety of diseases.

In 1990, Elion was awarded the Medal of Honor from the American Cancer Society. In 1991 she received the National Medal of Science and was elected to the National Academy of Sciences. Burroughs Wellcome gave Hitchings and Elion each $250,000 to donate to a charity of their choice in recognition of their contributions to science. Elion gave the money to Hunter College for scholarships for women in chemistry and biochemistry.[14] Also in 1991, she was elected to the National Inventors Hall of Fame, the first woman to be so honored.[15]

The greatest reward for Elion is hearing from the many patients whose lives have been saved through the drugs she discovered.[16] In retirement Elion keeps an active schedule, traveling, writing, teaching at Duke University School of Medicine, and serving on the boards of many cancer and health-related organizations.[17]

Notes

1. Sharon Bertsch McGrayne, *Nobel Prize Women in Science* (New York: Carol Publishing Group, 1993), pp. 285–286.

2. Ibid., p. 287.

3. Ibid., p. 288.

4. Katherine Boutoin, "Nobel Pair," *New York Times Magazine* 137 (January 29, 1989): 28.

5. McGrayne, *Nobel Prize Women,* p. 289.

6. Gertrude Elion, personal communication with Marilyn Parrish.

7. McGrayne, *Nobel Prize Women,* p. 291.

8. Gertrude Elion, personal communication with Marilyn Parrish.

9. McGrayne, *Nobel Prize Women,* p. 291.

10. Boutoin, "Nobel Pair," p. 28.

11. McGrayne, *Nobel Prize Women,* p. 296.

12. Marguerite Halloway, "The Satisfaction of Delayed Gratification: Profile of Gertrude Belle Elion, 1988 Winner of the Nobel Prize for Physiology/Medicine," *Scientific American* 265, no. 4 (October 1991): 40.

13. McGrayne, *Nobel Prize Women*, p. 300.

14. Ibid., p. 303.

15. Halloway, "The Satisfaction," p. 44.

16. McGrayne, *Nobel Prize Women*, p. 303.

17. Halloway, "The Satisfaction," p. 44. The author would like to thank Gertrude Elion for her kind assistance in providing information and reviewing this essay.

Bibliography

American Men and Women of Science, 1995–1996. New Providence, N.J.: Bowker, 1995.

Boutoin, Katherine. "Nobel Pair." *New York Times Magazine* 137 (January 29, 1989): 28.

Elion, G.B. "The Biochemistry and Pharmacology of Purine Analogs." *Federation Proceedings* 26 (1967): 898–904.

———. "The Purine Path to Chemotherapy." *Science* 244 (1989): 41–47.

Elion, G.B., et al. "Selectivity of Antiherpetic Agent, 9–(2–Hydroxyethoxymethyl) Guanine." *Proceedings of the National Academy of Sciences* 74 (1977): 5716–5720.

Halloway, Marguerite. "The Satisfaction of Delayed Gratification: Profile of Gertrude Belle Elion, 1988 Winner of the Nobel Prize for Physiology/Medicine." *Scientific American* 265, no. 4 (October 1991): 40.

McGrayne, Sharon Bertsch. *Nobel Prize Women in Science*. New York: Carol Publishing Group, 1993.

MARILYN MCKINLEY PARRISH

SANDRA MOORE FABER

(1944–   )

Astronomer

Birth	1944
1966	B.A., physics, Swarthmore College
1967	Married Andrew Faber
1972	Ph.D., astronomy, Harvard University
1972–77	Assistant Professor/Assistant Astronomer, Lick Observatory, University of California, Santa Cruz
1977–79	Associate Professor/Associate Astronomer, Lick Observatory, University of California, Santa Cruz

1979–	Professor/Astronomer, Lick Observatory, University of California, Santa Cruz
1985	Elected to National Academy of Sciences
1986	Heineman Prize, American Astronomical Society
1991	Darwin Lecturer, Royal Astronomical Society
1993	NASA Group Achievement Award, Hubble Space Telescope Wide-Field Planetary Camera Team

Sandra Faber is known for her research on the origin of the universe and of galaxies in particular. She has discovered scaling laws—correlations between galaxies' features that enable astronomers, having measured some features, to predict others. One, known as the Faber-Jackson law, holds that bigger elliptical galaxies have stars that are orbiting more rapidly. Faber's work has also been influential in convincing astronomers that much of the matter in the universe comes in the form of massive, invisible halos surrounding galaxies and that this "cold, dark matter" has played a determining role in the origin and development of galaxies and clusters of galaxies. During the 1980s, Faber joined in a collaboration that has since become famous as the Seven Samurai. The Seven Samurai found irregularities in the Hubble flow of galaxies and concluded that matter in the universe is clumped into immense concentrations that, by their gravitational attraction, perturb the smooth expansion of the universe.

Born in 1944, Sandra Moore decided early that her mother's domestic life seemed passive and dull compared to the more active, exciting life of her father, a World War II veteran, former civil engineer, and insurance executive. Unhappy at being barred from Little League and the Cub Scouts, she turned to collecting rocks and watching spiders. Soon she discovered that science was even more fun than baseball and that academic excellence was a smart girl's best revenge for exclusion. Her parents and teachers were supportive. She majored in physics and minored in mathematics and astronomy at Swarthmore College, then entered graduate school in astrophysics at Harvard in 1966. "The educational establishment seemed to me to be working actively in my behalf," she says; "I am an example of *how the system is supposed to work*, and for that I feel extremely lucky."[1] The first career compromise she made was one of her own choice, not one forced upon her: she went to Harvard for graduate school, even though at the time it did not offer the best opportunities in observational astronomy, because it was much closer to her future husband, Andrew Faber, who was still an undergraduate at Swarthmore. "That was," she has written in a recent autobiographical sketch, "the first of many career compromises that the two of us have made to stay together, and I have never regretted any of those choices."[2]

Her thesis on the spectrophotometry of elliptical galaxies got off to a bad start. In addition to the usual graduate student problems of refining a topic and coping with unreliable instrumentation, she had the terrible luck to fall off the observer's deck of the telescope on the very first night. She suffered a concussion and permanently injured her lower back. However,

> [t]he result was the first homogenous body of spectral data on elliptical galaxies showing both colors and absorption-line feature strengths. Sandra noticed that there was a relation between the spectrum and the size of the galaxy—big ellipticals were red and had strong absorption features, while small ones were blue and had weak absorption features. This was the first of many "scaling laws" for elliptical galaxies, some discovered by Sandra and many by others. These laws hold the clues to the formation of ellipticals, and explaining them still occupies much of Sandra's energy nearly 30 years later.[3]

In 1968, the Fabers, opposed to the Vietnam War, moved to Washington, D.C., so that Andrew could take a job doing underwater acoustical research at the Naval Research Laboratory. Sandra, still in graduate school, "became good at bumming—desk space, library privileges, and computer time," but she missed contact with other astronomers working on galaxies.

> Finally she settled in "paradise," at the Department of Terrestrial Magnetism, a branch of the Carnegie Institution of Washington in northwest Washington, D.C. At DTM's beautiful campus she punched the 75,000 computer cards needed to reduce her thesis photometry and rubbed shoulders with astronomers Vera Rubin and Kent Ford. These two, more experienced and highly imaginative, were inventing the observational study of dark matter in the Universe (though they didn't know it at the time) and also of irregularities in the expansion of the Universe. At DTM, Sandra absorbed the importance of forging boldly into unknown territory while at the same time maintaining total scientific integrity. Save for "thesis angst," it was a wonderful time.
>
> By great good luck, Sandra was finishing her thesis at just the time that Andy was ready to make a career change. He chose law, which, though neither realized it then, would make the future task of finding compatible career positions much easier. In another stroke of remarkably good luck, Andy was accepted to Stanford Law School, and Sandra got an Assistant Professor's post at the institution of her choice, the famous Lick Observatory at UC Santa

Cruz. Santa Cruz would prove a wonderful place to raise children, and the Observatory welcomed her with open arms—not a trace of discrimination did she feel as the first female staff member at Lick in its hundred-odd year history. The tenure-track Professorship freed her from the postdoc rat-race that young astronomers have to suffer today—another blessing.

This breather was needed, as Sandra, it turned out, was pregnant on her first day at work. Having postponed children in graduate school because she doubted her ability to have a family and a thesis at the same time, she and Andy decided that the time had come. She was 27. Looking back on this period, it was a good thing she had a tenured position, because it was *three years* before she published her next research project. Such a hiatus today would kill a promising career. During that time she had a baby, taught four new courses on subjects that she didn't know too well, and laid the ground-work for that all-important next paper. Fortunately it was a good one.[4]

The result of state-of-the-art spectroscopic observation done at Lick in collaboration with graduate student Robert Jackson was the paper that announced the Faber-Jackson scaling law mentioned above.[5]

The next few years saw an explosion of new detector technology, which Faber applied to the observation of hot stars and interstellar gas in galaxies. Having previously been mainly an observer, Faber began also to publish influential theoretical papers about the physical processes that might underlie the empirical relationships she and her colleagues were uncovering. In a paper coauthored with Jay Gallagher, Faber argued that galaxies are surrounded by massive halos.[6] This matter is invisible, but its presence is betrayed by its effect on the motion of matter that can be seen. (For a discussion of related work by Faber's mentor, Rubin, see the entry on **Vera Rubin** in this volume.) In two papers, one solo and one coauthored with Santa Cruz colleagues Joel Primack and George Blumenthal, Faber presented and developed the idea that this unseen matter has been responsible for the very formation of galaxies: small ripples in the density distribution of matter early in the history of the universe have resulted in hierarchical gravitational clustering, producing galaxies and clusters of galaxies with the properties we observe today.[7] "Though probably wrong in certain details," she writes, "the paper still stands as the current working paradigm for structure formation in the Universe."[8]

Faber's next major activity, the Seven Samurai (7S) collaboration, is now famous inside and outside the field of astronomy for identifying the Great Attractor, the nearest huge supercluster of galaxies. It is so big that the Local Supercluster containing our galaxy is just one of its suburbs, and its gravitational attraction accelerates everything in an immense region of space.

Faber describes the Samurai as:

> a lot milder than their name implies—a group of six observers and one theorist who set out, like Don Quixote, on a hopelessly misguided quest that ultimately yielded pay dirt. This is a recurrent theme in Sandra's career—she starts out doing one thing for good reasons, but reality turns out a very different way. In this case, the 7S thought they had discovered a way to determine the true shapes of individual elliptical galaxies, and they needed a homogenous catalog of accurate galaxy sizes and orbital speeds to properly calibrate their method. They set out on an eight-year project to collect data on 400 galaxies. When the dust settled, the original plan was dead—it simply did not work—but instead they had discovered a way to estimate the *distance* to every galaxy. With these distances they made a map of all the elliptical galaxies around us in space and noticed that the recessional speeds of galaxies (due to the expansion of the Universe) were not exactly as predicted from a smooth and uniform Hubble law. Rather, large patches of the Universe were moving away from us too slowly, while others were going too fast. The 7S had rediscovered what Rubin and Ford had announced years earlier: large mass concentrations perturbing the expansion of the universe in their neighborhood, pulling local matter into them and causing *irregularities in the Hubble flow*. This phenomenon has since matured to become one of the best ways to measure total mass density of the Universe, and hence whether it will continue to expand forever or one day recollapse. Sandra continues to work on this subject, acting as a data aide to other astronomers who are mathematically more gifted than she, since that is what the subject now requires.[9]

Faber has also been deeply involved in planning, securing funding for, and operating a number of significant astronomical instruments. She was one of the astronomers who started the project to build the 10–meter Keck telescope, currently the largest optical telescope in the world. She was part of the team responsible for the Wide-Field Planetary Camera for the Hubble Space Telescope, and she helped diagnose the spherical aberration of the telescope's famously flawed mirror and design the refurbishment plan that so triumphantly fixed it. She is currently working on the DEIMOS spectrograph for the second Keck telescope, now under construction.

Sandra Faber has received many honors: the Heineman Prize of the

American Astronomical Society; an honorary D.Sc. from Swarthmore College; distinguished lectureships at institutions including Lawrence Livermore Laboratories, Yale University, the Carnegie Institution of Washington, MIT, the Royal Astronomical Society of London, Pennsylvania State University, Stanford University, UC San Francisco, and the University of Michigan. She was elected to the National Academy of Sciences in 1985, and the American Academy of Arts and Sciences in 1989. For her work on the Wide-Field Planetary Camera of the Hubble Space Telescope, she shared a Group Achievement Award from NASA in 1993.

Since 1979, Faber has been a full professor at Lick Observatory, University of California at Santa Cruz. Her husband Andrew remains a lawyer and practices in nearby San Jose. They have two daughters, Robin and Holly.

Notes

1. Sandra Faber, in Alan Dressler, *Voyage to the Great Attractor: Exploring Intergalactic Space* (New York: Knopf, 1994), p. 35.

2. Sandra M. Faber, unpublished autobiographical sketch, July 12, 1995. This essay is derived largely from Faber's sketch; all subsequent quotations, including extracts, are taken from it.

3. Ibid. See S.M. Faber, "Variations in Spectral-Energy Distributions and Absorption-Line Strengths among Elliptical Galaxies," *Astrophysical Journal* 179 (1973): 731; and S.M. Faber, "Ten-Color Intermediate-Band Photometry of Stars," *Astronomy and Astrophysics Supplement* 10 (1973): 201.

4. Faber, unpublished autobiographical sketch.

5. See S.M. Faber and R.E. Jackson, "Velocity Dispersions and Mass-to-Light Ratios for Elliptical Galaxies," *Astrophysical Journal* 204 (1976): 668–683.

6. S.M. Faber and J.S. Gallagher, "Masses and Mass-to-Light Ratios of Galaxies," *Annual Reviews of Astronomy and Astrophysics* 17 (1979): 135–187.

7. S.M. Faber, "Galaxy Formation via Hierarchical Clustering and Dissipation," in *Astrophysical Cosmology* (Proceedings of the Vatican Study Week in Cosmology and Fundamental Physics, 1982), eds. H.A. Bruck, G.V. Coyne, and M.S. Longair (Vatican City: Specola Vaticana, 1983), p. 191; S.M. Faber, G.R. Blumenthal, J.R. Primack, and M.J. Rees, "Formation of Galaxies and Large Scale Structure with Cold Dark Matter," *Nature* 311 (1984): 517–525.

8. Faber, unpublished autobiographical sketch.

9. Faber, unpublished autobiographical sketch. See A. Dressler, S.M. Faber, D. Burstein, R.L. Davies, D. Lynden-Bell, R.J. Terlevich, and G. Wegner, "Spectroscopy and Photometry of Elliptical Galaxies: IV. A Large-Scale Streaming Motion in the Local Universe," *Astrophysical Journal Letters* 313 (1987): L37–L42; D. Lynden-Bell, S.M. Faber, R.L. Davies, A. Dressler, D. Burstein, R.J. Terlevich, and G. Wegner, "Spectroscopy and Photometry of Elliptical Galaxies: V. Galaxy Streaming towards the New Supergalactic Centre," *Astrophysical Journal* 326 (1988): 19.

Bibliography

Dressler, Alan. *Voyage to the Great Attractor: Exploring Intergalactic Space*. New York: Knopf, 1994.

Lightman, Alan, and Roberta Brawer. "Sandra Faber," in *Origins: The Lives and Worlds of Modern Cosmologists*. Cambridge, Mass.: Harvard University Press, 1990.

"Sandra Faber." *Omni* 12 (July 1990): 62–64, 88–92.

JOANN EISBERG

CATHERINE CLARKE FENSELAU

(1939–)

Chemist

Birth	April 15, 1939
1961	A.B., chemistry, *magna cum laude*, Bryn Mawr
1962–80	Married to Allan H. Fenselau
1965	Ph.D., chemistry, Stanford University
1967–87	Johns Hopkins School of Medicine, Dept. of Pharmacology: Instructor (1967–69); Assistant Professor (1969–73); Associate Professor (1973–82); Professor (1982–87)
1972–77	Research Career Development Award, National Institutes of Health
1973–89	Editor-in-Chief, *Biomedical and Environmental Mass Spectrometry*
1985	Garvan Medal, American Chemical Society; married Robert J. Cotter
1987–95	Professor and Chair, Dept. of Chemistry, University of Maryland, Baltimore County
1989	Maryland Chemist Award, American Chemical Society
1990–	Distinguished University Scholar, University of Maryland
1992	Merit Award, National Institutes of Health

Catherine Clarke Fenselau. Photo courtesy of Catherine Clarke Fenselau.

| 1993 | Pittsburgh Spectroscopy Award |
| 1995– | Interim Dean of the Graduate School and Associate Vice President for Research, University of Maryland, Baltimore County |

Dr. Catherine Clarke Fenselau, pioneer in mass spectronomy, was born and raised in York, Nebraska, where by the time she was in tenth grade she knew she was interested in science. She spent her high school years trying to sort out which science she wanted to study. An older family friend advised her to look beyond Nebraska for college and suggested she consider the "Seven Sisters" colleges. She decided to attend Bryn Mawr College, where she graduated with an A.B. in chemistry. At the time, she had no particular feeling about a women's college; but she realizes today as she looks back on it that women's colleges provide a special opportunity for serious-minded young women. Her interest in chemistry was strengthened by her opportunity to do research at Bryn Mawr in the summers following her sophomore and junior years. She found the experience very valuable and encourages all her undergraduate students to do research somewhere during their career.

During the Christmas break of her senior year, Catherine explored graduate schools. She looked at Minnesota, Stanford, and Berkeley. Although the climate may have had some influence on her decision, she feels she made the right one in choosing Stanford for her graduate work. She did her doctoral work with Carl Djerassi, who was a good mentor. In his laboratory she began her work in mass spectrometry, a new field for organic chemists.

Catherine did postdoctoral work at Berkeley with Melvin Calvin, who had about 120 people in his group at the time. She interacted most with A.L. Burlingame at Berkeley and at the NASA Space Science Laboratory, where she worked on the Lunar Landing Project. She helped develop mass spectrometry to look for biochemical markers in the moon rocks.

Catherine had married while she was a student, and she and her husband joined the faculty of the Johns Hopkins University Medical School. During her tenure as assistant professor, she had two sons. She considered the professional risks to be less and her career to be secure at that time. After all, the endowment for the Hopkins Medical School had been established by a group of Quaker women in Baltimore who said that the medical school must take women students, and there were women on the faculty when Catherine went there.

Being pregnant in a medical school in the early 1970s probably was not as unusual as it would have been in a chemistry department. Yet Fenselau realizes that Hopkins did not take her totally seriously. After the equal opportunity laws came into effect, a study was done on salaries. The next year, she suddenly received a 25 percent raise without any explanation. She was the first woman to become a full professor in a preclinical department at Hopkins.

Catherine Fenselau was a pioneer and a leader in the development of mass spectrometry. She believes that leadership is always easier to achieve in a new field and that women are well accepted as mass spectrometrists in this country. She thinks that "analytical chemistry is an exciting way to come into science. You are always useful and hireable."[1] With her colleagues, she used gas chromatography/mass spectrometry to identify the human metabolite of the anticancer drug cyclophosphamide (which is responsible for its cytotoxicity) and developed an assay to monitor its level in blood. Her collaborative studies of laetrile, a controversial anticancer remedy, helped to establish its legal definition. She may be the only Garvan Medalist whose work has been discussed in

Penthouse magazine.[2] Her interest in drug metabolism led her to the study of the metabolism of clofazamine, an antileprosy drug, at the request of the World Health Organization.

In a speech to the Women Chemists luncheon at the National American Chemical Society meeting in Miami in April 1985, Catherine talked about "Feminism and the Garvan Medal." When asked how she could accept an award intended for women only, she pointed out that there were at the time approximately 17,000 women in the American Chemical Society as compared with 7,300 organic chemists and 5,300 polymer chemists. (The latter two groups also have their own prizes.) She believed that receipt of the Medal would alert those in the chemical profession to women's presence and competence in chemistry.[3] Fenselau listed in her speech the objectives of a "feminist chemist" as equal salaries, opening of career opportunities that remain closed, and achievement of those "extra" professional rewards such as advisory positions, bonuses, and officerships in companies. She said she subscribes to the advice of M. Carey Thomas (former president of Bryn Mawr): set high standards for personal excellence and pursue goals with energy, persistence, and tact. To ensure success, she has also always followed the maxim: "Never let anyone lift heavy solvent bottles for you!"[4]

In discussing her move to the University of Maryland—Baltimore County (UMBC), Fenselau indicated that she did not see future personal growth at Hopkins and that she was very keen to get back into chemical education. She finds the undergraduates absolutely delightful. "Students keep us honest and keep us young." She was the first woman in the chemistry department at UMBC on a permanent basis, and two more have since been hired.

Fenselau is frequently invited to lecture in Europe, Asia, and South America. Extensively involved as an editor, she indicates that being an editor has made her name better known. As the founding editor of *Biomedical and Environmental Mass Spectrometry*, she was able to follow the instructions she received when she went to Hopkins to "exploit mass spectrometry in the service of biomedical research." She has found that being an editor enables her to know many people and their work. She views journals as projects of the entire scientific community. Fenselau has been associate editor of *Analytical Chemistry* since 1990 and currently serves on a number of editorial advisory boards including those for the *Journal of the American Society of Mass Spectronomy, Pharmaceutical and Biomedical Analysis, Drug Metabolism and Disposition, Mass Spectronomy Reviews*, and the *Journal of Mass Spectronomy*.

Fenselau's trips to Japan and England as a visiting professor have also given her an appreciation of the opportunities and approaches used in the United States. The Japanese medical school she visited had been endowed by a woman. While she was visiting there she gave lectures and talked to medical and graduate students, who were very interested in the American approach to both medicine and science. In England, her hosts were physical chemists. She spent a great deal of time talking about how to expand mass spectrometry into biochemistry. She found the role of women chemists to be very limited and not very visible in England.

In 1978, Fenselau received a grant from the National Science Foundation to establish a Mass Spectrometry Regional Center for the Middle Atlantic. The Center provided analytical service for high molecular weight materials (1,500-10,000 Daltons). The materials were biological, geochemical, and synthetic organic materials.

Her current research with students and collaborators is directed toward the interactions of drugs with proteins and the elucidation of the primary structures of proteins that have been modified by covalent reaction with anticancer agents, anti-inflammatory agents, and their metabolites. She likes to have a group of about twelve coworkers and students in her laboratory. She has several permanent Ph.D. staff to work with the instruments and to help provide instruction to graduate students. She normally adds one or two undergraduates to do research in the summer. The primary tool she uses to characterize biopolymer structures is mass spectrometry, the 1990s having been an exciting time for the development of mass spectrometry. Computerization has made it easier for people to use. New instrumental techniques such as electrospray and laser desorption have made it possible to provide molecular weights of intact and modified macromolecules.

When asked whether women do science differently than men, Fenselau suggests that their approach to competitiveness is different; women often worry more about making contributions to society in their research and less about working on the most competitive problem. She is certain that women manage departments or colleges differently: she has found that women generally try to do more to nurture their faculty, nominate them for awards, and generally view their role as one of responsibility for their co-workers. As an editor, she thinks that women as a whole give fairer reviews than men.

Although Fenselau has undertaken many activities as a chemist, editor, and mother, she does not regard herself as a "super woman." She admits that she has not done everything perfectly. She has always had laboratory lieutenants at work and someone to help at home. She received early encouragement from Carl Djerassi in her work and had a supportive chemist husband. However, her sons have no desire to be chemists. She said they decided that chemists work too hard. She hopes they will find careers for which they have the passion that she has for chemistry, and she hopes for grandchildren who might become chemists.

Notes

1. Unless otherwise noted, all quotations in this essay are from an interview with Catherine Fenselau by Nina M. Roscher on December 5, 1994.

2. See Linda M. Sweeting, "Catherine Clarke Fenselau 1985 Garvan Medalist," *Women Chemists Newsletter* (March 1985): 1.

3. Camille Peplowski McQueen and Margaret Cavanaugh, "Fenselau Addresses Women Chemists," *Women Chemists Newsletter* (July 1985): 1.

4. Ibid.

Bibliography

[Cover Feature] *Applied Spectroscopy* 47, no. 3 (March 1993).

Fenselau, Catherine. "Applications of Mass Spectrometry," in *Physical Methods in Modern Chemical Analysis*, Vol. 1. Edited by T. Kuwana, pp. 103–187. New York: Academic Press, 1978.

———. "Gas Chromatography Mass Spectrometry: A Report on the State of the Art." *Applied Spectroscopy* 28 (1974): 305–318.

———. "The New Mass Spectrometry: Desorption Ionization." *Chemtech* 18 (1988): 616–619.

Fenselau, C., and S. Pallante-Morell. "Mass Spectrometry in Cancer Research: 1969–1994," in *Therapeutic Aspects and Analytical Methods in Cancer Research*. Edited by E. Constantin, pp. 71–84. Strasbourg, France: AMUDES, 1994.

Fenselau, Catherine, M.M. Vestling, and R. Cotter. "Mass Spectrometric Analysis of Proteins." *Current Opinion in Biotechnology* 4 (1993): 14–19.

McQueen, Camille Peplowski, and Margaret Cavanaugh. "Fenselau Addresses Women Chemists." *Women Chemists Newsletter* (July 1985): 1.

Sweeting, Linda M. "Catherine Clarke Fenselau 1985 Garvan Medalist." *Women Chemists Newsletter* (March 1985): 1.

NINA MATHENY ROSCHER

MARY FIESER

(1909–   )

Chemist

Birth	May 27, 1909
1930	B.S., chemistry, Bryn Mawr
1931	A.M., organic chemistry, Radcliffe College

Mary Fieser. Photo reprinted by permission of Bryn Mawr College.

1932	Married Louis F. Fieser on June 21
1944	First edition of *Organic Chemistry* published
1959	Appointed research fellow in chemistry, Harvard University
1967	Volume 1 of *Reagents* published
1971	Garvan Medal, American Chemical Society
1994	Volume 17 of *Reagents* published

The names of Mary Fieser and her husband Louis are quite familiar to organic chemists. Some may have learned from their textbooks, but all have used their classic and continuing reference series on reagents. The history and practice of organic chemistry would be incomplete without the work of the Fiesers.

Mary (Peters) Fieser was born in Atchison, Kansas, in 1909 to Robert J. and Julia (Clutz) Peters. Her mother, one of seven children, graduated from Goucher College with a degree in English. Her father, the son of the president of Midland College, was a professor of English at Midland and subsequently at what was to become Carnegie Mellon University. Fieser grew up in a home that was a model of learning, work, and service to others, as exemplified by her father's brief position as the Pennsylvania secretary of labor and her mother's career as the owner and manager of a bookstore in Harrisburg, Pennsylvania. The Peterses encouraged their daughters to read and excel academically. Mary and her younger sister, Ruth, grew up in Harrisburg, where they attended a private girls school in preparation to enter college. Upon graduation from secondary school Mary entered Bryn Mawr College, where, inspired by her female family doctor, she was a premedical major. Her sister also attended Bryn Mawr, where she majored in mathematics.

Mary's change of major from premed to chemistry was influenced by her teacher and future husband, Louis F. Fieser. A Williams College graduate and Harvard Ph.D., Fieser was a chemistry instructor at Bryn Mawr from 1925 to 1930. It was during this time that he became well known as an organic chemist for his work on the synthesis of quinones, obtained by oxidizing quinic acid, which is present in some plants and berries. Because of the interest his work garnered while at Bryn Mawr, he was offered an assistant professorship at Harvard University in 1930, where he quickly rose through the ranks to become a full professor in 1937 and Sheldon Emery Professor of Organic Chemistry in 1939.

Upon graduation from Bryn Mawr in 1930, Mary accompanied Fieser to Harvard, where she was registered as a student at Radcliffe while she worked toward her master's degree in chemistry in Fieser's laboratory. Her experimental master's work in organic chemistry was concentrated in the study of naphthols and their synthetic conversions to quinones. In 1931 she obtained, as one of less than five women, her A.M. degree in organic chemistry from Radcliffe College. On June 21, 1932, she married L. F. Fieser and in 1933 they published their first collaborative paper, "A Synthesis of Phthaloylnaphthol," the fifth in a series of papers on the condensations and ring closures of naphthalenes.[1]

Mary never attempted to obtain a Ph.D. in chemistry. She determined that she would have greater access to the world of chemistry by working with her husband. This pragmatic approach was mutually beneficial for both Fiesers. By choosing to collaborate with her husband, Mary had access to ample funding and a laboratory at Harvard, a place virtually inaccessible to other women scientists of the time. For L. F. Fieser, this collaboration alone yielded fifteen original research papers in prestigious chemistry journals and seventeen well-received books. Unfortunately, however, Mary's excellent laboratory work and prolific authorship did not yield her the professional recognition from Harvard that she deserved. Harvard never provided her with a salary, and it was twenty-nine years before she was honored with the title of Research Fellow in Chemistry. But in 1971, Mary was honored by the American Chemical Society when she was awarded the Garvan Medal for her outstanding contributions to chemistry.

From 1933, the date of her first publication, to 1948, when she published "Absorption Spectroscopy and the Structure of Disterols," Fieser's laboratory work was directed toward the synthesis and structural examination of quinones and steroids.[2] As a researcher in her husband's laboratory, she contributed to work on the synthesis of vitamin K—a prothrombogenic medicinal; lapinone—an antimalarial; and cortisone—a potent steroid anti-inflammatory.

In 1944, while still fully involved in original research in organic chemistry, the Fiesers published the first edition of *Organic Chemistry*, a textbook for undergraduate students.[3] From 1947 to 1956, Mary published six more research articles with L. F. Fieser and began in earnest the work for which she has gained the most recognition, the dissemination of organic chemistry in the form of textbooks for students of chemistry and reference books for professional organic chemists. Her initiation as a textbook author was somewhat serendipitous. The Fiesers' first text, *Organic Chemistry*, was initially envisioned as a book to be authored solely by L. F. Fieser with help in information collection from his wife. However, due to the sheer magnitude of the information gathered by Mary Fieser, it was determined that she would be a co-author. In total, Mary contributed twenty chapters to *Organic Chemistry* on the applied aspects of organic chemistry (i.e., petroleum, rubber, microbiological processes), and it was these chapters that were considered to be the most novel feature of the text. In addition, she was primarily responsible for the indexing and data collection of this, and presumably future, texts. Notably, since the publication of their first textbook until the time of L. F. Fieser's death in 1977, the identities of the couple as textbook authors was almost completely blurred. During L. F. Fieser's lifetime Mary was not acknowledged as a sole force behind any text except in the *Style Guide for Chemists*, which was based on a set of "grammar, style and rhetoric" notes by Fieser for contributing authors to the serial *Organic Reactions*.[4]

The Fiesers' impact on organic chemistry throughout the twentieth century, from the publication of their first text up to the present-day serial publication of *Fieser and Fieser's Reagents for Organic Synthesis (Reagents)*, was a result of their strong desire to present the most current chemical information available. When new techniques (e.g., conformational analysis) or theories (e.g., the molecular orbital theory) demanded a fresh perspective, the Fiesers would willingly produce new editions of their texts. With this in mind, they would reinvestigate earlier work (their own and others') from a modern perspective. They would "reread most of the early papers, and scan the newer ones, with an eye for historical, personal and experimental items of interest."[5] In doing so, they almost entirely rewrote earlier editions of their own texts.

The Fiesers' reference works are notable for their up-to-date and definitive treatment of organic chemistry and have consistently been considered especially useful for experimental organic chemists. Throughout all their texts, they cited actual examples with complete bibliographies from the chemical literature rather than theoretical cases, as is often found in organic chemistry texts. In addition, they carefully decided what to include and exclude from their books. For example, in the first volume of *Reagents* they state that "the preparation of this book has involved repeated study of the question of excess reagent."[6] Not surprisingly, the Fiesers' bias toward their own areas of research is borne out in their textbooks. For example, in the first edition of *Organic Chemistry* they note that the "treatment of aromatic chemistry is somewhat more extensive than the relative importance of this branch of the subject alone would justify."[7] In the third edition of the same text, they sought to take into account advances from 1950 to 1956 in the structure and total synthesis of important "alkaloids, isoprenoids, tropolones, steroids and antibiotics."[8]

The Fiesers' texts and monographs have been well received for their concern for their reading public. They always paid great attention to details of potential importance, from the relative minutia of using Arabic numbers rather than Roman numerals to the elimination of nonessential technological details in order to save space and restrict book length in the third edition of *Organic Chemistry*. Their appealing writing style was both professional and personal. Their personal touches ran from the inclusion of 454 brief biographical sketches in *Organic Chemistry* to the trademark pictures of their Siamese cats in the preface of almost all their texts up until 1973. Their sense of collegiality was also evident in their abundant acknowledgment of their proofreaders and editors, who ranged from graduate students at Harvard to renowned organic chemists. In addition, they presented themselves as open and willing to learn from fellow chemists, inviting others to "submit corrections, comments, suggestions and contributions."[9]

The Fiesers' seminal work in organic chemistry is the serial *Fieser and Fieser's Reagents for Organic Chemistry*. The first volume of this serial was published in 1967 as a response to the enthusiasm of the readers of *Ex-*

periments in Organic Chemistry, which included a 49–page review of re-agents for organic synthesis. The structure of *Reagents*, a format now familiar to organic chemists worldwide, is easy to access, concise, and readable. It lists reagents alphabetically and provides relevant methods for reagent preparation and purification, pertinent examples of their use in organic synthesis, and all relevant references. From the first volume onward, *Reagents* has been considered a "Bible" for organic chemists.[10] In keeping with their previous reference works, the Fiesers worked diligently to present the state of the art of organic chemistry in *Reagents*.

In total, *Reagents* serves as an extremely useful reference work for practitioners of organic chemistry and as a treatise on the progress of synthetic organic chemistry over the last thirty years. Mary has continued to publish *Reagents* steadily since L.F. Fieser's death, an accomplishment for which generations of organic chemists are very grateful. In 1994, at the age of 85, she published Volume 17.

Notes

1. Mary Fieser and L.F. Fieser, "Condensations and Ring Closures in the Naphthalene Series. V. Synthesis of Phthaloylnaphthol," *Journal of the American Chemical Society* 55 (1933): 3010–3018.

2. S. Rajagopalan, L.F. Fieser, and Mary Fieser, "Absorption Spectroscopy and the Structure of Disterols," *Journal of Organic Chemistry* 13 (1948): 800–806.

3. Fieser and Fieser, *Organic Chemistry* (Boston: D.C. Heath, 1944).

4. Fieser and Fieser, *Style Guide for Chemists* (New York: Reinhold, 1960), p. iii.

5. Fieser and Fieser, *Steroids* (New York: Reinhold, 1959), p. iii.

6. Fieser and Fieser, *Reagents for Organic Synthesis*, Vol. 1 (New York: John Wiley and Sons, 1967), p. v.

7. Fieser and Fieser, *Organic Chemistry*, p. iii.

8. Fieser and Fieser, *Organic Chemistry*, 3rd ed. (New York: Reinhold, 1956), p. iii.

9. Fieser and Fieser, *Reagents*, Vol. 1, p. vi.

10. S. Pramer, "Mary Fieser: A Transitional Figure in the History of Women," *Journal of Chemical Education* 62 (1985): 190.

Bibliography

Brooks, C.J.W. "Louis Fieser." *Nature* 270 (1977): 768–769.

Costa, A.B. "Louis Frederick Fieser," in *Dictionary of Scientific Biography*, Vol. 17. Edited by F.L. Holmes, pp. 291–295. New York: Scribner's, 1990.

Pramer, S. "Mary Fieser: A Transitional Figure in the History of Women." *Journal of Chemical Education* 62 (1985): 186–191.

TAMI I. SPECTOR

WILLIAMINA PATON FLEMING
(1857–1911)
Astronomer

Birth	May 15, 1857
1879–81	Part-time Assistant, Harvard College Observatory
1881–98	Assistant, Harvard College Observatory
1899–1911	Curator of Astronomical Photographs, Harvard College Observatory
1906	Honorary Member, Royal Astronomical Society
1906?	Honorary Fellow in Astronomy, Wellesley College
1911	Guadalupe Almendaro Medal and honorary membership, Astronomical Society of Mexico; Honorary Member, Société Astronomique de France; lunar crater named in her honor
Death	May 21, 1911

When new technologies allowed the classification of stars from photographic plates, it was Williamina Fleming who created that system of classification. The discoverer of 79 stars and numerous other celestial phenomena, Fleming was a celebrated and world-renowned astronomer.

Williamina Paton was born in Dundee, Scotland, in 1857, the daughter of noted craftsman Robert Stevens and his wife, Mary Walker Stevens. She was educated in public schools and became a student teacher at age 14, continuing as a teacher until her marriage to James Orr Fleming on May 26, 1877. The couple emigrated to Boston in December 1878, but Williamina was abandoned by her husband early the next year. Desperately in need of work and pregnant, she became a maid in the household of Edward C. Pickering, the director of the Harvard College Observatory, who became noted for the hiring of women as computers.[1] There are two versions of what happened next: (1) in 1879 Pickering offered her a part-time position as a copyist and computer at the Observatory because he was "struck by her obviously superior education and intelligence," or, (2) a male assistant proved to be unsatisfactory and "in a huff Pickering is reported to have said . . . that he believed his housekeeper could do a better job."[2] She returned to Scotland in the fall of 1879 to deliver her son, whom she named Edward Pickering Fleming (and who eventually attended MIT and became a mining engineer), and then returned to Boston. In 1881 she was hired as a permanent member of the Observatory staff.

An important turning point in Williamina's career occurred with the commencement of the Henry Draper Memorial project on stellar spectra. After it was discovered in the 1800s that different stars had different spectra, several classification schemes were devised. The most useful of these was the work of Father Angelo Secchi, a Jesuit priest and astronomer who observed 4,000 stars and classified 500 according to five main types. With the advent of spectral photography by Henry Draper, the spectra of thousands of stars could be easily studied and classified. Such was the goal of the Henry Draper Memorial at the Harvard College Observatory (HCO), which was funded by Draper's widow, Anna Palmer Draper. Beginning in 1886, the women computers at the HCO followed a very tedious routine to carry out the work—first determining the identity of a star on a plate and then calculating its position. Using the sun's spectrum as a standard, they measured the positions of the spectral lines in each spectrum and then assigned an approximate spectral class to the star. The initial work was done mainly by Nettie Farrar, who decided to leave the HCO at the end of the year to be married. Pickering assured Mrs. Draper that he had an able replacement in hand, saying that Farrar was "instructing Mrs. Fleming who has assisted me, and who will, I think, take her place satisfactorily."[3] This is perhaps one of the greatest understatements in all of science, as Fleming did more than merely live up to Pickering's expectation and instead became one of the best-known women on the HCO staff.

In 1886, Williamina took over the Henry Draper Memorial project and was responsible for the examination, indexing, classification, and general care of the growing number of photographic plates at the HCO. In 1898 the Harvard Corporation formally recognized her position and gave her the official title Curator of Astronomical Photographs, the first official HCO appointment made to a woman. She supervised the women's work and interviewed new applicants as well. According to her contemporary, **Annie J. Cannon**, "her skill in administration was of great use in the numerous long pieces of routine work the observatory has carried through, and much of her time was spent on the proof of the numerous annals of the observatory."[4]

In her work, Fleming had found Secchi's classes to be too crude to explain all the subtleties in the spectra. She thus devised a system of spectral classes ranging from A to Q (omitting I) by the complexity of the spectrum lines and bands and the strength of the spectral lines due to hydrogen. Some of these classes were later abandoned as redundant or because of flaws in the plates. The first *Henry Draper Catalogue* was published in 1890 (Volume 27 of the HCO *Annals*) and contained the classified spectra of 10,351 stars, covering most of the stars visible to the naked eye. Pickering also published an introduction and discussion of the catalogue as Volume 26 of the *Annals* (1891). He gave Fleming credit for her tremendous work, as she had carried out the bulk of the classifications herself and had supervised the preparation of the catalogue; however, as was customary, her name is not listed as co-author of either Volume 26 or 27. However, Fleming became widely known throughout the astronomical community. An article from 1898 on a conference at Harvard describes a paper by Fleming that was read by Pickering (as was the custom):

The contribution of Mrs. Fleming, read by Mr. Pickering, Director, on "Stars of the fifth type in the Magellanic Clouds," continued some important statements in reference to the stars having spectra consisting mainly of bright lines, designated as Fifth type. Of these stars, which all lie in the Milky Way and in the Magellanic Clouds, 92 have been discovered during the last 14 years. In conclusion Professor Pickering said that Mrs. Fleming had omitted to mention that of these 79 stars nearly all had been discovered by herself, whereupon Mrs. Fleming was compelled by a spontaneous burst of applause to come forward and supplement the paper by responding to questions elicited by it.[5]

Fleming also studied stars of peculiar spectra and established that certain classes of variable stars and novae could be discovered from photographic plates by their spectra. In such a manner she discovered 10 novae, 59 gaseous nebulae, and 94 Wolf-Rayet stars. She also discovered 222 long-period variables by their peculiar spectra and set up a sequence of comparison stars to monitor her variables, leading to a paper entitled "A Photographic Study of Variable Stars" (1907). British astronomer H. H. Turner stated that "many astronomers are deservedly proud to have discovered one variable, and content to leave the arrangements for its observation to others: the discovery of 222 [long-period variables], and the care of their future on this scale, is an achievement bordering on the marvelous."[6]

Fleming was invited to participate in the Congress of Astronomy and Astrophysics at Chicago's Columbian Exposition in 1893, where she presented the paper entitled "A Field for Women's Work in Astronomy" that includes the widely quoted statement:

While we cannot maintain that in everything woman is man's equal, yet in many things her patience, perseverance and method make her his superior. Therefore, let us hope that in astronomy, which now affords a large field for woman's work and skill, she may, as has been the case in several other sciences, at least prove herself his equal.[7]

Fleming was considered "somewhat exceptional" because she was a woman in astronomy, and "in putting her work alongside that of others, it would be unjust not to remember that she left her heavy daily labours at the observatory to undertake on her return home those household cares of which a man usually expects to be relieved. She was fully equal to the double task."[8]

In 1906 Fleming became an honorary member of the Royal Astronomical Society, the sixth woman and the first American woman so honored. She was a charter member of the Astronomical and Astrophysical Society and an honorary member of the Société Astronomique de France. Wellesley College made her an Honorary Fellow in Astronomy. Also, a lunar crater has been named in her honor.

Fleming worked at the HCO for thirty years until her death from pneumonia at age 54. She left much work uncompleted. Indeed, "by the death of Mrs. Williamina Paton Fleming, astronomy has suffered an almost irreparable loss."[9] "Stars of Peculiar Spectra" was published posthumously in 1911, as was "Spectra and Photographic Magnitudes of Stars in Standard Zones."[10] Williamina Fleming was also missed by her colleagues at the HCO. She was described by Annie J. Cannon as

possessed of an extremely magnetic personality . . . fond of people and excitement, there was no more enthusiastic spectator in the stadium for the football games, no more ardent champion of the Harvard eleven. Industrious by nature, she was seldom idle . . . as much at home with the needle as with the magnifying eyepiece . . . she was never too tired to welcome her friends at her home or at the observatory, with that quality of human sympathy which is sometimes lacking among women engaged in scientific pursuits.[11]

Notes

1. For more information on the hiring of women at the HCO, see Dorrit Hoffleit, *The Education of American Women Astronomers before 1960* (Cambridge, Mass.: AAVSO, 1994); Bessie Zaban Jones and Lyle Gifford Boyd, *The Harvard College Observatory: The First Four Directorships, 1839–1919* (Cambridge, Mass.: Harvard University Press, 1971); Pamela E. Mack, "Straying from Their Orbits: Women in Astronomy in America," in *Women of Science: Righting the Record,* eds. G. Kass-Simon and Patricia Farnes (Bloomington: Indiana University Press, 1990); Margaret W. Rossiter, *Women Scientists in America: Struggles and Strategies to 1940*

(Baltimore: Johns Hopkins University Press, 1982).

2. Jones and Boyd, *The Harvard College Observatory*, p. 392; Hoffleit, *The Education of American Women Astronomers before 1960*, p. 20.

3. Jones and Boyd, *The Harvard College Observatory*, p. 235.

4. Annie J. Cannon, "Mrs. Fleming," *Scientific American* 104 (1911): 547.

5. Harriet Richardson Donaghe, "Photographic Flashes from Harvard Observatory," *Popular Astronomy* 6 (1898): 483.

6. H.H. Turner, "Obituaries," *Monthly Notices of the Royal Astronomical Society* 72 (February 1912): 262.

7. Mrs. M. Fleming, "A Field for Women's Work in Astronomy," *Astronomy and Astrophysics* 12 (1893): 689.

8. Turner, "Obituaries," p. 263.

9. William E. Rolston, "Mrs. W. P. Fleming," *Nature* 86 (1911): 453.

10. *Annals of the Harvard College Observatory* 56, no. 6, and 71, no. 2, respectively.

11. Annie J. Cannon, "Williamina Paton Fleming," *Astrophysical Journal* 34 (1911): 316.

Bibliography

Bailey, Solon I. *The History and Work of Harvard Observatory, 1839 to 1927*. New York: McGraw-Hill, 1931.

Cannon, Annie J. "Mrs. Fleming." *Scientific American* 104 (1911): 547.

———. "Williamina Paton Fleming." *Astrophysical Journal* 34 (1911): 314–318.

———. "Williamina Paton Fleming." *Science* 33 (1911): 987–988.

Donaghe, Harriet Richardson. "Photographic Flashes from Harvard Observatory." *Popular Astronomy* 6 (1898): 481–487.

Fleming, Mrs. Mina. "A Field for Women's Work in Astronomy." *Astronomy and Astrophysics* 12 (1893): 683–689.

Fleming, Williamina. "A Photographic Study of Variable Stars Forming a Part of the Henry Draper Memorial." *Annals of the Harvard College Observatory* 47, pt. 1 (1907).

———. "Spectra and Photographic Magnitudes of Stars in Standard Zones." *Annals of the Harvard College Observatory* 71, no. 2 (1911).

———. "Stars Having Peculiar Spectra." *Annals of the Harvard College Observatory* 56, no. 6 (1911).

Jones, Bessie Zaban, and Lyle Gifford Boyd. *The Harvard College Observatory*. Cambridge, Mass.: Harvard University Press, 1971.

Rolston, William E. "Mrs. W.P. Fleming." *Nature* 86 (1911): 453–454.

Turner, H.H. "Obituaries." *Monthly Notices of the Royal Astronomical Society* 72 (1912): 261–264.

KRISTINE M. LARSEN

HELEN M. FREE

(1923–)

Chemist

Birth	February 20, 1923
1944	B.S., College of Wooster, OH; began working as a chemist at Miles Laboratories in Elkhart, IN
1947	Married Alfred H. Free
1975	Published *Urinalysis in Clinical Laboratory Practice*
1976	Professional Achievement Award, American Society for Medical Technology
1978	M.A., Central Michigan University
1980	Garvan Medal, American Chemical Society; Distinguished Alumni Award, College of Wooster
1990	President, American Association for Clinical Chemistry
1993	President, American Chemical Society
1995	First recipient of the Helen M. Free Public Outreach Award

Helen M. Free has advanced the field of diagnostic chemistry by developing clinical laboratory tests that are effective and easy to use. In addition to making contributions to medical laboratory techniques, she has worked to promote chemistry and other sciences among the general public.

The daughter of Daisy (Piper) and James Summerville Murray, Helen Murray was born in Pittsburgh, Pennsylvania, in 1923. She attended the College of Wooster in Ohio and graduated with honors with a B.S. in chemistry in 1944. Free later earned an M.A. in laboratory management from Central Michigan University in 1978.[1]

After graduating from the College of Wooster, Free took a job as a chemist at Miles Laboratories (makers of Alka Seltzer) in Elkhart, Indiana. She worked her way up through the company, holding such positions as new products manager, director of clinical laboratory reagents, and director of marketing services of the research products division. Free currently serves as a consultant in professional relations of the diagnostics division of Miles Laboratories.[2] She is also an adjunct professor at Indiana University, South Bend.

As a result of her research involving clinical and medical testing methods, Free developed clinical laboratory diagnostic methods that have become standard throughout the world. According to polymer scientist Raymond B. Seymour, Free's research "led to the modification of convenient tablet tests for urinalysis and to the introduction and development of easy dip-and-read tests for various urinary constituents."[3] With the development of these methods, many clinical and medical tests have been simplified. As a result of her research Free has filed seven U.S. patents concerning clinical chemistry. She has authored and co-authored over two hundred publications, including multiple publications co-authored with her husband Alfred Free. Her book *Urinalysis in Laboratory Practice* has become a standard in the field.

Highly active in professional associations, Helen M. Free has contributed her time and energy to a number of scientific organizations. She was elected president of the American Association for Clinical Chemistry in 1990. She has been active in the American Chemical Society since 1945 and has served on numerous committees and boards at the national level, including the Committee on Public Affairs and Public Relations and the Women Chemists Committee. Her activity in the organization reached a high point when she was elected to serve as president of the American Chemical Society for 1993.

Public outreach has been a major focus of Free's work in professional organizations. As chairperson of the American Chemical Society (ACS) Board Committee on Public Affairs and Public Relations in 1994, she challenged all ACS members to become involved in communicating the positive and enjoyable aspects of science to students, teachers, parents, legislators, journalists, and other groups. She explained, "Public outreach is not a luxury. Only a small percentage of Americans know the basic facts of science. Many teachers, intimidated by a subject they know little about, avoid or abbreviate science lessons. In this climate, it is easy for people to accept pseudoscience and the claims of interest groups with antiscience agendas."[4] Free has demonstrated her commitment to public outreach by teaching grade-school students and working with teachers to advance science education in the classrooms.

The scientific community and the general public have honored Free with numerous awards. In 1967 she and her husband Alfred were awarded the Honor Scroll Award by the Chicago chapter of the American Institute of Chemists. Free was awarded the Garvan Medal in 1980 by the American Chemical Society. (The award recognizes distinguished service to chemistry by women chemists who are U.S. citizens.) In 1993 she was named YWCA Woman of the Year. Free was the first recipient of the Helen M. Free Public Outreach Award, named in her honor.

Helen Murray Free is married to Alfred Free, a fellow chemist whom she first met while working at Miles Laboratories. The couple married in 1947. They have six children, three boys and three girls.[5] With her ground-breaking work in diagnostic chemistry as well as her tireless efforts to promote the field of chemistry to women and the general public, Helen Free serves as a positive role model for women—not only in science but in all professions.

Notes

1. Raymond B. Seymour, "Helen Murray Free (1923–   )," in *Women in Chemistry and Physics: A Biobibliographic Sourcebook* (Westport, Conn.: Greenwood Press, 1993), p. 201.
2. Ibid., p. 202.
3. Ibid.
4. Helen M. Free, "Changing Public Perceptions," *Chemical and Engineering News* 14 (November 1994): 44.
5. Seymour, "Helen Murray Free (1923–   )," p. 201.

Bibliography

"ACS 1980 National Award Winners Announced [Garvan Medal]." *Chemical and Engineering News* 10 (September 1979): 62–80.
Free, Alfred H., and Helen M. Free. "Self Testing, an Emerging Component of Clinical Chemistry." *Clinical Chemistry* 30 (June 1984): 829–838.
————. *Urinalysis in Clinical Laboratory Practice.* Cleveland: CRC Press, 1975.
————. *Urodynamics: Concepts Relating to Urinalysis.* Elkhart, Ind.: Ames, 1972.
Free, Helen M. "Urine Glucose Tests for Diabetic Patients with Impaired Vision." *Diabetes Care* 1 (January–February 1978): 14–17.
"Meet Helen and Al." *American Journal of Chemistry Membership Newsletter* 14 (December 1989): 2.
Modern Urine Chemistry. Edited by Helen M. Free. Elkhart, Ind.: Ames Division, Miles Laboratories, 1986.
Seymour, Raymond B. "Helen Murray Free (1923–   )," in *Women in Chemistry and Physics: A Biobibliographic Sourcebook.* Edited by Louise S. Grinstein, Rose K. Rose, and Miriam H. Rafailovich. Westport, Conn.: Greenwood Press, 1993.
Who's Who in Technology. 6th ed. Detroit, Mich.: Gale Research, 1989.

HEATHER MARTIN

WENDY LAUREL FREEDMAN

(1957–)

Astronomer

Birth	July 17, 1957
1979	B.S., astronomy and astrophysics, University of Toronto; Robertson Award
1983	Henri Chretien Award, American Astronomical Society
1984	Ph.D., astronomy, University of Toronto; Ontario Scholarship; Amelia Earhart Fellowship; Carnegie Postdoctoral Fellowship
1985	Married Barry Francis Madore
1987–	Staff, Observatories of the Carnegie Institute of Washington in Pasadena, CA
1992–	Member, Extragalactic Distance Scale Key Project

Wendy Freedman and her colleagues working on the Hubble Space Telescope Project are concerned with a fundamental and exciting question: How old is the universe? Their answer to this question, still many calculations and a year or two away, may transform our understanding of how the universe evolved.

Wendy Freedman was born in Toronto, Canada, the oldest of three children of Dr. Harvey Freedman, a psychiatrist, and writer Sonya Freedman. She attended Cedarvale Public School until 1970 and graduated from Vaughan Road Collegiate Institute in 1975. Although she was interested in astronomy as a child, an interest stimulated and encouraged by her father, Freedman did not consider it a career possibility until the end of her first year at the University of Toronto. A twelfth-grade physics teacher had so inspired her that by the end of her high school years she had developed an interest in both biology and physics and applied to the university with the intention of pursuing a career in biophysics.[1]

This inspiration may well have carried over into her first professional work in astronomy. A 1987 biographical note mentioned that "her early work on galaxy evolution models, incorporating stochastic self-propagating star formation in differentially rotating disks, was soon found to have more immediate applications in biology and chemistry than in astronomy."[2]

Freedman has garnered several educational prizes. She was an Ontario

Wendy Laurel Freedman. Photo courtesy of Wendy Laurel Freedman.

Scholar in high school and was awarded the Robertson Award in 1979, along with her bachelor of science degree in astronomy and astrophysics, from the University of Toronto. In 1983 she received the Henri Chretien Award from the American Astronomical Society; and in 1984, when she earned her doctorate, she also received an Ontario Scholarship and an Amelia Earhart Fellowship.

While at the University of Toronto, Freedman met Barry Francis Madore, who collaborates with her in many of her publications and whom she married in 1985. Their daughter, Rachel, was born in 1987, and their son, Daniel, in 1988. Spending time with these three people is what she says she most enjoys doing whenever she can break away from the constraints of her demanding, exciting career.

Upon graduation from the University of Toronto in 1984, Freedman became a Carnegie postdoctoral fellow; in 1987 she was the first woman to join the staff of the Observatories of the Carnegie Institute of Washington in Pasadena, California. In 1988, Freedman reported on the first empirical test to see if different amounts of metal found in nearby galaxies has an effect on the cepheid period-luminosity (PL) relation, the relation widely accepted as one of the most accurate means of determining distances to nearby late-type galaxies. Using this relation she presented a new distance determination for IC 1613, an intrinsically faint, highly resolved, irregular galaxy.

In December 1992, Freedman became one of the fourteen members of the Extragalactic Distance Scale Key Project, led by the late Marc Aaronson. The primary aim of this team was to use the Hubble Space Telescope (HST) to observe about two dozen nearby galaxies in order to measure cepheid distances useful for calibrating a number of secondary distance methods. They began with observing M81 in the Virgo cluster and found 30 new cepheids, which they used to calibrate a new distance to that galaxy.

Freedman was well prepared for her role as a member of this Hubble team. By 1992 she had participated in numerous studies resulting in the publication of papers of photometric observations involving the entire range of morphological types of galaxies: dwarf ellipticals (NGC 147, NGC 185, NGC 205, Leo I), early-type spirals (M31), late-type spirals (M33), and irregulars (NGC 3109, NGC 6822, IC 1613). She had calibrated distances to other galaxies by monitoring primary distance indicators of cepheids, a group of variable or pulsating stars, once referred to as the rods of the universe; RR lyrae, a group of variable blue giants, often called cluster variables; and the position of the tip of the first-ascent red-giant branch.

Freedman had also published her observations and distance calculations based on various secondary distance indicators, those methods based on calibrations using the distance scale already established by the primary distance indicator, usually cepheids. Secondary indicator techniques are based either on properties of certain types of bright objects within the galaxies or on characteristics of the galaxies themselves. Most of Freedman's observations have taken place at Palomar Observatory, Las Campana, Chile; at Canada-France-Hawaii Telescope on 14,000–foot Mauna Kea in Hawaii; and at the Multiple Mirror Telescope Observatory at Kitt Peak in Arizona.

At the present time, there are three principal investigators on the HST Project. Freedman has primary responsibilities for the scientific analysis and publication of the observations. Robert Kennicutt from the University of Arizona is in charge of budget and scheduling of observations, and Jeremy Mould of Mount Stromlo Observatory in Australia coordinates the writing of the observing proposals. There are currently 27 team members, including postdoctorates and students.

In the fall of 1994, Freedman and her colleagues announced a new distance to the galaxy M100, a member of the Virgo cluster of galaxies, based on HST observations. From this measurement they calculated a Hubble constant of 80 kilometers per second per megaparsec. (The Hubble constant is the recession velocity of the galaxy divided by its distance.) Their findings indicate a cosmos between 8 and 12 billion years old, which poses a predicament because it makes the universe appear younger than its oldest stars, now believed to be 16 billion years old. This possibility brings up some very perplexing questions, which Freedman will be trying to answer in her future work.

But Freedman and her team do not stand alone. Studies by other astronomers have come up with a series of measurements that point to a relatively young universe. Her team's observations are considered to be more definitive because they are based on observations done using the Hubble's extraordinarily clear vision.

In November 1995, Freedman reported that she and her team had obtained data for nine galaxies: M81, M100, M101 (2 separate fields), NGC 925, NGC 3351, NGC 7731, NGC 3621, NGC 4414, and NGC 1365 (from the Fornax cluster). She noted that they had made great progress but still had about half the originally proposed sample to observe. Freedman explained that she and her team were still working on improving their calibrations and their data reduction algorithms. It would be a few months yet before they felt they could make a firm statement regarding how closely their findings for the Hubble constant from these latest observations compare with their earlier reported findings from observing the cepheids in M81 and M100 in the Virgo cluster. Freedman believed that the Fornax observations were going to be very exciting in this respect.

Wendy Freedman and her team now expect to make their final observations in 1997, and, allowing for one additional year to analyze their data completely, they expect to publish their final conclusions sometime in 1998. As a part of the key project, Freedman and her colleagues plan to make the HST data available to the entire astronomical community. Data for the cepheids they have found will be stored in addition to that for the other stars in the frames (tens of thousands of stars), excluding only those stars with measurement errors exceeding a certain limit, since such data would not be of much use to others.

The Carnegie Institution is undertaking the building of a large (6.5M) telescope at its observatory at Las Campanas, Chile. When it is finished, Freedman has two new large cosmology projects planned. She is scheduled to begin work on these projects in mid-1998.

Dr. Freedman states that she likes to work on problems that are amenable to several different observational tests; that is, they can be done in a variety of ways so that ultimately a quantitative estimate of the reliability of the results can be obtained. She is interested in projects that can be well developed over a long period of time so that she can do them as well as possible.

Notes

1. Unless otherwise noted, the information in this essay is based on e-mail interviews of Wendy Freedman by Ella N. Strattis on September 13 and 29 and November 16, 1995.

2. Wendy Freedman and Barry Madore, "Self-Organizing Structures," *American Scientist* 75 (1987): 252.

Bibliography

Cowen, R. "Hubble Constant: Controversy Continues." *Science News* (April 1, 1995): 198.

Flamsteed, Sam. "Crisis in the Cosmos." *Discover* (March 1995): 66–77.

Freedman, Wendy. "Distance to the Virgo Cluster Galaxy M100 from Hubble Space Telescope Observations of Cepheids." *Nature* 371 (1994): 757–762.

———. "The Expansion Rate and Size of the Universe." *Scientific American* 267 (1992): 54.

Freedman, Wendy, and Barry Madore. "The Cepheid Distance Scale." *Publications of the Astronomical Society of the Pacific* 103, no. 667 (1991): 933–957.

———. "Self-Organizing Structures." *American Scientist* 75 (1987): 252.

Freedman, Wendy, J.R. Mould, et al. "Limits on the Hubble Constant from the HST Distance of M100." *Astrophysical Journal* 449 (1995): 413.

Freedman, Wendy, et al. "The Hubble Space Telescope Extragalactic Distance Key Scale Project. I. The Discovery of Cepheids and a New Distance to M81." *Astrophysical Journal* 427, no. 2, pt. 1 (1994): 628–655.

Lemonick, Michael D., and J. Madeleine Nash. "Unraveling Universe." *Time* 145 (March 6, 1995): 76–84.

Wilford, John Noble. "Astronomers Debate Conflicting Answers for the Age of the Universe: Hubble Telescope's Data Have Upset Old Certainties." *New York Times* 144 (December 27, 1994): B7 (N).

ELLA N. STRATTIS

CATHARINE D. GARMANY
(1946–   )
Astronomer

Birth	March 6, 1946
1966	B.S., Indiana University

Catharine D. Garmany. Photo courtesy of Ken C. Abbott, Principal Photographer, Office of Public Relations, University of Colorado at Boulder.

1968	M.A., University of Virginia
1971	Ph.D., University of Virginia
1971–73	Research Associate, Dept. of Astronomy, University of Virginia
1976	Annie J. Cannon Award
1976–84	Research Associate, Joint Institute for Laboratory Astrophysics
1981–	Associate Professor Attendant Rank, Dept. of Astrophysical, Planetary, and Atmospheric Sciences, University of Colorado

1984–90	Senior Research Associate, Joint Institute for Laboratory Astrophysics
1985–	Fellow, Center for Astrophysics and Space Astronomy, University of Colorado
1990–	Fellow, Joint Institute for Laboratory Astrophysics; Director, Sommers-Bausch Observatory and Fiske Planetarium; Research Professor, Dept. of Astrophysical, Planetary, and Atmospheric Sciences, University of Colorado

Catharine D. (Katy) Garmany makes discoveries about the most massive, luminous, and hottest stars in the galaxy. These are the O- and B-type stars, which are very rare and quite unlike the ordinary star, such as our sun. O- and B-type stars are short-lived by stellar standards, eventually exploding as bright supernovas and throwing off matter into space. They frequently form in loose groupings of several thousand stars called *OB associations*, which are not bound gravitationally. "In the hierarchy of physically bound groupings of stars," Garmany states, "an OB association is the closest thing to nothing that is still something." Their association is a result of the fact that "the time it takes for them to be pulled apart by galactic tidal forces is comparable to the lifetimes of the most massive stars."[1] These stars are important, Garmany explains, because the matter they eject into space when they explode as supernovas provides "the heavier elements without which life on earth could not have formed." "There would be no planets like earth" without it.[2]

Garmany is also an astronomer with a past. "I think it is increasingly evident," she said, "that women who go on in science have had some fact in their past that has pushed them along that path." There were several such "semi-critical points" in her own past. She was born Catharine Doremus on March 6, 1946, in New York, New York. She has two younger brothers. Although she feels her parents did not actively encourage her, they did not especially discourage her either. "They never suggested a particular role for me as a girl," Garmany remembers. She became interested in astronomy in the fourth or fifth grade. Her father, who was a copy editor, brought home books from work, and she remembers reading all the astronomy books written by award-winning children's author Franklyn M. Branley. Growing up in the New York metropolitan area also provided unique opportunities. For example, her mother took her to visit the Hayden Planetarium, which, then as now, put on some fabulous sky shows.[3] Also, she was accepted into what is "perhaps the world's most famous secondary school for the teaching of science and math," the Bronx High School of Science.[4] "There was a special atmosphere there," Garmany said, "and it was assumed that the students would go to college *and* continue in science." She made lifelong friends in high school with women who did continue in science, eventually getting their doctorates in chemistry and biology. After high school Garmany went to Indiana University, where she earned a B.S. in astrophysics in 1966. For graduate study she went to the University of Virginia, receiving M.A. and Ph.D. degrees in astronomy in 1968 and 1971.[5] Garmany's dissertation reported the results of observations of the OB association III Cepheus, carried out over a three-year period at the Kitt Peak National Observatory in Arizona.[6] In 1970 she married George P. Garmany Jr., from whom she is now divorced. She has two sons: Rick and Jeff, born in 1974 and 1980 respectively.[7]

Another "semi-critical" point in her career came in 1976, when Garmany received the Annie J. Cannon Award in astronomy. Established by astronomer **Annie Jump Cannon** in 1933, this award was initially given to women "who had rendered distinguished service to astronomy."[8] A few years before Garmany received it, the award was transferred from the American Astronomical Society to the American Association of University Women and began to be granted for "promise" rather than for distinguished service.[9] The Cannon Award was a tremendous boost to Garmany's career. She wrote to the award committee in 1977, "If future recipients can be helped as much as I have, I believe this Award will be one of the most valuable ones in the field of Astronomy."[10] "We're dealing with the issue of self-respect," Garmany asserts. "Young women who enter science begin with low self-esteem. And the ones who leave sciences feel that they are not doing well enough, when, in fact, they are doing as well as the men." In Garmany's own case, after finishing graduate school she applied for jobs being quite willing to do the most menial tasks in her field and work on a volunteer basis.[11] Soon after she received the Cannon Award, Garmany was offered a postdoctoral research associate position at the Joint Institute for Laboratory Astrophysics (JILA) in Colorado and was asked to teach an astronomy class at the University of Colorado. She believes the publicity associated with the Cannon Award may have been a factor in securing those jobs.[12] Today, Garmany is a fellow of JILA and the Center for Astrophysics and Space Astronomy, University of Colorado. She is also director of the Sommers-Bausch Observatory and Fiske Planetarium on the University of Colorado campus and research professor at the University of Colorado.[13] The planetarium, she says, "is physically impressive, large and well equipped." As director, Garmany oversees the mission, which is to support instruction, give shows to the public, and develop programs for public school groups. She also oversees graduate students who teach labs in the observatory.[14]

Garmany attributes her own success to "wanting it so badly that nothing can make you give up." Although she would certainly not urge every woman to pursue a career in science, she would like to "pass the magic on to other people, that is, the true pleasure of figuring something out, not necessarily for the first time ever, but the first time for themselves." "This," she believes, "should be shared."[15]

Notes

1. Catharine D. Garmany, "Stellar Associations, OB-Type," in *The Astronomy and Astrophysics Encyclopedia*, ed. Stephen P. Maran (New York: Van Nostrand Reinhold, 1992), p. 825.

2. Catharine Garmany, telephone conversation with Janet Owens, July 12, 1995.

3. Ibid.

4. Kevin McKean, "Whiz Kids in the Fast Lane," *Discover* 4, no. 5 (May 1983): 21.

5. Garmany, telephone conversation.

6. Catharine D. Garmany, "A Spectroscopic Study of the OB Association III Cepheus," Ph.D. dissertation, University of Virginia, 1971.

7. Garmany, telephone conversation.

8. Dorrit Hoffleit, "Cannon, Annie Jump," in *Notable American Women, 1607–1950: A Biographical Dictionary*, ed. Edward T. James (Cambridge, Mass.: Belknap Press of Harvard University Press, 1971), Vol. 1, p. 283.

9. Garmany, telephone conversation.

10. Catharine D. Garmany, letter to the AAUW Educational Foundation Programs, Annie J. Cannon Award Committee, August 15, 1977.

11. Garmany, telephone conversation.

12. Garmany, letter to AAUW.

13. Catharine D. Garmany, letter to Janet Owens, July 11, 1995.

14. Garmany, telephone conversation.

15. Ibid.

Bibliography

American Men and Women of Science, 1995–96, 8 vols. 19th ed. New Providence, N.J.: R.R. Bowker, 1994.

Fitzpatrick, Edward L., and Catharine D. Garmany. "The H-R Diagram of the Large Magellanic Cloud and Implications for Stellar Evolution." *Astrophysical Journal* 363 (November 1, 1990): 119–130.

Garmany, Catharine D., and P.S. Conti. "Mass Loss in O-Type Stars: Parameters Which Affect It." *Astrophysical Journal* 284 (September 15, 1984): 705–711.

Garmany, Catharine D., Peter S. Conti, and C. Chiosi. "The Initial Mass Function for Massive Stars." *Astrophysical Journal* 263 (December 15, 1982): 777–790.

Garmany, Catharine D., Philip Massey, and Joel William Parker. "The OB Association LH 58 in the Large Magellanic Cloud." *Astronomical Journal* 108 (October 1994): 1256–1265.

Garmany, Catharine D., et al. "Mass Loss Rates from O-Stars in OB Associations." *Astrophysical Journal* 250 (November 15, 1981): 660–676.

JANET OWENS

MARGARET JOAN GELLER

(1947–)

Astronomer

Birth	December 8, 1947
1970	A.B., University of California, Berkeley
1972	M.A., Princeton University
1974–76	Fellowship, theoretical astrophysics, Center for Astrophysics
1975	Ph.D., Princeton University
1976–80	Research Associate, Harvard University
1980–83	Assistant Professor, Harvard University
1983	Astrophysicist, Smithsonian Astrophysical Observatory
1988–	Professor of Astronomy, Harvard University
1990	MacArthur "Genius" Award; Newcomb-Cleveland Prize, American Academy of Arts and Sciences

Margaret Joan Geller has become one of the most widely respected cosmologists of the late twentieth century. The daughter of a crystallographer, she was born in Ithaca, New York, while her father was a postdoctoral fellow at Cornell University. He later took a job at Bell Laboratories in Morristown, New York, where Margaret and her younger sister spent their youth.[1] Margaret has cited her father's influence as a vital factor in her career path: "He bought me a lot of toys that especially had to do with geometric things. I used to build solid shapes and play with all kinds of toys that really improved my 3–D perception. . . . I was trained by my father in that regard, whether I was going to become a scientist or not."[2] This ability to visualize the world in three dimensions has been an asset to her work on the large-scale structure of the universe. Margaret developed an acute interest in mathematics as a child and found her elementary school boring. Her parents allowed her to study algebra on her own at home instead of attending school on many occasions.

Margaret learned, even as a young child, that society was not always accepting of a girl's interest in mathematics and science: "I was somewhat aware that there was something funny about being a little girl interested in mathematics. I got messages from my teachers. One in particular gave me a very hard time, which I think was one of the reasons I didn't like to go to school too much."[3] Fortunately, she found more intellectual stimulation and encouragement in her father's laboratory, where she learned that "science was an exciting thing to do and that there were people who really enjoyed doing it and that it could be rewarding."[4]

In the 1960s the Geller family moved to Los Angeles, where Margaret finished high school and became an undergraduate in physics at the University of California at Berkeley from 1966 to 1970. When she was considering graduate schools, Charles Kittel, one of her professors and a respected physicist, suggested that she pursue a field that was beginning to open up and would provide excitement for years after Margaret attained her Ph.D.[5] She applied to Berkeley for solid-state physics and to Princeton for astrophysics, finally deciding to attend the latter. Margaret has described a lack of support while attending Princeton: "I was the second woman to be admitted to the graduate program in physics, and a number of the faculty members made it very clear they didn't approve of me."[6] She chose James Peebles, a cosmologist whose specialties include the early universe and the clustering of galaxies, as her thesis advisor. Her Ph.D. thesis, "Bright Galaxies in Rich Clusters: A Statistical Model for Magnitude Distributions," explored a particular statistical model of the luminosity patterns of the brightest members of galaxy clusters. As Margaret explained in her thesis,

> Resolution of the debate on the appropriateness of the various models is fundamentally important for the discussion of the origin and evolution of galaxies and clusters of galaxies. Equally important, the magnitudes [brightnesses] and redshifts [motion] of first-ranked galaxies in rich clusters are currently being used in setting cosmic distance scales and in attempting to determine whether, for big bang cosmologies, the universe is open, flat, or closed.[7]

Margaret was a postdoctoral fellow at the Harvard-Smithsonian Center for Astrophysics in 1974–1976 and was afterwards hired as a research associate. She also spent eighteen months as a senior visiting fellow at the Institute of Astronomy at Cambridge University in England. In 1980, Margaret had to choose between an untenured faculty position at Harvard and a tenured position at the University of Michigan.[8] She chose to remain at Harvard and eventually became the second female tenured astronomy professor in Harvard's history, the first having been **Cecilia Payne-Gaposchkin**.

The 1980s marked the beginning of Margaret's fruitful, ongoing collaboration with John Huchra, also at Harvard, in the area of redshift surveys. One of the ultimate goals of cosmologists is to map the structure of matter in the universe by studying how galaxies are arranged in clusters, superclusters, and any larger features. By understanding the density of material in the universe, the ultimate fate of the universe (whether it will expand forever and at what rate) can be ascertained. By utilizing Hubble's law, the relationship between the distance and recessional velocity of a galaxy (due to the expansion of the universe), cosmologists can make maps of the structure of the visible universe. The velocities are derived from the Doppler shift of the galaxy's spectral lines, and the distance can later be inferred from Hubble's law. Such studies of large-scale structure are called redshift surveys.[9] There are subtleties in this methodology, such as the large uncertainty in the exact relationship between distance and velocity (Hubble's Constant) and the effect of the motions of galaxies within a cluster or supercluster (the so-called non-Hubble Flow). The ultimate result of these studies was expected to show that galaxies (and hence matter) are arranged in an overall homogeneous fashion, the same homogeneity that is found in the cosmic background radiation (the "echo" of the Big Bang).

The first Harvard-Smithsonian Center for Astrophysics (CfA) redshift survey was published by Huchra and others in 1983. Margaret Geller, John Huchra, and graduate student Valerie de Lapparent conducted an extension to the CfA survey, which was first published in 1986. A map of galaxies visible in a long, thin strip of the sky was turned into a "pie wedge" by supplying the redshift data to show structure at various distances. Instead of showing a trend toward smoothness on the large scale, the distribution of galaxies looked "like a slice through the suds in the kitchen sink: it appears that the galaxies are on the surfaces of bubble-like structures with diameters $25–50h^{-1}$ Mpc. This topology poses serious challenges for current models for the formation of large-scale structure."[10] Geller and Huchra found similar structures in other adjacent and nonadjacent strips in the sky, proving that such structures appeared commonplace in the nearby universe.

The most startling discovery was the "Great Wall," a chain of galaxies stretching across the entire length of the survey on the order of 500 million by 200 million by 15 million light years in extent—the largest coherent structure yet seen in the universe.[11] Geller and Huchra's survey remains ongoing and has become the largest survey of its kind. It has been extended to the southern hemisphere and contains over 11,000 galaxies. Collaboration with L. Nicolaci da Costa has produced an additional survey including more than 3,500 galaxies, leading to coverage of more than one-third of the sky.[12] These surveys have demonstrated similar structure in the southern hemisphere, including a "Southern Wall" similar to the Great Wall.

Margaret Geller has utilized the redshift survey data to probe other aspects of galaxies and galactic clusters including the structure and evolution of galactic clusters, X-ray properties of galaxy clusters, and the luminosity distribution of galaxies; and she has attempted to reconcile the observed structure in the CfA surveys with theoretical models of the geometry of the universe and the proposed types of dark matter.[13] In 1990 she received a MacArthur Fellowship and the Newcomb-Cleveland Prize of the American Academy of Arts and Sciences for her important contributions to cosmology. Margaret has also been involved with making cosmology accessible to the general public through two collaborative films with cinematographer Boyd Estus: the award-winning 8–minute film of the CfA results, "Where the Galaxies Are" (1991); and "So Many Galaxies . . . So Little Time" (1992), a documentary that shows a general audience "how a scientific group works."[14]

Margaret Geller has made a significant impact on the study of the structure of the universe and continues to probe the frontier with an explorer's heart:

> The unanswered questions in cosmology are profound. I often feel that we are missing some fundamental element in our attempts to understand the large-scale structure of the universe. . . . Mapping the universe will undoubtedly keep us busy, awed, and fascinated for a long time to come. I often ask myself what we will learn about large-scale structure during my lifetime. There will be surprises, answers to old questions, and the uncovering of new puzzles. At every stage we will think we understand, but at every stage there will be nagging doubts in the minds of those who wonder.[15]

Notes

1. Donald Goldsmith, *The Astronomers* (New York: St. Martin's Press, 1991), p. 92.

2. Alan Lightman and Roberta Brawer, *Origins* (Cambridge, Mass.: Harvard University Press, 1990), p. 360.

3. Ibid.

4. Ibid., p. 362.

5. Gloria Skurzynski, "Beyond the Stars: A Profile of Cosmologist Margaret Geller," *Cricket* 21 (March 1994): 53–56.

6. Michael D. Lemonick, *The Light at the Edge of the Universe* (New York: Princeton University Press, 1993), pp. 66–67.

7. Margaret Joan Geller, "Bright Galaxies in Rich Clusters: A Statistical Model for Magnitude Distributions," Ph.D. thesis, Princeton University, 1974.

8. Goldsmith, *The Astronomers*, p. 93.

9. For a clear and detailed explanation of redshift surveys, see Margaret Geller, "Mapping the Universe," in *Bubbles, Voids and Bumps in Time: The New Cosmology*, ed. James Cornell (Cambridge: Cambridge University Press, 1989), pp. 50–72.

10. Valerie de Lapparent, Margaret J. Geller, and John P. Huchra, "A Slice of the Universe," *Astrophysical Journal* 302 (1986): L1.

11. Margaret J. Geller and John P. Huchra, "Mapping the Universe," *Science* 246 (1989): 897–903.

12. L. Nicolaci da Costa et al., "A Complete Southern Redshift Survey," *Astrophysical Journal* 424 (1994): L1–L4; L. Nicolaci da Costa et al., "The Power Spectrum of Galaxies in the Nearby Universe," *Astrophysical Journal* 437 (1994): L1–L4.

13. For example, see Massimo Ramella et al., "The Birthplace of Compact Groups of Galaxies," *Astronomical Journal* 107 (1994): 623–628; Ian P. Dell'Antonio, Margaret J. Geller, and Daniel G. Fabricant, "X-Ray and Optical Properties of Groups of Galaxies," *Astronomical Journal* 107 (1994): 427–447; R.O. Marzke, J.P. Huchra, and M.J. Geller, "The Luminosity Function of the CfA Redshift Survey," *Astrophysical Journal* 428 (1994): 43–50; Changbom Park et al., "Power Spectrum, Correlation Function, and Tests for Luminosity Bias in the CfA Redshift Survey," *Astrophysical Journal* 431 (1994): 569–585.

14. Skurzinski, "Beyond the Stars: A Profile of Cosmologist Margaret Geller," p. 56.

15. Geller, "Mapping the Universe," pp. 71–72.

Bibliography

da Costa, L. Nicolaci, M.J. Geller, et al. "A Complete Southern Sky Redshift Survey." *Astrophysical Journal* 424 (1994): L1–L4.

de Lapparent, Valerie, Margaret J. Geller, and John P. Huchra. "A Slice of the Universe." *Astrophysical Journal* 302 (1986): L1–L5.

Fabian, A.C., Margaret J. Geller, and Sandor Szalay. *Large Scale Structures in the Universe.* Sauverny-Versoix: Geneva Observatory, 1987.

Geller, Margaret J. "Mapping the Universe: Slices and Bubbles," in *Bubbles, Voids and Bumps in Time: The New Cosmology.* Edited by James Cornell. Cambridge: Cambridge University Press, 1989.

Geller, Margaret J., and John P. Huchra. "Mapping the Universe." *Science* 246 (1989): 897–903.

Lightman, Alan, and Roberta Brawer. *Origins.* Cambridge, Mass.: Harvard University Press, 1990.

Skurzynski, Gloria. "Beyond the Stars: A Profile of Cosmologist Margaret Geller." *Cricket* 21, no. 7 (1994): 53–56.

Vogeley, Michael S., Changbom Park, Margaret J. Geller, and John P. Huchra. "Large-Scale Clustering of Galaxies in the CfA Redshift Survey." *Astrophysical Journal* 391 (1992): L5–L8.

KRISTINE M. LARSEN

ELLEN GLEDITSCH

(1879–1968)

Nuclear Chemist

Birth	December 29, 1879
1902	Degree in pharmacology
1902–07	Studied at Oslo University
1907–12	Worked with Marie Curie in Paris
1912	*Licencée ès Sciences*, La Sorbonne, Paris
1914	Worked at Yale University; D.Sc., *honoris causa*, Smith College
1926–29	President, International Federation of University Women (IFUW)
1929–46	Professor of Chemistry, Oslo University
1948	Honorary Doctorate, University of Strasbourg, France
1962	Honorary Doctorate, La Sorbonne, Paris, France
Death	June 5, 1968

When Ellen Gleditsch was appointed professor of chemistry in 1929, she was the second woman to attain a professorial chair at Oslo University, Norway. Her appointment had met with considerable resistance, as it was regarded particularly unwise to place a woman in a chair of such importance as chemistry.

At the height of the controversy, which reached the local press, a letter arrived from Paris from Gleditsch's mentor and friend, **Marie Curie**. Although reluctant to get involved in what she saw as an internal struggle, Curie responded to a request from Kristine Bonnevie, professor of biology at Oslo University, and recommended her former student to the position.

Despite her many accomplishments and rewards, Gleditsch is today virtually unknown abroad as well as in her native Norway. Her life story sheds light on the horrendous barriers encountered by women scientists and on the courage and perseverance with which they fought for their right to do science.[1]

Ellen Gleditsch, born in 1879, was the oldest in a family of ten children. She excelled in her studies, particularly in mathematics and science, but the exam that would have qualified her for university entry was closed to girls. She trained instead as a pharmacist. With a degree in pharmacology, she was permitted to take courses at Oslo University. There she met a chemist, Eyvind Bødtker, who suggested she continue her studies in Paris. During a study leave, he personally visited the laboratory of Marie Curie and gave her the message that "my talented assistant wishes to work for you, solely out of love for science, not to gain a degree."[2] Mme. Curie was short of space in her laboratory, but Bødtker assured her that "Mademoiselle Gleditsch is so small and slight she will not take up much room."[3] As Curie needed a chemist, Ellen was offered space in the laboratory. She soon became Curie's personal assistant, and a friendship developed that later came to include Curie's daughters, Irène and Eve. The correspondence between Gleditsch and the Curie family continued uninterrupted until the death of Irène, spanning almost fifty years.

Upon her arrival at the Curie laboratory in 1907, Gleditsch was given the tedious task of separating and purifying radium salts for the production of the pure metal, which had been discovered by Curie in 1898. Curie informed Gleditsch of the conditions for exemption from laboratory fees in a letter sent to Oslo:

> If you would take on this work (recrystallization of barium and radium salts), which will only take up part of your time, and which is of general benefit for the laboratory, I could exempt you from the fees because of these favours. You can at the same time work on another problem of greater interest, which could lead to new results.[4]

In Paris, Gleditsch undertook a series of studies of radium and uranium content in radioactive minerals. In 1909, when her first publications on radioactive minerals appeared, the concept of isotopes was unknown. It was unclear where in the Periodic Table the newly found "elements" would fit in. Painstaking analytical study of the composition of minerals and the relationship between various radioactive components eventually led to the solution of the puzzle of nuclear decay and radioactive series.

In 1890, Ernest Rutherford had observed a new parameter, characteristic of a radioactive species: its half-life. When Gleditsch returned to Oslo in 1912, she continued research started in Paris on the determination of the half-life of radium, an area of study that was to lead to her most fruitful scientific work. However, she soon was being hampered by the lack of proper equipment. She contacted Dr. Boltwood at Yale, the leading authority on radiochemistry in the United States, and Dr. Lyman, at Harvard, for permission to come and work in their laboratories. The events and correspondence leading up to Gleditsch's arrival at Yale shed light on the bias at the time against women doing science alongside men.

Lyman pointed out that so far no woman had ever set foot in the physics laboratory at Harvard. Boltwood's response was equally typical of contemporary attitudes. In his letter, he politely cautioned her against making a hasty decision about coming to Yale before she could see for herself if "the facilities I can offer are sufficient to make you feel it is worth your while."[5] However, in a letter written to Rutherford the following day, the tone was very different:

> I have a piece of news that will interest you. Mlle. Gleditsch has written that she has a fellowship from the American-Scandinavian Foundation (I never heard of it before!) and wishes to come and work with me in New Haven!! What do you think of that? I have written to her and tried to ward her off, but as the letter was unnecessarily delayed in being forwarded to me, I am afraid she will be in New York before I get there. Tell Mrs. Rutherford that a silver fruit dish will make a very nice wedding present!!![6]

Mrs. Mary Rutherford replied in the same mocking tone, "Are you engaged to the charmer yet, I forget who she was?"[7]

Gleditsch did indeed arrive in New York before Boltwood was able to "ward her off," and it appears they managed to establish an amicable working relationship. At Yale, Gleditsch completed her work on the half-life of radium—her reported value of 1,686 years was to stand for 35 years, when it was adjusted to 1,620 years.

Upon her return to Norway, Gleditsch expanded her work with radioactive minerals. She developed analytical procedures for the isolation of numerous radioactive substances found in Norwegian minerals and, using the new concept of radioactive series, carried out accurate age determinations of minerals.[8] But to seek inspiration and reprieve from professional isolation in Oslo, Gleditsch regularly visited and worked at the Curie laboratory in Paris. During Marie Curie's visit to South America in 1920, she entrusted to Gleditsch the running of the "Institute du Radium."

Gleditsch's research group at Oslo University included numerous women who were drawn to the new field of radiochemistry. Before the occupation of her own country, she provided a safe place for researchers fleeing Nazi forces in central Europe. One of them, Tibor Graf, brought with him the technology of Geiger-Müller counters and proceeded to build an apparatus for Gleditsch's laboratory. Together they became pioneers in the use of this nondestructive analytical technique.

Gleditsch continued active research after her retirement. Of her many honors, the one she probably cherished most was an honorary doctoral degree at the Sorbonne in 1962. Gleditsch's list of publications, which has around 150 entries, includes scientific papers (a few co-authored with her sister Liv), monographs on current topics, and biographies of famous scientists. She wrote several textbooks. One on radiochemistry was co-authored by Eva Ramstedt, a Swedish nuclear scientist. Her last article, about the Swedish chemist Carl Wilhelm Scheele, was published the year of her death, 1968, when she was 88 years old.

Despite her impressive research and list of scientific publications, Gleditsch's greatest contributions to science were (1) as an educator of the public at large through numerous articles, books, and popular lectures, on radio and at the university; and (2) as a mentor for her many students and assistants. She took a personal interest in her students: "If you have a problem—go to Ellen," was a saying among the students. Her single status allowed her to take an active part in their lives and they became her extended family. She promoted their careers by securing places for them in laboratories abroad. Traveling and working in another country was in her view the best way to broaden one's horizons and to gain valuable experience.

Above all, Ellen Gleditsch was a role model for young women who flocked to the university to study science, often inspired by tales about Marie Curie. From 1926 to 1929 she served as the president of IFUW (International Federation of University Women) and traveled extensively in the United States to promote its cause. She was a staunch supporter for equal access to education for all. Once asked what young women should fight for, she answered, "That their parents should understand that it is equally important that girls as well as boys get an education, and that is regardless of the fact whether the girls will marry or not."[9] At the same time, she was the first one to acknowledge the difficulties facing women who wanted to combine family and a scientific career, and she herself chose to remain single.

> Their (women's) presence in the home is often required. A woman who wants to be a researcher has to reconcile two opposing demands. The research requires first and foremost a tranquil atmosphere, opportunity to think in peace and quiet, and to concentrate on a particular problem. Material worries, concern for a husband or children who are left at home without adequate help or care, will kill all chances of a first rate effort.[10]

Less than a week before her death, Gleditsch gave a dinner party in her home for some of her former students, all women, and expressed her delight that their education and training had made them valuable citizens from Norway to Tanzania. When she died of a stroke in 1968, she was 88 years old.

Notes

1. The material for this essay is taken mainly from Torleiv Kronen and Alexis C. Pappas, *Ellen Gleditsch: Et liv i forskning og medmennesklighet* [Ellen Gleditsch: A Life in Research and Humanitarian Work] (Oslo: Aventura, 1987). The book has not been translated into English. It includes the extensive correspondence between Gleditsch and the Curie family. Direct quotes have been translated from Norwegian or French by the contributor, Anne-Marie Kubanek, who wishes to thank Grete Grzegorek for assistance with translations.

2. Boedtker to M. Curie, June 23, 1907, in Kronen and Pappas, *Ellen Gleditsch*, p. 208.

3. Kronen and Pappas, *Ellen Gleditsch*, p. 34.

4. M. Curie to Gleditsch, July 26, 1907, in ibid., p. 34.

5. Boltwood to Gleditsch, May 1, 1913.

6. Boltwood to Rutherford, September 12, 1913, in *Rutherford and Boltwood: Letters on Radioactivity*, ed. Lawrence Badash (New Haven: Yale University Press, 1969), p. 285.

7. Mary Rutherford to Boltwood, October 6, 1913, in *Rutherford and Boltwood*, p. 286.

8. The technical content of this essay is taken from a special publication issued by the Chemical Institute of the University of Oslo, written (in Norwegian) by Alexis C. Pappas for the 100th anniversary of the birth of Ellen Gleditsch. The contributor is indebted to Professor Pappas for the use of that communiqué.

9. Kronen and Pappas, *Ellen Gleditsch*, p. 157.

10. From a speech by Gleditsch at the IFUW Fourth Congress, Amsterdam, 1926, quoted in Kronen and Pappas, *Ellen Gleditsch*, p. 112.

Bibliography

Gleditsch, E. *Contributions to the Study of Isotopes*. Oslo: J. Dybwad, 1925.

———. "The Life of Radium." *American Journal of Science* 41 (1916): 112–116.

———. *Radioaktivitet og grundstofforvandling*. Oslo: O. Norlis Forlag, 1924.

Gleditsch, E., and E. Ramstedt. *Radium og de radioaktve processer*. Kristiania: Aschehoug Forlag, 1917.

Kronen, Torleiv, and Alexis C. Pappas. *Ellen Gleditsch: Et liv i forskning og medmennesklighet* [Ellen Gleditsch: A Life in Research and Humanitarian Work]. Oslo: Aventura, 1987.

ANNE-MARIE WEIDLER KUBANEK

JENNY PICKWORTH GLUSKER

(1931–)

Crystallographer, Cancer Researcher

Birth	June 28, 1931
1953	B.A., chemistry, Somerville College at Oxford
1957	D.Phil., chemistry, Somerville College at Oxford
1967	Assistant Member, Institute for Cancer Research; Associate Member, Institute for Cancer Research
1969–	Adjunct Professor of Biochemistry and Biophysics, University of Pennsylvania, Philadelphia
1978	Philadelphia Section Award, American Chemical Society
1979	President, American Crystallographic Association; Garvan Medal, American Chemical Society
1979–	Senior Member, Institute for Cancer Research
1985	D.Sc., College of Wooster, OH
1987–	Editor, *Acta Crystallographica*
1995	Fankuchen Award, American Crystallographic Association

Jenny Pickworth Glusker travels the world and the world travels like-wise to her laboratory in the pursuit of crystallographic research. Her research has made her a recognized, prize-winning authority on cancer-causing chemicals, and her presentations and publications have helped to further the understanding of her chosen field around the globe.

Jenny was born in Birmingham, England, in 1931. She is the oldest in her family, having a brother, John, and sister, Marjorie, who continue to live in England. Initially she was discouraged from studying chemistry by her parents, Dr. Frederick Alfred Pickworth and Dr. Jane Wylie Stocks, who were both physicians and who would have preferred her to continue in the family tradition (which included, on her mother's side of the family, the doctor to four czars of Russia, Dr. James Wylie). Jenny nevertheless enrolled in Somerville College at Oxford University to study chemistry with **Dorothy Hodgkin** as her tutor.

To obtain honors for her B.A. degree, Jenny worked with infrared spec-troscopy in the laboratory of Sir Harold Thompson, who believed that women did not belong in the laboratory because they were prone to create fires and floods.[1] It was there where she met her future husband,

Donald L. Glusker, an American Rhodes Scholar, who was doing his D.Phil. studies. For her own D. Phil. work, Jenny returned to Hodgkin's laboratory and worked on an important problem: determining the structure of vitamin B_{12}. Her work helped to clarify the most difficult crystallographic problem to be resolved up to that time. Dorothy Hodgkin, her mentor, was awarded the Nobel Prize for her own contributions in determining the structures of difficult molecules including penicillin, vitamin B_{12}, and insulin. Glusker has been actively doing crystallographic research, which involves clarifying the structures of molecules at an atomic level, since her student days.

After completing her thesis work in 1956, Jenny and Don Glusker decided to move to the United States and get married. This decision did not please her parents. Jenny's father made sure that for their first year together she had enough money to purchase a ticket for a return trip to England, should she feel the need to go. In the United States, Jenny and Don both held postdoctoral positions at Cal Tech for one year. Jenny worked with Robert Corey and Linus Pauling. When they had completed their postdoctoral experience, they faced the increasingly common problem of professional couples today—where to live so both could find work close to each other. Philadelphia offered the best possibility. Jenny worked in A. Lindo Patterson's laboratory, and Don joined the industrial firm of Rohm and Haas.

Lindo Patterson's laboratory was at the Institute for Cancer Research of the Fox Chase Cancer Center. It was one of the leading crystallographic laboratories in the world. Patterson was very supportive of women scientists (his wife was a respected biochemist), and Jenny continued to work part-time while her three children, Ann, Mark and Katherine, were small. She worked on the structures of various small molecules of biological interest, such as citrates. These are small molecules that are involved in the body's degradation of metabolic products known as the citric acid cycle. The elucidation of the citric acid cycle was a major achievement and required the efforts of numerous investigators. Glusker's proposal of a mechanism for an enzyme in this cycle, aconitase, was important in understanding the cycle on a molecular level.[2]

Patterson died unexpectedly in 1966, and Glusker took over as head of the laboratory at the Institute for Cancer Research. She continues to work there and has pursued further investigations on vitamin B_{12} derivatives.[3] She has become a leading authority on chemical carcinogenesis based on the structure determinations of various chemical carcinogens that she has carried out.[4] In this area, she has also performed calculations from the beginning on simple aromatic hydrocarbons that act as models for polycyclic aromatic hydrocarbons. In addition, she has studied many antitumor agents that inhibit chemical carcinogenesis.[5]

Jenny's use of the Cambridge Structural Data Base, which contains

results of all small molecule crystal structures, to determine basic information about chemical behavior has led to many research publications
that include, for example, describing where metal ions bind in proteins
and what role they play in the structure.[6] She has become increasingly
interested in macromolecular compounds. The clarification of xylose
isomerase, an enzyme that isomerizes sugars, and its mechanism was
based on the extremely accurate data for this enzyme collected in her
laboratory.[7]

She continues to be extraordinarily active, surprising everyone with
her productivity, and she enthusiastically embraces new problems of interest. She has been adjunct professor of biochemistry and biophysics at
the University of Pennsylvania since 1969. She is a master at teaching
crystallographic techniques and, because of her organized and clear presentations, is sought after as a lecturer for audiences ranging from elementary schoolchildren to postgraduate students. She served as the chair
of Summer School in Crystallography in Tianjin, China, in 1988 and as
chair and organizer of the Winter School in Crystallography in Bangkok,
Thailand, in February 1990. She has organized and was chair for the
"Asian Region Seminar in Crystallography in Molecular Biology" held
in Madras, India, in December 1993; and she was director of the workshop on "X-Ray Crystallography" at the 13th Biennial Conference on
Chemical Education in August 1994 at Bucknell University, Pennsylvania.

Jenny's ability to write and speak clearly on a subject considered difficult by many has led her to be a co-author or editor of many books
dealing with the history of crystallography and the impact of crystallographic results. She has written two textbooks in crystallography: *Crystal
Structure Analysis: A Primer* (with Ken Trueblood)—considered to be one
of the best introductions to the area, it is widely used, has been translated
into Polish and Russian, and is now in its second edition; and *Crystal
Structure Analysis for Chemists and Biologists* (with M. Lewis and M. Rossi).

Jenny has been awarded many honors for her research accomplishments, including both the Philadelphia Section Award and the Garvan
Medal from the American Chemical Society. She was elected president
of the American Crystallographic Association and was given a D.Sc. from
the College of Wooster. For her research and her teaching of crystallography, she was awarded the Fankuchen Award by the American Crystallographic Association (ACA) in 1995 (jointly with Ken Trueblood).

Jenny's dedication and sense of responsibility to science led to her
involvement in many other capacities. She founded the ACA newsletter
and was its editor for fifteen years. In this capacity, knowledge was
spread not only among the American members but to other crystallographic associations worldwide, and it became a model for the other
societies' own newsletters. She has served on study sections for the Na-

tional Institutes of Health, the American Cancer Society, and the Damon Runyan–Walter Winchell Foundation. She has been editor of *Acta Crystallographica* since 1987 and was instrumental in the founding of a new journal for macromolecular compounds (Section D).

Apart from all these accomplishments that are recorded, there is another important aspect to Jenny's career as a scientist. Many have benefited from her ability to act as a good mentor, and through her own example she has served as an extraordinary role model for many women scientists. Jenny has the ability to inspire an independent curiosity in those fortunate enough to have worked with her. Her legendary generosity and enthusiasm extend from helping former students get started in setting up their own laboratories, to the invitations she has given to scientists in countries whose laboratory experiences were limited. Her laboratory is a Mecca for crystallographers from other countries and a truly international place, with visitors from England, Australia, Poland, Germany, Japan, India, Italy, and Israel, just to name a few. This exchange of ideas has a beneficial effect on all participants.

Jenny is also generous at home. She and Don regularly entertain guests. A perusal of their guest book finds everyone from Nobel Prize winners to undergraduate students. The combination of her research accomplishments, her integrity, her caring attitude, and her ability to remain modest, sincere, and forthright makes Jenny Glusker an outstanding and rare scientist.

Notes

1. Donald Glusker, personal communication with Miriam Rossi.

2. J.P. Glusker, "Mechanism of Aconitase Action Deduced from Crystallographic Studies of Its Substrates," *Journal of Molecular Biology* 38 (1968): 149; J.P. Glusker, "Citrate Conformation and Chelation: Enzymatic Implications," *Accounts of Chemical Research* 13 (1980): 345.

3. J.P. Glusker, "Vitamin B_{12} and the B_{12} Enzymes," *Vitamins and Hormones* 50 (1995): 1.

4. J.P. Glusker, "X-Ray Crystallographic Studies on Carcinogenic Polycyclic Aromatic Hydrocarbons, "in *Polycyclic Aromatic Hydrocarbons and Cancer*, eds. P.O.P. Ts'o and H. Gelboin, vol. 3 (New York: Academic Press, 1981), p. 61; J.P. Glusker, "X-Ray Analyses of PAH Metabolite Structures," in "Polycyclic Aromatic Hydrocarbons and Cancer," ed. R. Harvey, *ACS Symposium Proceedings* (Washington, D.C.: American Chemical Society, 1985), p. 125.

5. See *Plant Flavonoids in Biology and Medicine, Biochemical, Pharmacological and Structure-Activity Relationships*, eds. V. Cody, E. Middleton Jr., and J.R. Harborne (New York: Alan R. Liss, 1986).

6. C.W. Bock, A.K. Katz, and J.P. Glusker, "Hydration of Zinc Ions: A Comparison with Magnesium and Beryllium Ions," *Journal of the American Chemical Society* 117 (1995): 3754.

7. H.L. Carrell, B.H. Rubin, T.J. Hurley, and J.P. Glusker, "X-Ray Crystal Struc-

ture of D-xylose Isomerase at 4 A Resolution," *Journal of Biological Chemistry* 259 (1984): 3230.

Bibliography

"Garvan Medal." *Chemical and Engineering News* 56 (September 11, 1978): 43–44.

Glusker, J.P., and K.N. Trueblood. *Crystal Structure Analysis: A Primer.* New York: Oxford University Press, 1972.

———. *Crystal Structure Analysis: A Primer.* 2nd ed. New York: Oxford University Press, 1985.

Glusker, J.P., M. Lewis, and M. Rossi. *Crystal Structure Analysis for Chemists and Biologists.* Mannheim and Boca Raton: VCH, 1994.

Hodgkin, D.C., J. Kamper, M. Mackay, J. Pickworth, K.N. Trueblood, and J.G. White. "Structure of Vitamin B_{12}." *Nature* 173 (1956): 64–66.

Rose, Rose K., and Donald L. Glusker. "Jenny Pickworth Glusker (1931–   )," in *Women in Chemistry and Physics: A Biobibliographic Sourcebook.* Edited by L.S. Grinstein, R.K. Rose, and M.H. Rafailovich. Westport, Conn.: Greenwood Press, 1993.

Rossi, M., J.P. Glusker, L. Randaccio, M.F. Summers, P.J. Toscano, and L.G. Marzilli. "The Structure of a B_{12} Coenzyme: Methylcobalamin Studies by X-Ray and NMR Methods." *Journal of the American Chemical Society* 107 (1985): 1729–1738.

MIRIAM ROSSI

MARIA GOEPPERT-MAYER
(1906–1972)
Nuclear Physicist

Birth	June 28, 1906
1930	Ph.D., physics, University of Göttingen; married Dr. Joseph Edward Mayer
1931–39	Volunteer Associate, Johns Hopkins University
1939–46	Lecturer in Chemistry (unpaid), Columbia University and (1942–45) Sarah Lawrence College
1946–60	Physicist, SAM Laboratories, Columbia University; Senior Physicist, Argonne National Laboratory; Professor of Physics (unsalaried), University of Chicago, and Enrico Fermi Institute for Nuclear Studies

1960–72	Professor of Physics, University of California, La Jolla
1960	D.Sc., Russell Sage College
1961	D.Sc., Mt. Holyoke College, Smith College
1963	Nobel Prize in Physics, jointly with Hans Jensen and Eugene Wigner
1968	D.Sc., University of Portland
1970	D.Sc., Ripon College
Death	February 20, 1972

Women scientists have always encountered great difficulties in getting recognition for their contributions. The Nobel Prize certainly offers an instructive example of this imbalance. More than three hundred men have received the prize in the sciences, but just nine women have been so honored. To date, only two women have won the Nobel Prize in physics—Marie Curie in 1903 and Maria Goeppert-Mayer sixty years later.

Maria Göppert was born in Kattowitz (after World War I renamed Katovice) in Upper Silesia in 1906, the only child of Friedrich and Maria (Wolff) Göppert. In 1910 her father, a pediatrician, received an appointment as professor at the University of Göttingen, where he also directed a children's hospital. Maria grew up in an academic environment. As the sixth generation of university professors in his family, Friedrich Göppert belonged to the well-educated professional class, which, until 1933, set the tone socially in small German university towns like Göttingen. Professor Göppert held quite advanced views on child rearing. He encouraged self-confidence and an inquisitive mind-set, which fostered his daughter's intellectual disposition. He assumed as a matter of course that she would attend a university, even though at that time the admittance of women was still quite rare in Germany.

Maria went to the public elementary school for girls until age 15. At that point, no further public education was available for girls. Luckily, however, Göttingen did have a three-year private school for girls, the Frauenstudium. Founded by a group of suffragettes, this small school prepared girls for the examination that all students had to pass before admission to the university. Maria attended this school, but she was not able to finish the whole curriculum because the school fell victim to the rapidly increasing inflation in post–World War I Germany. She decided to take the exams one year early and, to the teachers' surprise, she passed. In spring of 1924, she enrolled at the University of Göttingen in mathematics. Originally she planned on obtaining a teaching certificate in mathematics, but soon she became bored with some of the required courses and looked for a more challenging curriculum.

Göttingen in the 1920s was the leading center of research for mathematics and physics. The roster of professors, visiting lecturers, and graduate students of that period seems like a "Who's Who" of future Nobel laureates and leading thinkers in nuclear physics and quantum mechanics. Maria Göppert knew most of these scientists socially, since her parents kept a hospitable house and the pretty, intelligent girl was a great favorite with visiting professors. David Hilbert, the great mathematician, was a neighbor of the Göpperts. He had invited Maria to a lecture he was giving on atomic physics while she was still attending high school, and she had become intrigued by the subject. When she debated about changing her field of study, the physicist Max Born invited her to come to his seminar on quantum physics. She was hooked. In a later interview she explained her change of majors: "This was wonderful. I liked the mathematics in it. . . . Mathematics began to seem too much like puzzle-solving. . . . Physics is puzzle-solving, too, but of puzzles created by nature, not by the mind of man. . . . Physics was the challenge."[1]

The exciting new field of quantum physics attracted the brightest students from all over Europe and America to Göttingen, among them Paul Dirac, Robert Oppenheimer, Enrico Fermi, Johann von Neumann, Linus Pauling, Leo Szilard, and many more. Maria became friends with most of them. No one could have imagined that many of them would meet again and work together in the United States only a few years later.

When Professor Friedrich Göppert died in 1927, Maria decided that she would continue the family tradition and pursue a doctorate. She switched to physics. In 1928 she was awarded a fellowship for study abroad, spending a semester at Cambridge University and attending lectures by Ernest Rutherford. After her return, Max Born accepted her as a doctoral student and she started doing research for her thesis. He and Werner Heisenberg were developing the Born-Heisenberg theory of quantum mechanics at that time. Quantum mechanics, a branch of physics, deals with the structure and behavior of the atom. Very complicated mathematical calculations were necessary to describe the movements of electrons around the nucleus of the atom. Maria's background in mathematics enabled her to join this ground-breaking research effort from the very beginning.

Maria's mother had begun renting out rooms to students, and in 1929 Joseph Edward Mayer became a boarder at the Göppert house. Two years older than Maria, he had earned a Ph.D. in chemistry at the University of California at Berkeley and had come to Göttingen on a Rockefeller grant to study crystal theory with Max Born and James Franck. He and Maria fell in love and were married on January 18, 1930.

Joe Mayer encouraged Maria to finish her degree, and shortly after the wedding she received her doctorate. For her dissertation she had performed the calculations to support the probability of double photon emissions from electrons orbiting the nucleus of an atom. Thirty years later, her theoretical treatment was borne out experimentally by laser technology. After her graduation, the Mayers took a steamer to the United States and arrived in Baltimore on April 1, 1930. Johns Hopkins University had offered Joe Mayer a position as assistant professor in chemistry.

Maria Goeppert-Mayer (she had Americanized the spelling of her maiden name) had known that as a woman she would never be appointed as professor at a German university, but she hoped that the United States would offer her such a possibility. American universities might be more liberal than German ones in admitting women students, but most institutions strenuously resisted hiring women professors. During the Depression, nepotism rules were especially stringently enforced and until the 1960s were used as a convenient tool to keep women out. In addition, Maria's specialty of quantum mechanics was practically unknown as yet in the United States, and the physics department at Johns Hopkins held to a traditional approach teaching classical mechanics, theory, and engineering.

Since the department showed no interest, Maria decided she should do research on a volunteer basis. Rather reluctantly, the university administration assigned her an office in the attic of the science building. She received a very small salary for helping a physics professor with his German correspondence, and she was given a number of titles such as "voluntary assistant" or "research assistant," but she was not listed by name in the catalogue. The Mayers stayed at Johns Hopkins for ten years, but Maria's position never changed. She worked at first with the chemist Karl Herzfeld, a fellow German, and used her expertise in quantum mechanics to investigate the structure of organic compounds and of dyes. Her application of quantum mechanics to chemical physics led to ground-breaking research and to her most important publications while she was a volunteer researcher. Until 1937, Maria spent her most of her summers in Göttingen and worked with Max Born.

In 1933 the Mayers had their first child, their daughter Marianne. Maria stayed home with the baby for one year, but then she wanted to get back to research. The year 1933 assumed added importance for her, because it marked the beginning of the exodus of Jewish scientists from Germany. Many of Maria's friends and mentors emigrated to the United States. Max Franck arrived to teach at Johns Hopkins University in 1935. Edward Teller accepted a position at George Washington University in nearby Washington, D.C. Maria finally had a chance again to discuss her research with leaders in the field of nuclear physics. By then, she was teaching some graduate courses in the physics department (although she was still not listed by name in the catalogue) and had acquired a doctoral student, Robert G. Sachs, with whom she wrote her first article on nuclear physics entitled "Calculations on a New Neutron-Proton Interaction Potential." She also taught a course in statistical mechanics together with her husband and afterwards co-authored *Statistical Mechanics*, an advanced textbook published in 1940 that became a classic and went through numerous editions. The book had been started as a project for Maria while she was pregnant with her second child. Her son, Peter, was born in 1938, but the book did not appear until two years later.

The year 1938 brought another change for the Mayers. Johns Hopkins University did not grant Joe tenure, effectively firing him. Maria suspected that her position (unofficial though it was) and her attic office had created resentment in the College of Physical Sciences, and she felt guilty on her husband's account. Columbia University and the University of Chicago immediately offered Joe Mayer positions at double his old salary. He decided to go to Columbia, where he joined the chemistry department headed by Harold Urey, a Nobel Prize winner in 1934 for his discovery of heavy water. Unfortunately, this move did not improve Maria's professional position. Columbia's physics department was as opposed to women faculty members as Johns Hopkins had been. She did not get an appointment. Fortunately, however, Harold Urey was aware of her contribution to *Statistical Mechanics*, and he offered her a lectureship in the chemistry department as well as an office. When Sarah Lawrence College, a women's school in New York, asked Maria Goeppert-Mayer to teach an interdisciplinary science course in December 1941, it was her first proper academic faculty job.

Columbia University provided a challenging research environment and congenial friends for the Mayers. Harold Urey, Enrico Fermi (who had joined the faculty at Columbia around the same time), and their wives became close friends of the Mayers. The three families settled in the small town of Leonia in New Jersey, and other scientists followed their example. During the first few years in Leonia, Maria stayed mostly at home because Columbia University would not give her any opportunities.

As for so many other women, World War II proved a turning point for Maria. Joseph Mayer started working on weapons research at the Aberdeen Proving Grounds in Maryland and came home only one day a week. With Joe gone so much, Maria accepted an offer from Sarah Lawrence College and taught interdisciplinary science courses off and on until the end of the war. Scientists, especially physicists, played a major role in the war effort from the very beginning. The U.S. government decided in 1941, after promptings from prominent physicists, to develop an atomic weapon. Research projects were set up at various locations under the code name "Manhattan Project." Harold Urey now headed a part of this secret research project at Columbia, where he engaged in work on uranium isotopes. This project was known as the Substitute Alloy Materials project, or SAM. He offered Maria a job with SAM, which she accepted on a part-time basis. She soon supervised a group of scientists and felt "here I was suddenly taken seriously, considered a good scientist."[2] She worked on uranium isotope separation by photochemical actions. The published results of this work enhanced her reputation among the top scientists involved in the Manhattan Project.

Edward Teller had been part of the Manhattan Project from its inception, and at his invitation Maria worked at Los Alamos for a few months in 1945. She had already done work at Columbia under Teller's direction on the effect of extremely high temperatures on the properties of matter and radiation, and she continued this research at Los Alamos. The hectic pace of the war years took their toll on Maria's health. She had two operations and contracted pneumonia, but she would not give in to health problems. After the conclusion of the Manhattan Project, she went back to teaching at Sarah Lawrence College.

After the end of the war, the University of Chicago set up the interdisciplinary Institute of Nuclear Studies and offered jobs to many of the scientists of the Manhattan Project. Joe Mayer accepted a full professorship and joint appointment in the chemistry department and the Institute. In February 1946, the Mayers moved to Chicago. Harold Urey, Enrico Fermi, James Franck, and Edward Teller also accepted positions at the University of Chicago. The University welcomed Maria Goeppert-Mayer as well and appointed her associate professor, but without a salary because of rigidly enforced nepotism rules. She did, however, get an office and participated fully in university activities without encountering the resentment she had faced at Johns Hopkins and at Columbia. In very short order Chicago became the place to be for atomic physicists, just as Göttingen had been in the 1920s.

Fortunately, Robert Sachs, Maria's first graduate student, was by then in a leading position at the new Argonne National Laboratory located just outside Chicago. He offered her a half-time appointment as senior physicist. Since the Argonne Laboratory specialized in nuclear physics, Maria focused her research efforts on that field, even though she had not worked in that area since having written two papers during the 1930s. Edward Teller had become interested in the possible origins of the chemical elements, and he asked Maria to collaborate with him because the research involved complex mathematics, which was Maria's area of expertise.

She investigated atoms of the same element, but with varying numbers of neutrons, so-called isotopes. Some isotopes are more prevalent than others, because their nuclei are very stable and do not decay radioactively. Stable isotopes accumulate because they do not break down into other elements. No satisfactory theoretical explanations for the occurrence of these stable isotopes had been put forward. Teller suggested that Maria investigate this phenomenon. She became intrigued by the problem and started the research project that was to earn her the Nobel Prize. Experimental physicists at Chicago had access to a cyclotron and could supply her with empirical data. She methodically collected statistics on elements with stable nuclei. When she painstakingly sifted through her data, she concluded that all the stable nuclei had in common certain numbers of protons or neutrons. She came up with the numbers 2, 8, 20, 28, 50, 82, and 126. A physicist friend, Eugene Wigner, called this sequence "magic numbers" in jest, but Maria liked the term and used it in her publications.

In the meantime, Edward Teller had become involved in the development of nuclear weapons and had lost interest in the joint research project, but Maria went on collecting data. She discussed her findings with Enrico Fermi. In 1948 she published a paper suggesting that the particles inside the atomic nucleus orbited in layers "like the delicate shells of an onion with nothing in the center."[3] Shell models had been proposed and then abandoned by other physicists, but Maria collected additional data and added the "magic numbers." She still, however, did not have a satisfactory theory in terms of quantum mechanics for the behavior of the particles in the nucleus. When she discussed her project with Fermi one day, he asked casually, "What about spin-orbit coupling?" In a moment of exhilarating intensity, Maria intuitively saw the solution fall into place. She immediately worked out the calculations to explain the magic numbers and to prove the important effect of spin-orbit coupling inside the nucleus.

Spin-orbit coupling means that protons and neutrons orbit inside the nucleus while at the same time spinning around their axis, some clockwise and some counter-clockwise. Because slightly less energy is needed to spin in one of the two directions, this small energy differential accounted for the magic numbers shared by stable isotopes. Joseph Mayer urged his wife to publish her results immediately, but she hesitated, especially since she had found out that a team of German physicists had come to similar conclusions and she did not want to preempt their publication. Eventually she published two papers detailing her theory in April 1950. Later the same year, the Mayers visited German universities on behalf of the State Department to report on scientific research. Maria visited Hans Jensen, the German physicist who had also developed a version of the shell theory. They immediately took to each other and decided that instead of competing against each other, they would work together on a book outlining the shell theory.

As it turned out, Maria did most of the writing because Jensen tended to procrastinate. Her name appears on the title page as first author. *El-*

ementary Theory of Nuclear Shell Structure came out in 1955. After its publication, both Maria and Jensen were elected to the National Academy of Sciences. But along with these honors came difficult experiences for Maria. Her great friend, Enrico Fermi, died of cancer in 1954; and in 1956 she lost her hearing in one ear. With Fermi's death, the University of Chicago and the Institute of Nuclear Studies lost much of their attraction. Teller had left earlier and Harold Urey had accepted a position at the University of California—San Diego, in La Jolla. Urey invited the Mayers to join him, so in 1959 they moved to San Diego. The University of California offered Maria her first paid professorship. After more than twenty years of volunteer work in physics, barely tolerated by the universities where she had tried to eke out a professional career, she finally became a full professor.

It almost was too late. In October 1960, Maria suffered a stroke that affected her speech somewhat and paralyzed her left arm. She enjoyed teaching and still engaged in research, but her health problems forced a slower pace. Ever since Goeppert and Jensen had published their research about the structure of the shell model, other scientists had suggested their work for the Nobel Prize. Maria was not very hopeful, and with each passing year the possibility seemed more remote.

Early in the morning of November 5, 1963, a newsman called the Mayers' house from Stockholm with the news that she and Hans Jensen had won half the Nobel Prize for their work on the shell model. Eugene Wigner, an old friend from the University of Göttingen, had been awarded the other half for his work on the atomic nucleus. Maria Goeppert-Mayer thus became the first American woman to be awarded a Nobel Prize in physics. This achievement is all the more impressive because she reached the pinnacle of her field without an academic appointment, working mostly on a volunteer or part-time basis. Recognition came late in her career but did not change the pattern of her life. Always modest about her achievements, she put the honor in perspective when she said in an interview: "If you love science, all you really want is to keep on working. The Nobel Prize thrills you, but it changes nothing."[4]

True to her principles, she kept on with her research and teaching, although her health began to fail seriously. In 1968 she had to have a pacemaker. The last of her many papers was written jointly with Hans Jensen and appeared in 1965. She died of a pulmonary embolism in San Diego in 1972.

Notes

1. Joan Dash, *A Life of One's Own: Three Gifted Women and the Men They Married* (New York: Harper and Row, 1973), p. 252.

2. Ibid., p. 294.

3. Sharon Bertsch McGrayne, *Nobel Prize Women in Science: Their Lives, Struggles and Momentous Discoveries* (New York: Carol Publishing Group, 1993), p. 196.
4. Ibid., p. 200.

Bibliography

American Men of Science: A Biographical Dictionary. Edited by James McKeen Cattell et al. Editions 1–11. Lancaster, Pa.: Science Press, 1906–1970.

Bailey, Martha J. *American Women in Science: A Biographical Dictionary.* Santa Barbara, Calif.: ABC-CLIO, 1994.

Dash, Joan. *A Life of One's Own: Three Gifted Women and the Men They Married.* New York: Harper and Row, 1973.

Fermi, Laura. *Atoms in the Family.* Chicago: University of Chicago Press, 1954.

Haber, Louis. *Women Pioneers of Science.* New York: Harcourt, Brace, Jovanovich, 1979.

Hall, Mary Harrington. "Maria Mayer: The Marie Curie of the Atom." *McCall's* 91 (July 1964): 38–39.

Herzenberg, Caroline L., and Ruth H. Howes. "Women of the Manhattan Project." *Technology Review* 35 (November/December 1993): 32–40.

Kass-Simon, G., and Patricia Farnes. *Women of Science: Righting the Record.* Bloomington: Indiana University Press, 1990.

Mayer, Maria Goeppert. "The Shell Model." *Science* 145 (September 1964): 999–1006. [Reprint of her Nobel lecture.]

McGrayne, Sharon Bertsch. *Nobel Prize Women in Science: Their Lives, Struggles and Momentous Discoveries.* New York: Carol Publishing Group, 1993.

Moritz, Charles. *Current Biography Yearbook 1964.* 25th Annual Cumulation. New York: H.W. Wilson, 1964.

Opfell, Olga S. *The Lady Laureates: Women Who Have Won the Nobel Prize.* 2nd ed. Metuchen, N.J.: Scarecrow, 1986.

Rempel, Trudy D. "Maria Gertrude Goeppert Mayer," in *Women in Chemistry and Physics: A Biobibliographic Sourcebook.* Edited by Louise S. Grinstein, Rose K. Rose, and Miriam H. Rafailovich. Westport, Conn.: Greenwood Press, 1993.

Rossiter, Margaret W. *Women Scientists in America: Struggles and Strategies to 1940.* Baltimore: Johns Hopkins University Press, 1982; 1983 printing.

Sachs, Robert G. "Maria Goeppert Mayer." *Biographical Memoirs of the National Academy of Sciences* 50 (1979): 311–328.

The Who's Who of Nobel Prize Winners. Edited by Bernard S. Schlessinger and June H. Schlessinger. Phoenix: Oryx Press, 1986.

World Who's Who in Science: A Biographical Dictionary of Notable Scientists from Antiquity to the Present. Edited by Allen G. Debus. Chicago: Marquis Who's Who, 1968.

Zuckerman, Harriet. *Scientific Elite: Nobel Laureates in the United States.* New York: Macmillan, 1977.

IRMGARD WOLFE

GERTRUDE SCHARFF GOLDHABER
(1911–)
Physicist

Birth	July 14, 1911
1935	Ph.D., physics, University of Munich
1935–39	Research Associate, Imperial College, University of London
1939	Married Maurice Goldhaber
1939–50	Research Physicist, University of Illinois
1950–79	Physicist, Brookhaven National Laboratory
1972	Elected to National Academy of Sciences
1972–74	Research Advisory Committee, National Science Foundation
1973–81	Report Review Committee, National Academy of Sciences
1974–77	Science Consultant, Arms Control and Disarmament Agency
1983–88	Visiting Scholar, National Advisory Committee on Science, Technology, and Society
1984–85	Visiting Scholar, Phi Beta Kappa
1984–87	Committee on Human Rights, National Academy of Sciences

Gertrude Goldhaber has been not only a respected researcher in nuclear energy but also an effective voice for women in science. Brookhaven National Laboratory recognized her contribution with an award in her name.

Gertrude "Trudy" Scharff Goldhaber was born in Mannheim, Germany, in 1911, the daughter of Otto Scharff and Nelly (nee Steinharter). As a 4–year-old, she first realized a fascination with numbers. Her family moved to Munich when she was 5, and it was there she grew up. As a teenager she decided to study physics and mathematics, even though at that time in Germany it was expected that girls studying such subjects would become schoolteachers.

Gertrude did not, however, desire to work in the field of education. Her father had suggested she study law so she could help him with litigation related to the family's century-old wholesale and retail food business. Instead, seeking to "understand what the world is made of," she entered the University of Freiburg.[1] She later attended universities in Zurich and Berlin prior to receiving a Ph.D. in physics from the University of Munich in 1935. She views her courses of study as fortuitous.

Being Jewish, Gertrude fled Germany with the rise of the Nazis. Physics, she discovered to her delight, was something "that could be done anywhere."[2] She moved to England and became a research associate in physics at the Imperial College of the University of London. In 1939 she married Maurice Goldhaber, a physicist from Austria with whom she later had two sons, Alfred and Michael. That same year she came to the United States, where she became a naturalized citizen in 1944.

While at the University of Illinois, Gertrude and her husband experimented with nuclear physics and made several notable discoveries regarding electrons and neutrons. In addition, she served as a consultant for the Argonne National Laboratory from 1948 to 1950. The couple later moved to Brookhaven National Laboratory, where he became director. As an associate at Brookhaven National Laboratory from 1950 to 1958, she was the first woman staff physicist. She later became a senior physicist. At Brookhaven, she, Maurice, and others studied aspects of nuclear energy. In addition, she continued to pursue her own wide-ranging research, which focuses on a variety of nuclear physics topics including fission, inertia, and electromagnetics.

Gertrude became involved in a considerable number of other activities besides her activities at Brookhaven. She has been a consultant with the Los Alamos National Laboratory since 1953. She was involved in a number of professional activities, including chairing the ad hoc panel on evaluation of nuclear data compilations for the National Science Foundation's Natural Resources Advisory Committee from 1972 to 1981. From 1972 to 1974 she was on the Research Advisory Committee of the National Science Foundation. She was also a member of the Board of Trustees of the Fermi National Accelerator Laboratory from 1972 to 1978 and on the Physics Visiting Committee at Harvard University from 1973 to 1977. Since 1982, Gertrude has been a member of the Education Advisory Committee of the New York Academy of Sciences.

In addition to her own extensive publications, she was active in reviewing and editing research. From 1973 to 1981 she was a member of the National Academy of Sciences report review committee. She was a member of the Editorial Committee of *Annual Reviews of Nuclear Science* for five years until 1977. Two years later she became an editorial board member for the *Journal of Physics G* and continued until 1980. She was also named as the North American representative to the *Europhysics Journal*, Institute on Physics, in Bristol, England, from 1978 to 1980.

Dr. Goldhaber's public service activities include having been on the President's Committee on the National Medal of Science from 1977 to

1979. During the Carter administration, from 1974 to 1977, she served as science consultant for the Arms Control and Disarmament Agency. She was also a member of the National Academy of Sciences Committee on Human Rights from 1984 to 1987.

Goldhaber has continued to be involved in academic activities. From 1983 to 1988 she was a visiting scholar for the National Advisory Committee on Sciences, Technology, and Society. She has also been an adjunct faculty member. She was at Cornell University from 1980 to 1982 and has been at Johns Hopkins since then.

Dr. Goldhaber has long advocated an expanded role for women in the natural sciences. In 1979 she was a founding member of Brookhaven's Women in Science. She has also been a member of the Committee on Education and Employment of Women in Science and Engineering since 1978, as well as a member of the forum committee of the Committee on Problems of Women in Physics, 1971–1972. In 1978 she became a member of the Committee for the Education and Employment of Women in Science and Engineering of the National Research Council Committee on Human Resources and continued until 1983. In honor of her life and work, the Brookhaven National Laboratory awarded the first Gertrude S. Goldhaber Prize in Physics, which was sponsored by Brookhaven's Women in Science project in 1992. In 1984 and 1985 she toured eight U.S. colleges on behalf of Phi Beta Kappa, talking about her own work and the role of women in science. She frequently speaks on the problems faced by women scientists, particularly relating to pay inequity, layoffs, and balancing family life with professional responsibilities.

Her leisure-time interests include the history of science, tennis, swimming, hiking, literature, and art as well as the latest results in neuroendocrinology. She currently lives near Brookhaven in Bayport, New York.

Notes

1. Lawrence Van Gelder, "Woman of Science Shares Her Heart," *New York Times* (July 29, 1984): Sect. 21, 2.
2. Ibid.

Bibliography

"All in the Family." *Baltimore Evening Sun* (April 30, 1991): A3.
American Men and Women of Science. 19th ed. New Providence, N.J.: R.R. Bowker, 1994.
Bailey, Martha J. *American Women in Science.* Denver: ABC-CLIO, 1994.
Goldhaber, Gertrude, and D.D. Clark. *Experimental Studies of Nuclides Far from Stability with the TRISTAN II Fission Product Separator at the Brookhaven National Laboratory.* Final Report. Upton, N.Y.: Brookhaven National Laboratory, March 1983. DOE Report No. DOE/ER/10576–10.

O'Neill, Lois Decker. *Women's Book of World Records and Achievements*. New York: Anchor/Doubleday, 1979.

Symposium on the History of Nuclear Physics, May 18–21, 1977. Minneapolis: University of Minnesota, 1977. 17 video cassettes.

Van Gelder, Lawrence. "Woman of Science Shares Her Heart." *New York Times* (July 29, 1973): Sect. 21, 2.

Weeks, Dorothy W. "Women in Physics Today." *Physics Today* 13 (August 1960): 22–23.

"Women in Science." *Science Teacher* 40 (1973): 15.

DANIEL LIESTMAN

MARY LOWE GOOD

(1931–)

Chemist

Birth	June 20, 1931
1950	B.S., University of Central Arkansas (Arkansas State Teachers College)
1952	Married Bill J. Good
1953	M.S., University of Arkansas, Fayetteville
1954–58	Instructor and Assistant Professor of Chemistry, Louisiana State University (LSU), Baton Rouge
1955	Ph.D., inorganic and radiochemistry, University of Arkansas, Fayetteville
1958–63	Associate Professor of Chemistry, LSU, New Orleans
1963–74	Professor of Chemistry, University of New Orleans
1973	Garvan Medal, American Chemical Society; Distinguished Alumnae Citation, University of Arkansas
1974–78	Boyd Professor of Chemistry, University of New Orleans
1978–80	Boyd Professor of Materials Science, LSU, Baton Rouge
1980–91	National Science Board, Oversight Committee of National Science Foundation (appointed by President Carter, 1980; by President Reagan, 1986)
1980–85	President, Inorganic Division, International Union of Pure and Applied Chemistry (IUPAC); Vice President–Director of Research, Universal Oil Products, Inc.

Mary Lowe Good. Photo courtesy of Public Relations, LSU.

1985–86	President–Director of Research, Signal Research Center (which became Allied Signal Research and Technology Laboratory)
1986–88	President—Engineered Materials Research, Allied Signal
1987	President, American Chemical Society
1987–93	President, Zonta International Foundation
1988–93	Senior Vice President—Technology, Allied Signal
1991	President's Council of Advisors on Science and Technology (PCAST) (appointed by President Bush)
1991	Charles Lathrop Parsons Award, American Chemical Society
1993–	Commerce Department Under Secretary for Technology (appointed by President Clinton)

Mary Lowe Good—award-winning chemist and public servant—was born in 1931 in Grapevine, Texas. Her inspiration and role models were her parents, both of whom were college-educated public school teachers. Her mother was an English and mathematics teacher as well as a school librarian. Her parents were very supportive when Mary began her postsecondary schooling.

Mary attended Arkansas State Teachers College (now the University of Central Arkansas) and graduated with a degree in chemistry in 1950, having changed her major from home economics after her introduction to chemistry in a required freshman course. Having graduated with high scholastic standing, she moved on to the University of Arkansas, Fayetteville, to begin graduate studies. Here she met and married Bill J. Good, a graduate student in the physics department. Two major milestones occurred in 1953: her first son, Billy, was born, and she was awarded her master of science degree in chemistry. Two years later she received her Ph.D. in inorganic and radiochemistry.

Dr. Good's first professional position was as an instructor in the chemistry department of Louisiana State University (LSU) in Baton Rouge. Her friend and mentor, Virginia Rice Williams, smoothed the way for the 22–year-old instructor with a young child and a career to develop. Dr. Williams was a faculty member in biochemistry (a division of the chemistry department) and helped Good navigate through the organization and power structure of the chemistry department at the time. (Dr. Williams's contributions to LSU were honored with a classroom and library building named after her in 1976.) In 1956, Good was promoted to assistant professor, a tenure track position. In the spring of 1958 her second son, James, was born and her husband Bill's one-year postdoctoral position at LSU was completed, which followed the awarding of his Ph.D. in the physics department.

These events and a political/cultural change heralded a turn in Dr. Good's career. New Orleans in 1958 was the largest city in the United States without a public institution of higher education. Mary and Bill Good were invited to be a part of the effort to establish an LSU campus in New Orleans. With two academic careers and a promotion to consider, the Goods became members of the faculty of Louisiana State University at New Orleans, Mary in the chemistry department (now as an associate professor with tenure) and Bill in the physics department. Homer Hitt, who was to become chancellor of this new public education institution and had recruited the Goods, proved to be another positive and long-term influence in Mary's life and career.

Dr. Good's research and teaching progressed steadily, and in 1963 she was promoted to professor. Good had done some radiochemistry as part of her doctoral research and now pursued this research interest. At LSU in New Orleans she pioneered the development of the new technique of ruthenium Mössbauer spectroscopy, a method for studying the local chemical and physical properties of chemical compounds by absorbing gamma radiation from a fixed source. Rudolf Mössbauer earned the Nobel Prize in physics for this technique in 1961. Mary's years of research in this area were exciting in that she did not realize until she published her work in 1969 that Mössbauer had also been studying ruthenium compounds for their physical properties after doing his initial groundbreaking work using iron compounds.[1] Dr. Good was among the first to find a practical application of the technique in basic chemical research.

The national attention garnered by Good's research into Mössbauer spectroscopy, the inorganic chemistry of the later transition metals, and forays into bioinorganic chemistry and catalysis led to her receipt of the Garvan Medal of the American Chemical Society in 1973. (This award is given to one American woman chemist annually; it was established by Francis and Mabel Garvan in 1937.) The following year she was given the highest honor a researcher at Louisiana State University can earn: the title of Boyd Professor. This systemwide distinguished professorship remains a lifetime position; in 1978, Dr. Good moved back to the Baton Rouge campus, taking up her Boyd chair in the materials science department. By then, Good's research support was nearly $200,000 per year and supported a large, diverse research group.

In 1980, Good was offered opportunities that were to change the course of her life and professional career. First, she was appointed by President Carter to the National Science Board, the oversight committee of the National Science Foundation. She was also elected president of the Inorganic Division of the International Union of Pure and Applied Chemistry. In an interview just after her appointment to the National Science Board, she stated that she was pleased with her appointment for two reasons:

> first, for the "personal satisfaction" of being selected from a group of "able and well-known scientists" and second . . . because the "president was willing to make appointments from areas of the country and from institutions which are not necessarily considered to be the educational elite. . . ." "That's important," she added, "because science must be diversified and supported throughout the country."[2]

Most significant for the remainder of her scientific career was Good's decision to accept an offer to leave academics and become the vice president—director of research of Universal Oil Products, Inc., in 1980. When asked what precipitated this change, she replied that it was "a challenge I just couldn't turn down. I had done an awful lot of the things I wanted to do in the academic community. I was taken in by the opportunity to try."[3] UOP labs was an "old, highly regarded research center, doing some of the most fascinating chemistry in the country." In her new position, she directed the research and development efforts of several hundred people in this high-tech company.

For twelve years, Good held several positions as vice president and president in a number of companies; but in her own words, "all of those years, I essentially worked for the same company."[4] She considers those changes a unique mirror of the 1980s when the climate of business and industry in general was one of mergers and buyouts. Throughout all the changes and workforce reductions Good remained at the top of her profession, always carrying a title and position relating to research and technology.

Good has been very active in professional organizations, serving on committees and holding elected offices in local and national societies. She was elected president of the American Chemical Society for 1987, representing 150,000 chemists nationwide and directing the organization in the difficult economic climate of the mid-1980s. For five years she held the post of president of the Zonta International Foundation, a philanthropic arm of Zonta International, a multinational organization dedicated to improving the status of women and encouraging high ethical standards in business. Among other funds, the foundation awarded Amelia Earhart fellowships to women in graduate studies in aerospace science.

In June 1990 it was said about Good: "If you really want to get something done in chemistry or . . . in science in general . . . , one of the better ways to go about it is to persuade Mary L. Good to serve on the cognizant committee, preferably to chair it."[5] Service on committees, advisory boards, national panels, and elected offices have enabled her to leave a mark on the practice and policy of science and the education of future scientists. In recognition of her outstanding public service, the American Chemical Society named her the Charles Lathrop Parsons Award recipient for 1991. By then, Good's experience in academics, industry, organizational management, and leadership had established her as an expert and trusted professional. New challenges were brought to her in 1991 when President Bush appointed her to the President's Council of Advisors on Science and Technology, an elite group of knowledgeable and trusted scientists who guide and shape U.S. scientific policy.

In 1993, President Clinton asked Good to make one more career move into government service. She is now under secretary for technology in the Commerce Department, where one of her first projects was the Clean Car Initiative. The goal of this joint project between the government and the Big 3 automakers is to develop an automobile that will average 82 miles per gallon. Good's ongoing challenge is to establish a basis for civilian technology within the government structure that can help lead the country in its quest for a competitive economy in the twenty-first century.

Mary Lowe Good has been awarded eighteen honorary doctorates, the most recent from Louisiana State University in May 1995. She has been particularly pleased with her Distinguished Alumnae certificate and honorary doctorate from the University of Arkansas for their home-town quality and the doctorate from Duke University because of her admiration for its academic excellence. After she leaves government service, Good has indicated that she would like to work with some companies on technology policy issues, perhaps by serving on corporate boards or by advising technology development projects for developing countries of importance to the United States.[6]

Notes

1. C.A. Clausen III, R.A. Prados, and M.L. Good, "Moessbauer Effect Parameters in Ruthenium Compounds," *Journal of the Chemical Society D* 20 (1969): 1188–1189.

2. Elizabeth Hansen, "Dr. Good Adds New Job to Busy Schedule," *State-Times* [Baton Rouge, Louisiana], July 25, 1980.

3. Telephone conversation between Mary Lowe Good and Eileen Stanley, July 5, 1995.

4. Ibid.

5. "Mary Good Wins ACS's Parsons Award for Public Service," *C & E News* (June 11, 1990): 16–19.

6. Mary Lowe Good to Eileen Stanley, personal correspondence, July 12, 1995.

Bibliography

"Allied-Signal's Mary Good Analyzes New Threats to Chemical Profession." *Chemical and Engineering News* (September 8, 1986): 7–14.

Brown, B.H. "Mary Lowe Good Fills Roles of Scientist, Industry Spokesperson." *Industrial Research and Development* (October 1982): 155–156.

Cavanaugh, Margaret A. "Mary Lowe Good (1931–)," in *Women in Chemistry and Physics: A Biobibliographic Sourcebook.* Edited by Louise S. Grinstein et al., pp. 218–229. Westport, Conn.: Greenwood Press, 1993.

Frame, Phil. "Mary Good, Director of Super Car Project, Faces Formidable Task of Uniting Government, Auto Industry." *Automotive News* (October 25, 1993): nn–41.

Hansen, Elizabeth. "Dr. Good Adds New Job to Busy Schedule." *State-Times* [Baton Rouge, Louisiana], July 25, 1980.

Heylin, Michael. "Mary Good Wins ACS's Parsons Award for Public Service." *Chemical and Engineering News* (June 11, 1990): 16–19.

Mathews, Pam. "Mary Good: Chemists Have a Responsibility." *American Press* [Lake Charles, Louisiana], April 30, 1987.

EILEEN HORN STANLEY

JEANETTE GRASSELLI (BROWN)

(1929–)

Analytical Chemist

Birth	August 4, 1929
1950	B.S., *summa cum laude*, Ohio University
1958	M.S., Western Reserve University
1982–84	National Science Foundation, Advisory Committee for Analytical Chemistry
1985	Distinguished Service Award, Society for Applied Spectroscopy
1986	Garvan Medal, American Chemical Society
1987–88	White House Initiative on Historically Black Colleges and Universities, Science and Technology Advisory Committee
1987–89	U.S. Department of Energy, Energy Research Advisory Board
1988–91	National Institute of Standards and Technology, Visiting Committee
1989	Ohio Women's Hall of Fame
1990–	International Women's Forum; Ohio Academy of Sciences
1990–94	Smithsonian, "Science in American Life" Exhibition Advisory Board
1991	Ohio Sciences and Technology Hall of Fame
1992–95	Chair, U.S. National Committee, International Union of Pure and Applied Chemistry
1992–	Chair, Board of Trustees, Cleveland Scholarships Program
1993	Fisher Award in Analytical Chemistry, American Chemical Society
1995	National Research Council/Committee on Women in Science and Engineering

Jeanette Grasselli Brown has made major contributions to the advancement of science and to the advancement of women in science. During the course of her career, she has received numerous awards for her contributions in developing new problem-solving techniques in analytical chemistry. Her research has resulted in practical ways to solve real-life

Jeanette Grasselli (Brown). Photo courtesy of Jeanette Grasselli Brown.

problems, from identification of contaminants in gasoline, to solving structures of new plastics, to analyzing pollution problems in the environment.

Grasselli worked her way up the ranks of Standard Oil of Ohio (Sohio) to become the first woman director of corporate research, Environmental and Analytical Sciences. During her 38–year career with Standard Oil, Grasselli was responsible for developing many innovative applications for molecular spectroscopy, which are now at the forefront of industrial practice. "Under her direction, the use of in-situ analyses, computerized spectroscopy, and systematized data retrieval and storage have been major advances."[1] One of Grasselli's lifelong goals has been to bridge the gap between research and practical applications and between academia and industry.

After graduating from Ohio University in 1950, Grasselli was hired to work on a problem-solving team for Standard Oil of Ohio (now BP of America). Spectrometry had been around since the end of the last century, but not until World War II was the electronics developed to make the instrumentation necessary to solve complex problems. Grasselli was put in charge of a new instrument called an infrared spectrometer and told "to see what she could do with it."[2] This led her to become one of the foremost contributors to infrared and Raman spectrometry of the century.

Spectroscopy is an analytical technique whereby the interaction of electromagnetic radiation with matter is measured. The measurements of these interactions between energy and matter provide information that allows scientists to plot a unique graphical representation of a chemical called a spectrum. With the new instrumentation that Grasselli had, molecules could be positively identified and quantitatively analyzed in ways that had never before been possible. The methods are nondestructive and require only very small amounts of sample, making them powerful tools for providing scientists with information at the atomic and molecular levels. The data can then be used to solve chemical problems in industry, academia, or the environment. For example, scientists can test the environment to detect the effects of pollution and identify unknown chemicals in samples, which leads to the development of better processes, new and cleaner products, and a cleaner environment.

Always an avid mystery reader and a fan of Sherlock Holmes, Grasselli was fascinated by the "detective" aspects of her work. She was intrigued by the similarities of approach in problem-solving techniques at Standard Oil and what detectives do when they track down clues to solve crimes. An infrared spectrometer can measure what amounts to a "fingerprint" of all the types of molecules that make up our world.[3] Intrigued by the possibilities of this new instrument and the ways it might be used, she began consulting with the coroner's office in Cleveland to help with their work. They began to use infrared spectrometry to identify and analyze unknown substances in the crime lab, because any substance—whether it be fuel, air, water, soil, blood, hair, or fibers from clothing—can be subjected to spectroscopic analysis. The forensic analytical methods are the same as the analytical methods that Grasselli helped develop in industry and are now used widely with growing applications.

Although retired now from Standard Oil, Grasselli is still widely sought after as a speaker and consultant on a wide range of environmental and industrial problems. She frequently travels to Europe, especially eastern Europe, to teach about the uses of spectroscopy for soil, air, and water pollution problems. One thing Grasselli likes most about using spectroscopic analysis is that it is a noninvasive, nondestructive type of testing and analysis that does not further harm the environment.

Bringing the detective aspect into her work has helped Grasselli develop a problem-solving philosophy that is widely sought after by both private and public organizations. Besides getting specialized training, Grasselli believes that problem solvers must be able to work in teams, learn to communicate well, and utilize and cross-apply techniques from other disciplines when necessary. Her problem-solving approach involves combining the specialist's expertise with a broader, generalist's view of the total picture in which the problem is viewed as a part of the total puzzle. She combines an enthusiasm for discovering the unknown in science with showing how practical applications can be made.

Grasselli's education and upbringing helped her to develop this broader point of view. Born Jeanette Gecsy, Grasselli was the daughter of Hungarian immigrants. She grew up in a working-class, Hungarian neighborhood in Cleveland, Ohio, where hard work, initiative, and education were emphasized. Her parents made learning interesting and wanted both her and her younger brother to have a college education. She was in a high school college track program, planning to major in English, when she took her first course in chemistry and was immediately captivated by this fascinating subject. With the strong encouragement of her chemistry teacher, she switched her proposed college major to chemistry. Jeanette won a four-year scholarship to Ohio University, where she enjoyed learning everything from chemistry to Shakespeare to dance. Today, Grasselli's broad range of activities includes chairing the Cleveland Scholarship Programs Board (an organization that provides financial aid to inner-city students), chairing the Education Committee of the Cleveland Orchestra, and sitting on the board of five major corporations as well as the Ohio Board of Regents and the Foundation Board of Ohio University.

Grasselli has published 75 papers in scientific journals, written 9 books, and received 7 honorary Doctor of Science degrees. She uses the name Grasselli from her first marriage of twenty-eight years because it is the one under which she is known professionally. Grasselli loves to share her knowledge and enthusiasm for chemistry and has given over one hundred talks at scientific conferences, one hundred seminars for graduate students, and over five hundred talks to the general public. One of her favorite talks is a program she developed called Operation Super Sleuth. In this program, Grasselli has combined her burning interest in discovering the unknown in science with showing how practical applications can be made. She likes to describe how scientists use advanced analytical techniques to solve crimes. She uses cases from the Cleveland and FBI crime files to illustrate how problem-solving techniques have been used, ranging from simple cases of DNA typing to an investigation of the bombing of Pan Am Flight 103. Grasselli also likes to show students what a career in industry can be like, combining her enthusiasm for science with her interest in people and education to encourage high school students to study chemistry and pursue careers in industry.

One of the issues near and dear to her heart is the future of women in science. In 1989, Grasselli was elected to the State of Ohio's Women's

Hall of Fame; and in 1991, she was the first woman elected to the Ohio Science and Technology Hall of Fame. She states that working hard and rising to the top management ranks of a national corporation had many rewards. "My career has been challenging and exciting. Chemistry is fun because I could channel my inherent curiosity about everything into research plans and objectives aimed at discovery and innovation, hoping to provide products or processes that could benefit mankind."[4] However, she also recognizes that it has always been difficult for women who want a career and a family. This led Grasselli to champion recognition and support for women in the workplace. She was a defender of part-time work for women, equal salaries, and corporate child-care facilities long before they were national issues. She uses her life experience, broader perspective, and approach to problem solving to try to find sensible, socially constructive ways to integrate women into the workplace and into immensely rewarding careers in science and technology.

Notes

1. "Garvan Medal," *Chemical and Engineering News* 14 (October 1985): 48.
2. Edward Hochberg, "Jeanette Gecsy Grasselli," in *Women in Chemistry and Physics* (Westport, Conn.: Greenwood Press, 1993).
3. L. Wasnak, "Making It in a 'Man's' World," *Northern Ohio Business Journal* (April 1982): 57, 78.
4. Jeanette Grasselli Brown, personal communication with Francie Bauer, July 1995.

Bibliography

Eiss, M.I. "1986 Garvan Medalist—Jeanette G. Grasselli," in *Women Chemists*, Vol. 1. Washington, D.C.: ACSW, 1986.

Farkas, K. "Noted Scientist Retires to a Busy Life as a Teacher." *Cleveland Plain Dealer* (April 20, 1989): B2.

"Garvan Medal." *Chemical and Engineering News* 14 (October 1985): 48.

Hochberg, Edward. "Jeanette Gecsy Grasselli," in *Women in Chemistry and Physics*. Westport, Conn.: Greenwood Press, 1993.

Reid, S. "Chemist Finds Industrial Research Rewarding." *Chagrin Valley Times* (January 28, 1988): 24.

Wasnak, L. "Making It in a 'Man's' World." *Northern Ohio Business Journal* (April 1982): 57, 78.

FRANCIE BAUER

ARDA ALDEN GREEN
(1899–1958)
Biochemist

Birth	May 7, 1899
1921	B.S., University of California
1927	M.D., Johns Hopkins University School of Medicine; Fellow, Harvard University Medical School
1929	Research Associate, Harvard University Medical School; Tutor, biochemical sciences, Radcliffe College
1941	Research Associate in Pharmacology, Washington University, St. Louis, MO
1945	Staff Member, Research Division, Cleveland Clinic
1953	McCollum-Pratt Institute of Johns Hopkins
Death	January 22, 1958
1958	Posthumously awarded the Garvan Medal of the American Chemical Society

Arda Green not only contributed to the Coris' Nobel Prize–winning research with her work on enzymes, but her methods also influenced the eventual synthesis of RNA and DNA. And it was Arda Green who unlocked the secret of what makes fireflies glow.

Arda Green, one of the best examples of the phrase "the woman behind the man," was born in Prospect, Pennsylvania, in 1899. She graduated from the University of California at Berkeley with the highest honors in chemistry and honors in philosophy. For one year she pursued graduate studies in philosophy, then decided to go into medicine instead. She studied for two years at Berkeley until a professor, Herbert M. Evans, encouraged her to interrupt the standard medical curriculum for a year of research at Harvard University Medical School.

At Harvard, Green worked in the laboratory of Edwin J. Cohn. After finishing her time there, she had so impressed Cohn that he helped obtain for her a Leconte memorial fellowship upon her return to the University of California. After completing the fellowship she transferred to Johns Hopkins, where she received her M.D. degree in 1927. While there, she had the opportunity to work with biochemist Leonor Michaelis on the conductivity of electrolytes within membranes, and she published her first article.

After graduation she returned to Cohn's laboratory for two years as a National Research Council fellow. Green did her classic studies on the equilibrium relationship between oxygen and hemoglobin and the effect of hydrogen ion activity change by determining the solubility of hemoglobin in various ionic media, work that remained the standard in the field for decades. In fact, Linus Pauling based some of his views on the structure of hemoglobin on her work.

After her fellowship ended Green remained at Harvard, working with Cohn for an additional five years, then as a pediatrics research associate with C. F. McKhann. Her intense concentration and thoroughness enabled her to develop elegant methods for the isolation of blood proteins. With Irvine Page, Green co-authored the first paper to describe angiotensin, a substance that circulates in the blood and helps regulate blood pressure, kidney functions, and water and sodium balance. Of this period, Sidney Colowick noted that "her success in this field was in part due to her conviction that the purification of proteins was a sufficiently important goal in itself to deserve a lifetime of concentrated effort." He added that "more important was her ability to consider problems of protein purification in a logical manner—an ability based on her broad knowledge of their physiochemical properties."[1]

Green also did immunological investigations with McKhann on placental extracts to prevent measles. While carrying out this research and the hemoglobin research, she also found time to be a biochemical sciences tutor at Radcliffe College. By her example, she influenced several of her students there to go into the medical research field.

After a long tenure at Harvard, Green left for Washington University in St. Louis, Missouri, where she worked in the laboratory of Carl and **Gerty Cori** as a research associate in pharmacology. Within a year she was an assistant professor of biological chemistry. She developed a method for the isolation of phosphorylase, upon which rested much of the Coris' 1946 Nobel Prize–winning research on polysaccharide synthesis. "It is noteworthy that she developed this method in the short space of ten weeks, after three other investigators in the same laboratory had devoted three fruitless years to the same problem."[2] The Coris' research involved conversion of carbohydrates into energy, the mechanism of glucose-glycogen conversion, and the effects of hormones on this process.

Green did other research on muscle proteins while at Washington University, developing purification and characterization methods and crystallizing rabbit aldolase, the enzyme that works in the process of converting fructose to fuel for muscle activity. By virtue of being in this high-profile laboratory at the time, she influenced the course of enzymology in the United States as future biochemists, such as Severo Ochoa (the first to synthesize RNA) and Arthur Kornberg (the first to synthesize

DNA), learned her methods and logical approach. Ochoa and Kornberg shared the 1959 Nobel Prize in medicine.

Green then left the Coris for an opportunity to work again with Irvine Page, this time at the famed Cleveland Clinic. She joined the research team in the same year that Page became director of research there, indicating that she was specifically recruited to come aboard. She isolated and characterized a substance in the blood serum that constricted blood cells, which her research group named serotonin. Once serotonin was identified and could be studied directly, scientists found it was a mechanism in many biochemical processes, including circulation, pain, sleep, drug effectiveness, and psychological function. Her accomplishment, revolutionizing the understanding of how the central nervous system works, was hailed as a revolution in pharmacology. Green also collaborated with F. M. Bumpus on identifying the proteins involved in renal hypertension.

In 1953, Green joined W. D. McElroy at the McCollum-Pratt Institute of Johns Hopkins University. She began to make systematic efforts to purify the enzyme known as luciferase, the substance that allows fireflies to glow. Biochemists knew that luciferase was necessary, but not how it actually caused bioluminescence. In 1956, Green successfully crystallized the enzyme from firefly lanterns. Using the crystalline enzyme, Green proved that it would glow in the presence of oxygen, adenosinetriphosphate, magnesium, and its own substrate, luciferin. She then studied the reaction step-by-step, identifying an adenyl-luciferin intermediate and explaining the mechanism by which co-enzyme A promotes the emission of light.

The methods Green developed allowed systematic study of broader problems in catalysis involving similar substances, since they "established the basic similarity of universally occurring enzyme reactions for the activation of acetate and amino acids."[3] After publishing her studies, she began to work on the isolation of enzymes involved in bioluminescence of bacteria, describing its properties in 1955 and continuing to work to crystallize the protein. Her health, however, was failing. About this time, she began treatment at Sinai Hospital for breast cancer.

Arda Green was popular with her colleagues and "devoted herself at home to cooking, dressmaking, music, and entertaining. It was she who saw to it that no unattached members of the laboratory staff ever went unfed on Thanksgiving Day."[4] Her research teams recognized her remarkable ability to unravel the mysteries of the smallest of substances and even teased her about it. "Many have attributed her success in crystallizing proteins to her 'magic touch.' Some even suggested to her that her secret lay in some mysterious seeding effect of the ash falling from the cigarette which never left her mouth when she was working."[5]

In the autumn of 1957, Green learned that the American Chemical

Society would award her the Garvan Medal for her contributions to biochemistry at its annual meeting in 1958. The Garvan Medal, begun in 1937, was created to honor a woman scientist for distinguished service in the field of chemistry. The 1948 winner of the medal was her former collaborator, Gerty Cori, for enzymatic synthesis and reactions. The acknowledgment had to be sweet: "Arda Green did not always receive the recognition which she deserved, partly because she always worked in the shadow of men of great scientific reputation. The uniformly high quality of her research over the years makes it clear that she had a unique talent which was responsible for her success."[6]

Arda Green died at Sinai Hospital in January 1958 of carcinoma of the breast at the age of 57. She was to have been awarded the Garvan Medal in San Francisco in April of the same year. Her sister, Metta Claire Green Lewis, accepted the award on her behalf.

Notes

1. Sidney P. Colowick, "Arda Alden Green, Protein Chemist," *Science* 128 (September 5, 1958: 520–521.
2. Ibid., p. 520.
3. Ibid.
4. Ibid., p. 521.
5. Ibid.
6. Ibid., p. 520.

Bibliography

[ACS Award—Garvan Medal.] "Arda A. Green." *Chemical and Engineering News* 36 (April 28, 1958): 123.

Bumpus, F.M., Arda A. Green, and I.H. Page. "Purification of Angiotensin." *Journal of Biological Chemistry* 210 (September 1954): 281–286.

Colowick, Sidney P. "Arda Alden Green, Protein Chemist." *Science* 128 (September 5, 1958): 519–521.

[Deaths.] "Green, Arda Alden." *Journal of the American Medical Association* 166 (1958): 1762.

Green, Arda A., and F.M. Bumpus. "Purification of Hog Renin Substrate." *Journal of Biological Chemistry* 210 (September 1954): 287–294.

Green, Arda A., and W.D. McElroy. "Crystalline Firefly Luciferase." *Biochimica et Biophysica Acta—Amsterdam* 20 (April 1956): 170–176.

[Necrology.] "Arda A. Green." *Chemical and Engineering News* 36 (February 17, 1958): 121.

O'Neill, Lois Decker. *The Women's Book of World Records and Achievements*. Garden City, N.J.: Anchor Press/Doubleday, 1979.

KELLY HENSLEY

DOROTHY ANNA HAHN

(1876–1950)

Chemist

Birth	April 9, 1876
1899	B.A., chemistry, Bryn Mawr
1899–1906	Professor, chemistry, Pennsylvania College for Women, Pittsburgh
1906–07	Studied at University of Leipzig
1907–08	Studied at Bryn Mawr
1908–14	Instructor, Mt. Holyoke College
1913	First published article
1914–18	Associate Professor, chemistry, Mt. Holyoke College
1916	Ph.D., Yale University
1918–41	Professor, chemistry, Mt. Holyoke College
1941	Retired from Mt. Holyoke College
Death	December 10, 1950

Dorothy Hahn and **Emma Perry Carr** formed the team at Mount Holyoke College that allowed many women chemists of distinction to further their education and make significant contributions to the field of chemistry.

Dorothy Hahn was born in Philadelphia, Pennsylvania, in 1876, the younger of two daughters of Carl S. Hahn and Mary (Beaver) Hahn. Her mother was a native of Pennsylvania, but her father was born in Germany. Her father worked variously as a clerk or bookkeeper, a seller of artificial flowers, and a linguist. Little is known of Hahn's childhood. She attended Philadelphia Girls' High School for a time and graduated from Miss Florence Baldwin's School (now The Baldwin School) in Bryn Mawr, Pennsylvania. She attended Bryn Mawr College from 1895 to 1899. After graduation from Bryn Mawr in 1899, Hahn was hired as professor of chemistry and biology at Pennsylvania College for Women (later Chatham College), where she taught from 1899 to 1906. From 1904 to 1906, she was also professor of biology at Kindergarten College in Pittsburgh.

In 1906–1907, Hahn worked in the laboratory of Professor Arthur Hantzch at the University of Leipzig in Germany. Returning to the United

Dorothy Anna Hahn. Photo reprinted by permission of The Mount Holyoke Archives and Special Collections.

States, she did graduate study at Bryn Mawr under the organic chemist E. P. Kohler. In 1908, she became a faculty member at Mount Holyoke College in South Hadley, Massachusetts. With the exception of two leaves of absence, she remained on the faculty at Mount Holyoke until 1941.

Hahn attended Yale University on an AAUW (American Association of University Women) fellowship in 1915–1916, receiving her Ph.D. from Yale in 1916. At Yale she worked with Treat B. Johnson, an organic chemist. In 1922 she co-authored a book, *Theories of Organic Chemistry*, with Johnson. Because Johnson was a strong supporter of women in chemistry, she continued to send him her best students throughout the course of her career. As a result of her encouragement and inspiration, many students from Mount Holyoke obtained Ph.D.s in chemistry from Yale.

One of these students recalled: "a woman of great sincerity and strength of character, Miss Hahn possessed imagination, humor and a sense of realism—the latter tempered by her insight and sympathy. For many students, she opened a new horizon of what might and should be achieved."[1]

Hahn's first paper was published in 1913. It was a report of work done at Mount Holyoke using the newly developed ultraviolet spectroscopy method developed by Emma Perry Carr of Mount Holyoke. In this paper, written with one of her graduate students, Hahn confirmed the ring structure of hydantoins. Hydantoins are organic compounds closely related to naturally occurring compounds such as vitamin B_1.

In 1915, Hahn was one of the first to describe in English the relationship of electrons to chemical valence, a year before Gilbert Lewis and Irving Langmuir published their valence theories.[2] (Valence electrons orbit the outer shell of an atom and largely determine the atom's properties.) From 1913 to 1940 she was the author of twenty-two articles, most of them published in the *Journal of the American Chemical Society*. She was also the co-author of three books or monographs in addition to the one she wrote with Treat Johnson.

During Hahn's years at Mount Holyoke, Emma Perry Carr was the most prominent figure in the chemistry department, chairing it from 1913 to 1946. One of their colleagues recalled: "Miss Carr and Miss Hahn were both remarkable women, but they didn't see things in the same light. They agreed very peaceably to each run their own department. Miss Hahn ran the organic chemistry as a separate department."[3] For many years the course catalogues of the college reflected this split in the department, listing the course offerings in organic chemistry separately from the others.

Hahn's summers were often spent away from South Hadley. In 1917 and 1918 she worked in the research laboratories of the Barrett Company, researching coal tar products. From 1927 to 1948 she spent her summers at the Connecticut shore in Noank, Connecticut, in a cottage shared with Dean Margaret Morriss of Pembroke College in Rhode Island. During these summers, she became known as a skilled sailor.

After her retirement Hahn continued to live the rest of the year in South Hadley, sharing a house with another faculty member, Dorothy Foster, of the Mount Holyoke English department. Upon her death in 1950 and at her request, her ashes were scattered on a hillside facing Mount Tom and Mount Holyoke. Eighty of her former students organized the Dorothy Hahn Memorial Fund to furnish a seminar room in the new chemistry building at Mount Holyoke in her honor. Dorothy Hahn was known "within and outside the college for scholarship, research achievements, informed and lucid thinking, and gifted teaching."[4]

Notes

1. Alice G. Renfrew, "Dorothy A. Hahn, Scientist and Teacher," *Mount Holyoke Alumnae Quarterly* (February 1951): n.p.
2. George Fleck, "Mary Lura Sherrill," in *Women in Chemistry and Physics: A Biobibliographic Sourcebook*, eds. L.S. Grinstein, R.K. Rose, and M. Rafailovich (Westport, Conn.: Greenwood Press, 1993), p. 533.
3. Interview with Lucy Weston Pickett by Carole B. Shmurak, May 1990. (Tape in Mount Holyoke College Library/Archives.)
4. Renfrew, "Dorothy A. Hahn," n.p.

Bibliography

Hahn, Dorothy A., and Angie Allbee. "Saturated Delta-Ketonic Esters and Their Derivatives." *American Chemical Journal* 49 (1913): 171–179.

Hahn, Dorothy A., and M.E. Holmes. "The Valence Theory of I. Stark from a Chemical Standpoint." *Journal of the American Chemical Society* 37 (1915): 2611–2626.

Hahn, Dorothy A., and Treat B. Johnson. *Pyrimidines: Their Amino and Aminooxy Derivatives*. Monograph in *Chemical Reviews*, 1933.

———. "Researches on Hydantoins." *Journal of the American Chemical Society* 39 (1917): 1257–1266.

———. *Theories of Organic Chemistry*. New York: John Wiley, 1922.

Hahn, Dorothy A., and Margaret K. Seikel. "The Isomerization of Certain Saturated and Unsaturated Hydantoins." *Journal of the American Chemical Society* 47 (1936): 647–649.

Renfrew, Alice G. "Dorothy Anna Hahn," in *Notable American Women, 1607–1950: A Biographical Dictionary*. Edited by Edward T. James. Cambridge, Mass.: Belknap Press of Harvard University Press, 1971.

CAROLE B. SHMURAK

HEIDI HAMMEL

(1960–)

Astronomer

Birth	March 14, 1960
1982	B.S., earth and planetary science, Massachusetts Institute of Technology
1988	Ph.D., physics and astronomy, University of Hawaii

Heidi Hammel. Photograph by Donna Coveney.

1989	Team Member, NASA Voyager Imaging Science Team, Neptune Encounter
1990	NASA Group Achievement Award for Voyager Science Investigation
1990–	Principal Research Scientist, Massachusetts Institute of Technology
1994	NASA Science Investigations, Team Leader, NASA Hubble Space Visible/Near-UV Imaging Team, Comet Shoemaker–Levy 9 Collision with Jupiter; Vladimir Karapetoff Award (in recognition of contributions to science and education), Massachusetts Institute of Technology
1995	Klumpke-Roberts Award, Astronomical Society of the Pacific

Heidi Hammel fits the definition of an astronomer: she does extensive research, analyzes immense amounts of data, and publishes her findings in scientific papers. In fact, she has published over thirty such papers. But what she finds to be of probably equal importance to her purely scientific work is sharing her excitement of discovery in astronomy with ordinary people who may not have a scientific background. When she was head of the twelve-person Hubble Space Telescope observation team of the Comet Shoemaker-Levy 9 crash into Jupiter, the scene at Space Telescope Science Institute in Baltimore probably was closer to that of a ballpark after a victorious world series ballgame than a scientific observatory. After saying that she felt really sorry for Jupiter because it was being covered with "splotches, black eyes and bruises," she and the team began cheering and toasting each other with swigs from a champagne bottle.[1] Then she grabbed another bottle of champagne and ran upstairs to inform the three discoverers of the comet, Carolyn and Eugene Shoemaker and David Levy, of what was happening.

Hammel, born Sacramento, California, in 1960, was one of three children in a working-class family. When Heidi was 6 years old the family moved to Scranton, Pennsylvania. Her father worked for the state and her mother was a nurse. It was stressed to her as a child that "you can do anything you want."[2] She also grew up understanding the importance of and experiencing the joy of reading everything.

> I watched the usual high intellectual shows on television like "Batman" and "The Dating Game" . . . but my parents encouraged me to read books all the time. All kinds of books. Reading was an important part of our world. They also read to us a lot. There might have been an occasional "golden book" in the sciences, but they read all kinds of books. Science was just one thing you were exposed to like everything else.[3]

Hammel's first sky watching took place with a tiny toy telescope that was a gift from her parents, but she later moved on to the local planetarium in Harrisburg. Yet as a child she was not particularly interested in astronomy. A math wizard growing up, she ran for treasurer of her high school class. When her opponent made a cartoon of her with her head full of equations, she lost the election. "I went home and cried about it, but I learned it's what you do after you finish crying that makes a difference."[4] Losing the election for class treasurer in high school was not her only disappointment. Heidi Hammel had not considered applying to MIT. "I was thinking about schools in Pennsylvania," she recalls. "I even had a dormitory room assigned at Penn State."[5] A calculus teacher, however, encouraged her to apply to MIT. When she asked a male chemistry teacher for a letter of recommendation to MIT, he refused, saying she would never get in. And when she did get in, he said "it was only because you're a woman and they've got quotas to fill."[6]

It was in her sophomore year at MIT that Hammel's interest in astronomy first peaked. Like most first-year students, she mainly had to take required classes, which gave her a solid grounding in all subjects. But then she could sign up for electives and take any class that she thought interesting. Astronomy was one of them. Hammel was the only female in a four-person astronomy class that included two seniors and a graduate student. Feeling out of place, she considered dropping out but was persuaded by the instructor to stay. "One clear night (under the stars) made all the difference," she says.[7] She credits her astronomy professor for convincing her that "she was good enough to be an astronomer."[8] Hammel went on to receive her Ph.D. in physics and astronomy from the University of Hawaii at Manoa, writing her doctoral thesis on the clouds and structure of Neptune. She chose Hawaii because it had the best and biggest telescopes for what she was studying.

That was when Hammel had her first experience in explaining science to the public.

With Halley's comet's visit to the Earth in 1986 I spent a lot of time talking to elementary school children, travel agents, and the general public. I did get a lot of experience in talking to the public with the crash of Shoemaker-Levy 9 into Jupiter, but basically it had been something that I had been doing for years. So many people think that science is esoteric, that it is something you must study for years to understand. But science is a way of asking questions and exploring. It is a process. It is a way you ask the right question and how you find the right answer. In elementary school the emphasis should be on the diversity of science—how neat things are. While they need basic facts like the names of the planets, elementary school students should learn that science is something you do, not something that is read in a book. In high school students should be taught that science is a process, a way of thinking about things. All the memorization is just about terminology.[9]

Her first job after graduate school was at NASA's Jet Propulsion Laboratory in Pasadena, California, where she was a member of the imaging team for *Voyager 2*'s encounter with Neptune. Like many astronomers, she did not think the comet crash into Jupiter would be all that interesting, but she had applied for a grant with two other scientists to study the atmospheric waves from the comet. A NASA committee informed her that she had been selected to head the imaging team, and Hammel felt that it was because "they wanted someone young, with the stamina to deal with the situation."[10]

During the comet crash, which lasted from July 16 to July 22, 1994, Hammel became a bit of a media star. She used language to describe the crash that all could understand and shared her excitement on television stations and in many interviews. When she compared the blackened plumes on Jupiter to bruises and black eyes and said that she felt sorry for the planet, she was using language to which scientists are unaccustomed. She believes scientists need to loosen up. "I don't like to talk over people," she explains.

> I prefer to talk very simply and very plainly.... Even in science conferences you can either talk to the top 5 percent in your field who will know what you're talking about, or you can talk to the 95 percent of the people who are not necessarily that familiar with your work and just want to learn something. At scientific meetings the normal scientific style is so boring . . . I just get up there and have a good time.[11]

Since July 1995, Heidi Hammel has done quite a bit of traveling, lecturing about the comet crash to groups from scientists to second graders. She believes she has the perfect job because she finds scientific discovery so much fun. She continues to study the atmosphere and surface of Neptune, and she continues to analyze data from the comet crash.

> Planetary collisions (with other members of the solar system) happen all the time, but this was the first time in history we were able to observe it. Now we have hard data about what actually happens. We learned about the altitude and gravity of Jupiter, its clouds, its wind and its atmosphere. Before the crash all we had about Jupiter were theories. Now we have data to back it up. And we learned a whole lot about the strength of comets.[12]

In the spring, Hammel teaches an astronomy class at MIT. She says she may be starting to enjoy traveling less and less. "I am working to find the right compromise between traveling, teaching, and research. I would love to give talks to all the groups that ask. I just can't. I need to do my own research."[13] But teaching and doing research in astronomy are not the only things that occupy Hammel's life. She is a trained pitch percussionist, has performed with a Chinese opera performance group, and has had a recipe published on a home brew of beer. And she enjoys reading romance novels and science fiction.

Her advice to people considering a career in science is to "read everything in sight. Take classes in everything, including English, that is essential for good communication skills. Don't focus on any one subject too early. Take a broad variety of classes. I didn't decide what I wanted to be until I was in college."[14] Hammel feels that reading a lot as a child influenced her brother and sister also. Her brother is a lawyer and her sister is in the retail book business. Finally, her message to everyone is to "keep studying, do your math, work hard. . . . My job is such great fun. . . . It doesn't get better than this."[15]

When Hammel starts talking about Jupiter's atmosphere her voice lights up, she speaks in language everyone can understand, and it is obvious she is having a great time. Whether or not she makes great discoveries in astronomy, Heidi Hammel has already convinced the public that you need not have a Ph.D. in science to know "that our universe is a living and changing entity and that we are part of it."[16]

Notes

1. Kim McDonald, "The Comet Drama's Biggest Hit," *Chronicle of Higher Education* (July 27, 1994): A7.
2. Margery Eagan, "Stargazer Shining Example of Women's Bright Future," *Boston Herald* (July 21, 1994): 4.
3. Heidi Hammel and Natalie Kupferberg, telephone interview, September 14, 1995.
4. Eagan, "Stargazer," p. 4.
5. Hammel, telephone interview.
6. Eagan, "Stargazer," p. 4.
7. J. Kelly Beatty, "The Women Who Watch the Sky, They Touch the Future," *Parade* (July 16, 1995): 4–5.
8. McDonald, "The Comet Drama's Biggest Hit," p. A7.
9. Hammel, telephone interview.
10. McDonald, "The Comet Drama's Biggest Hit," p. A12.
11. Ibid.
12. Hammel, telephone interview.
13. Ibid.
14. Ibid.
15. Beatty, "The Women," p. 5.
16. Ibid.

Bibliography

Baines, K.H., and H.B. Hammel. "Clouds, Hazes and the Stratospheric Methane Abundance in Neptune." *Icarus* 109 (1994): 20.
Beatty, J. Kelly. "The Women Who Watch the Sky, They Touch the Future." *Parade* (July 16, 1995): 4–5.
Eagan, Margery. "Stargazer Shining Example of Women's Bright Future." *Boston Herald* (July 21, 1994): 4.
Hammel, H.B., and Natalie Kupferberg. Telephone interview. September 14, 1995.
Hammel, H.B., et al. "An Atmospheric Outburst on Neptune from 1986 through 1989." *Icarus* 99 (1992): 363.
———. "Collision of Comet Shoemaker-Levy 9 with Jupiter Observed by the NASA Infrared Telescope Facility." *Science* 267 (1995): 1277.

————. "Hubble Space Telescope Imaging of Neptune's Cloud Structure in 1994." *Science* 268 (1995): 1740.
————. "Impact Debris Particles in Jupiter's Stratosphere." *Science* 267 (1995): 1288.
McDonald, Kim. "The Comet Drama's Biggest Hit." *Chronicle of Higher Education* (July 27, 1994): A7, A12–A13.

NATALIE KUPFERBERG

ANNA JANE HARRISON
(1912–   )
Chemist

Birth	December 23, 1912
1933	B.A., chemistry, University of Missouri
1935	B.S., education, University of Missouri
1937	M.A., chemistry, University of Missouri
1940	Ph.D., physical chemistry, University of Missouri
1940–45	Instructor and Assistant Professor of Chemistry, Sophie Newcomb College
1945–47	Assistant Professor of Chemistry, Mt. Holyoke College
1947–50	Associate Professor of Chemistry, Mt. Holyoke College
1950–79	Professor of Chemistry, Mt. Holyoke College
1960–66	Chair, Dept. of Chemistry, Mt. Holyoke College
1969	Manufacturing Chemists Association Award in College Chemistry
1978	President, American Chemical Society
1979	Retired from Mt. Holyoke College
1982	American Chemical Society Award in Chemical Education
1983	President, American Association for the Advancement of Science
1989	Publication of *Chemistry: A Search to Understand*

Anna Jane Harrison carried on the tradition of teaching excellence and research in chemistry at Mount Holyoke College that was begun by **Emma Carr** and **Dorothy Hahn**. Harrison's interest in the development

Anna Jane Harrison. Photo reprinted by permission of The Mount Holyoke Archives and Special Collections.

of science policy led her to the National Science Board and the presidency of the American Association for the Advancement of Science as well as the American Chemical Society. She was the Society's first woman president.

Anna Jane Harrison was born in Benton City, Missouri, in 1912, the daughter of Albert Harrison and Mary Katherine (Jones) Harrison. Anna Jane, her older brother, and her parents were very much a part of the farming community. Anna Jane was only 7 years old when her father died, but her mother continued to manage the farm until 1960. Growing up in a rural community, Anna Jane and her brother learned to be independent and responsible. "We learned not to be limited by other people's expectations."[1] She had very little to do with science in grade school, although she remembers her teacher saying "we're going to do science" and sending the children home to find out about caterpillars. Consulting her father, she learned a lot about Caterpillar tractors. "Fortunately the teacher called on someone else."[2]

Harrison went to Mexico High School in Mexico, Missouri, where she became interested in the sciences because "the best teachers I had were in math and science; I just found it most stimulating."[3] After graduation she went to the University of Missouri, from which she received two degrees: a B.A. in chemistry in 1933 and a B.S. in education in 1935. She remembers that when she signed up for advanced physics courses, "semester after semester, the chairman of the department would explain to me that this was a very foolish thing for a girl to do. But he wouldn't stand in my way if I insisted."[4]

She spent a few years teaching in a one-room schoolhouse, the Sunrise School in Andrain County. Her students ranged in age from 5 to 21 and were in grades 1–6 and 9. Of her experience there, she said, "It may have been embarrassing to be teaching at a one-room schoolhouse, but it was important experience, since I was completely on my own. If I got into problems, I had to get myself out."[5] She returned to the University of Missouri for graduate work, earning her M.A. in 1937 and her Ph.D. in physical chemistry in 1940. Harrison was the first woman to earn a Ph.D. in chemistry at that institution, but she did not discover that fact until a few years afterward: "maybe it was better that I didn't know."[6]

In 1940 "the only places that were hiring women in chemistry were women's colleges," so Harrison went to Sophie Newcomb College, which was part of Tulane University.[7] She began as an instructor and was promoted to assistant professor in 1942. In 1943 she worked for the National Defense Research Council on the collection and detection of toxic smoke. In 1945, with the war over, she moved to Mount Holyoke College. She had met Emma Perry Carr and **Lucy Pickett** at meetings of the American Chemical Society and was eager to join them at Mount Holyoke. Indeed, she spent the rest of her teaching and research career there until her retirement in 1979.

Harrison was promoted to associate professor at Mount Holyoke in 1947 and to full professor in 1950. As a college teacher, she put emphasis on thinking and problem solving rather than on one right way to do things, allowing students to select the methods that were best for them.[8] She was noted for her lively sense of humor and for teaching the freshman chemistry course in a way that made the basic principles seem fascinating and the complex seem simple.

Throughout her career at Mount Holyoke, she was a member of the research team that established the structure of many organic molecules through spectroscopy. Much of her research concerned the absorption of organic compounds in the vacuum ultraviolet. An AAUW fellowship enabled her to go to Cambridge University in England to do research in flash photolysis. During a sabbatical in 1959–1960 she worked at the National Research Council in Canada on photolysis by high frequency ultraviolet light.

During the 1970s, Harrison became more concerned with the application of scientific knowledge to the solution of societal problems and the importance of scientists' participation in the development of national policies. From 1972 to 1978 she served on the National Science Board (NSB), the policy-making unit of the National Science Foundation. As a member of the NSB, she traveled to Antarctica to survey the scientific work being done at the American installation there. In 1971 she became the first woman to chair the Division of Chemical Education of the American Chemical Society. She served on the editorial board of the *Journal of Chemical Education* from 1959 to 1966 and from 1970 to 1973, chairing that board in 1960 and 1964–1965. She also served on the editorial boards of the *Journal of College Science Teaching, Chemical and Engineering News,* and *Science 80.*

In 1978, Harrison became the first woman to be elected president of the American Chemical Society in the 102–year history of that society. Speaking of her election, she said that she saw no particular significance in this, because she did not campaign as a woman; however, she hoped that she would not be the last woman to be president of the ACS.[9] She was elected president of the American Association for the Advancement of Science in 1983, the fourth woman to occupy that post. Between 1975 and 1990, Harrison received the honorary degree of D.Sc. from ten institutions, including Smith College, Williams College, Lehigh University, Worcester Polytechnic Institute, Mount Holyoke College, and the University of Missouri. She also received the Manufacturing Chemists Association Award in College Chemistry for outstanding teaching in 1969 and the American Chemical Society Award in chemical education in 1982.

Speaking of her travels in the Oregon wilderness and in South America, Harrison once said, "What I really like is to go places one isn't supposed to go."[10] In many ways this statement characterizes Anna Jane Harrison's entire life and career as a teacher, a researcher, and a spokesperson for the scientific community.

Notes

1. Anna J. Harrison, interview with Carole B. Shmurak, April 11, 1995.
2. Ibid.
3. Ibid.
4. *Boston Globe*, November 23, 1977. In Harrison papers, Mount Holyoke College Library Archives.
5. Interview, April 11, 1995.
6. Ibid.
7. Ibid.
8. Ibid.

9. *Chemical and Engineering News*, November 22, 1976. In Harrison papers, Mount Holyoke College Library Archives.

10. *Springfield Union* [Springfield, Massachusetts], November 5, 1970. In Harrison papers, Mount Holyoke College Library Archives.

Bibliography

Harrison, A.J. "Chemical Education and the Expectations of Society." *Journal of Chemical Education* 51 (1974): 569–571.

———. "Goals of Science Education." *Science* 217 (1982): 109.

———. "The Role of Chemical Education." *Journal of Chemical Education* 48 (1971): 719.

———. "Science, Engineering and Technology." *Science* 223 (1984): 543.

Harrison, A.J., and E.S Weaver. *Chemistry: A Search to Understand*. New York: Harcourt Brace, 1989.

Roscher, Nina Matheny. "Anna Jane Harrison," in *Women in Chemistry and Physics: A Biobibliographic Sourcebook*. Edited by L.S. Grinstein, R.K. Rose, and M. Rafailovich. Westport, Conn.: Greenwood Press, 1993.

CAROLE B. SHMURAK

MARGARET HARWOOD

(1885–1979)

Astronomer

Birth	March 19, 1885
1907	A.B., Radcliffe College
1916	M.A., University of California, Berkeley
1916–57	Director, Maria Mitchell Observatory in Nantucket, MA
1957	Medal for Distinguished Achievement, Radcliffe College
1962	Annie J. Cannon Prize in Astronomy
Death	1979

A dedicated scientist, Margaret Harwood devoted her life to the study of astronomy. She was born in 1885 in Littleton, Massachusetts. Her parents were Emelie Augusta (Green) and Herbert Joseph Harwood. Margaret received a bachelor of arts degree from Radcliffe College in 1907 and a master of arts degree from the University of California at Berkeley in 1916.

From 1907 to 1912, Harwood worked as a computer assistant at the Harvard College Observatory. She was an astronomical fellow at the Maria Mitchell Observatory in Nantucket, Massachusetts, from 1912 to 1916. In 1916 she became director of the Observatory and held the position until 1957. Perhaps Harwood was inspired in her work by the Observatory's namesake, **Maria Mitchell**, a trailblazing nineteenth-century scientist who is often cited as the "first American woman astronomer."[1] Harwood also worked as a staff member of the radiation laboratory at the Massachusetts Institute of Technology from 1944 to 1946.

Harwood pursued her commitment to astronomy through membership in a number of American and British astronomical societies. She was a fellow of the American Association for the Advancement of Science and held memberships in the American Association of Variable Star Observers and the Maria Mitchell Association of Nantucket. Harwood was also a member of the Royal Astronomical Society and the International Astronomical Union.

In 1917, Harwood discovered an asteroid, also known as a minor planet. However, male astronomer George H. Peters was credited with the discovery of Asteroid No. 886 and had the honor of naming it. Peters received credit for the discovery although he sighted the asteroid four days after Harwood discovered it.[2]

Although Harwood did not receive official credit for her discovery of the asteroid Washingtonia, she did receive recognition for her work as an astronomer. In 1962 she received the Annie J. Cannon Prize in Astronomy from the American Association of University Women and the American Astronomical Society. A grant given to women who are conducting astronomy research, the award helped support Harwood's research on variable stars.[3] Harwood also received a medal for distinguished achievement in her field from Radcliffe College.

Notes

1. Martha J. Bailey, *American Women in Science: A Biographical Dictionary* (Santa Barbara, Calif.: ABC-CLIO, 1994), p. 253.

2. Blaine P. Friedlander Jr., "Familiar Places Go to Outer Space," *Washington Post* (June 5, 1986): district weekly section, final edition.

3. Claire Walter, *Winners, the Blue Ribbon Encyclopedia of Awards*, rev. ed. (New York: Facts on File, 1982), pp. 447–448.

Bibliography

Bailey, Martha J. *American Women in Science: A Biographical Dictionary*. Santa Barbara, Calif.: ABC-CLIO, 1994.

Friedlander, Blaine P., Jr. "Familiar Places Go to Outer Space." *Washington Post* (June 5, 1986): district weekly section, final edition.

Walter, Claire. *Winners, the Blue Ribbon Encyclopedia of Awards*. Rev. ed. New York: Facts on File, 1982.

Who's Who of American Women, 1958–59. 1st ed. Chicago: Marquis Who's Who, 1958.

Who's Who of American Women, 1964–65. Chicago: Marquis Who's Who, 1964.

HEATHER MARTIN

CAROLINE HERSCHEL

(1750–1848)

Astronomer

Birth	March 16, 1750
1772	Moved to England to live with brother William
1781	William Herschel discovered planet Uranus
1782	Caroline and William gave last musical performance
1786	Discovered first comet on August 1
1798	Became first woman to publish in the *Philosophical Findings* of the Royal Astronomical Society with her *Index to Flamsteed's Observations of the Fixed Stars*
1822	Returned to Hanover, Germany
1828	Received Gold Medal by Royal Astronomical Society for her work *The Reduction and Arrangement in the Form of a Catalogue in Zones of All the Star Clusters and Nebulae Observed by Sir William Herschel*
1835	Became one of first two women elected honorary members of the Royal Astronomical Society (**Mary Somerville** is the other)
1838	Honorary Member of the Royal Irish Academy
1846	Awarded Gold Medal for Science by King of Prussia
Death	January 9, 1848

Caroline Lucretia Herschel was the first woman to discover a comet. She had served her older brother William as housekeeper and general assistant, in which capacity she learned astronomy from him. Her own work was limited, however, by social restrictions on women and by her position as his assistant. But in 1786 her brother's absence allowed her the first opportunity to step beyond the role of observer and recorder and prove what she could do on her own. Comets remained her field of expertise; William never discovered one. She sent an account of her discovery to the Royal Society, requesting that it be communicated to her "brother's astronomical friends."

> The employment of writing down the observations, when my brother uses the 20–feet reflector, does not often allow me time to look at the heavens. But as he is now on a visit to Germany, I have taken the opportunity of his absence to sweep in the neighborhood of the sun, in search of comets; and last night, the 1st of August, about 10 o'clock, I found an object very much resembling in colour, and brightness the 27 nebula of the Connoissance des Temps, with the difference, however, of being round. I suspected it to be a comet; but a haziness coming on, it was not possible to satisfy myself as to its motion till this evening. I made several drawings of the stars in the field of view with it, and have enclosed a copy of them, with my observations annexed, that you may compare them together.[1]

Born in Hanover, Germany, in 1750, Caroline was one of six children of a musician in the band of the Royal Hanoverian Foot Guards. Isaac and Anna Herschel raised their children in the difficult times in Prussia under Frederick II (the Great). Owing to poverty and the lack of opportunities and education for girls, as well as her position as the youngest daughter at home, Caroline worked as an unpaid servant. However, the fact that her father gave her secret violin lessons and showed her the stars probably offered Caroline some vision of possible accomplishments. Six years after her father died, when she was 22 years old, William had Caroline move to England. He had moved to England several years earlier to seek employment as a musician. When he became organist at Octagon Chapel at Bath, his financial position improved. With twin interests in music and astronomy and the financial wherewithal, he could afford and needed assistance. When William heard from a younger brother about Caroline's difficult life at home and about her good voice, William saw the opportunity to help her and himself. What Caroline thought of originally as two years away from home became fifty. She never saw her mother again, and the course of her life radically changed.

Caroline did become a successful soprano. But when William's interest turned completely to astronomy, she began training as an assistant. In an address to the Royal Astronomical Society on the occasion of Caroline's receiving an honorary medal, J. South, Esq., proclaimed:

she it was whose pen conveyed to paper his observations as they issued from his lips; she it was who noted the right ascensions and polar distances of the objects observed; she it was who, having passed the night near the instrument, took the rough manuscripts to her cottage at the dawn of day and produced a fair copy of the night's work on the following morning; she it was who planned the labour of each succeeding night; she it was who reduced every observation, made every calculation; she it was who arranged everything in systematic order; and she it was who helped him to obtain his imperishable name.[2]

William Herschel, called the father of stellar astronomy, had among his many achievements the discovery of the planet Uranus. Yet his daughter-in-law, Mrs. John Herschel, writes of Caroline that "she might have become a distinguished woman on her own account. . . . But the pleasure of seeking and finding for herself was scarcely tasted." She goes on to describe Caroline's attitude about her accomplishments: "in her later life she met with honour and recognition from learned men and learned societies; but the dominant idea was always the same—I am nothing, I have done nothing; all I am, all I know, I owe to my brother. I am only the tool which he shaped to his use—a well-trained puppy-dog would have done as much."[3]

After William died in 1822, Caroline returned to Hanover, Germany, to live and work for another twenty-six years. She continued to encourage the work of her nephew, John, and presented to him her revised version of William's catalogue of the nebulae in zones, for which she received the Gold Medal from the Royal Astronomical Society. She received other awards during this time, but her health prevented her from doing much work. Caroline died in 1848 at the age of 97. Her self-written epitaph depicts her feelings of both pride and humility:

The eyes of Her who is glorified were here below turned to the starry Heavens. Her own Discoveries of Comets, and her participation in the immortal Labours of her Brother, William Herschel, bear witness of this to future ages. The Royal Irish Academy of Dublin, and the Royal Astronomical Society of London enrolled Her name amoung their Members.[4]

Notes

1. Margaret Alic, *Hypatia's Heritage: A History of Women in Science from Antiquity through the Nineteenth Century* (Boston: Beacon Press, 1986), pp. 128–129.

2. Mrs. John Herschel, *Memoir and Correspondence of Caroline Herschel* (New York: D. Appleton and Company, 1876), p. 224.

3. Ibid., pp. vi, ix.
4. Ibid., p. 352.

Bibliography

Alic, Margaret. *Hypatia's Heritage: A History of Women in Science from Antiquity through the Nineteenth Century.* Boston: Beacon Press, 1986.

Ashton, Helen, and Katharine Davies. *I Had a Sister: A Study of Mary Lamb, Dorothy Wordsworth, Caroline Herschel and Cassandra Austen.* [Reprint of the 1937 edition published by L. Dickson, London.] Folcroft, Pa.: Folcroft Library Editions, 1975.

Clerke, Agnes M. *The Herschels and Modern Astronomy.* New York: Macmillan, 1895.

Herschel, Mrs. John. *Memoir and Correspondence of Caroline Herschel.* New York: D. Appleton, 1876.

Land, Barbara. *The Telescope Makers, from Galileo to the Space Age.* New York: Thomas Y. Crowell, 1968.

Lubbock, Constance A. *The Herschel Chronicle: The Life-Story of William Herschel and His Sister Caroline Herschel.* Cambridge: University Press, 1933.

Phillips, Patricia. *The Scientific Lady: A Social History of Women's Scientific Interests, 1520–1918.* New York: St. Martin's Press, 1990.

Ronan, Colin. *The Astronomers.* London: Evans Brothers Limited, 1964.

Schiebinger, Londa. *The Mind Has No Sex? Women in the Origins of Modern Science.* Cambridge, Mass.: Harvard University Press, 1989.

CAROL B. NORRIS

DOROTHY CROWFOOT HODGKIN
(1910–1994)
Crystallographer

Birth	May 10, 1910
1931	B.A., honors, Somerville College at Oxford University
1933	Research Fellow, Somerville College
1936	D.Phil., Cambridge University; Tutor and Fellow, Somerville College
1937	Married Thomas Hodgkin on December 16
1946	University Lecturer, Somerville College
1947	Fellowship of the Royal Society

1955	University Reader, Somerville College
1956	Royal Medal of the Royal Society
1960	Wolfson Research Professor of the Royal Society
1964	Nobel Prize in Chemistry
1965	Order of Merit
1970–89	Chancellor, Bristol University
1972–75	President, International Union of Crystallography
1976	Copley Medal of the Royal Society
1977–78	President, British Association for the Advancement of Science
Death	July 30, 1994

Dorothy Crowfoot Hodgkin will always be remembered for her many significant contributions to the structural analyses of important molecules and her active role in fostering an international scientific community that is sensitive to the needs of third world countries.

Dorothy Crowfoot was born in Cairo, Egypt, in 1910. Her father, John Winter Crowfoot, was an archaeologist in the service of the British government and worked for the Egyptian Ministry of Education. Her attraction to archaeology started with her visits to various sites with her parents as they worked. As a young girl, she also developed a love of chemistry and was encouraged in this by her mother, Molly Hood Crowfoot. In particular, she was influenced by a book written by William Henry Bragg, the Nobel Prize–winning scientist, in which he described the new science of X-ray crystallography that made it possible to discern the structure of molecules.

Dorothy went to study chemistry at Somerville College, Oxford University, in 1928. There she worked with H. M. Powell, a young demonstrator in the laboratory of Professor H. L. Bowman, on X-ray studies of the thallium dialkyl halides (Thallium is an element that has physical properties that look like lead.) In 1932 she went to Cambridge to work with J. D. Bernal for her doctorate degree. He oversaw a very active laboratory where her creative genius was allowed to flourish, and he helped her to develop her research methods and ideas. Among the many successful projects she studied during that fruitful time was the X-ray analysis of cholesteryl iodide, the iodide salt of cholesterol, a steroid molecule.[1] This was one of the first three-dimensional structural determinations to be performed and certainly one of the more complicated problems to be studied. Her results demonstrated the importance of X-ray analysis in chemistry because it clarified the complete chemical and stereochemical structure of this compound, thus removing any misconceptions regarding the structure of steroids. She and Bernal also pursued the earliest X-ray study of a protein, pepsin.

In 1937, Dorothy married Thomas Hodgkin, who was an educator and a historian. They had three children: Luke (1938), Elizabeth (1941), and Toby (1946).

After leaving Bernal's group Dorothy returned to work at Oxford, where she continued her crystallographic studies. She crystallized insulin in 1934 and took some preliminary X-ray data. She became involved with the study of penicillin, which was discovered in 1929 by Alexander Fleming and isolated in 1941 by Howard Florey and Ernst Chain. Once again, there were many misconceptions regarding its structure. Robert Robinson and John Cornforth felt very strongly that its chemical formula should not contain a β-lactam structure, a structural feature formed by the junction of a five-membered sulfur ring with a four-membered, nitrogen-containing ring. When Hodgkin's work, with Barbara Low, A. Turner-Jones, and Charles Bunn, on various penicillin derivatives showed that it did indeed contain the β-lactam ring, not only was the unusual chemical structure elucidated but the power of X-ray crystallography as a technique was shown and Hodgkin's reputation as an excellent scientist was established.[2] This work also introduced many of the now standard procedures for solving and refining structures.

Hodgkin's work on penicillin also pioneered the use of computers in crystallography. An IBM analog computer was used for help with the tedious Fourier transformations on the data that had been collected. Hodgkin subsequently went on to determine structures of other penicillin derivatives and related antibiotics, such as cephalosporin. Following these scientific victories she was admitted to the Royal Society of London in 1947 at 37 years of age, only the third woman to be elected.

Hodgkin went on to clarify the structure of vitamin B_{12}, which was the largest molecule studied up to that point. It contained over 100 atoms. While undergoing this structure determination, the use of digital computers as an aid in the many calculations required by this large molecule was pushed by Hodgkin's group. Crystallography became one of the earliest scientific disciplines to use computing power. The three-dimensional structure was finally shown by Hodgkin and **Jenny Pickworth Glusker** to contain a corrin ring system. This was similar to the macrocyclic porphyrin ring, the structure found at the active site of hemoglobin, but missing a bridging CH_2 group. (Hemoglobin is the oxygen-transporting protein in red blood cells.) Once again, the power of crystallography was shown because the corrin ring had not been predicted by classical organic chemical analytical techniques and, indeed, was identified first by crystallographic methods. In 1956 the complete structure of vitamin B_{12} was published.[3]

In the fall of 1964, Dorothy Hodgkin was the sole recipient of the

Nobel Prize in chemistry for the crystallographic structure elucidation of many important molecules. In 1965 she received the Order of Merit, the highest honor given to civilians in Britain, from Queen Elizabeth. The only other woman to have received this honor was Florence Nightingale in 1907.

Hodgkin continued her work on insulin. Its structure became clear in 1966 and its complete three-dimensional structure was announced in 1969 in a *Nature* article.[4] Insulin is a hexamer of six molecules linked by two zinc ions. As was the case with many of Hodgkin's scientific inquiries, determining the insulin structure proved to be difficult—it is a small protein and the fact that Hodgkin had first crystallized it thirty-four years earlier gives some clue as to the complexity of the project and her perseverance. Her work in protein crystallography anticipated many of the unique problems that arise from macromolecular data collection and structure determination.

Given such a full private and scientific life, it was a cruel twist of fate that Hodgkin for most of her life was plagued with an autoimmune degenerative disease, rheumatoid arthritis, that led to disfigurement of her hands and feet and, at times, was the source of much pain and suffering. The disease manifested itself first in 1934, when she was at the start of her brilliant career. Except for some remissions, she triumphed over her disease. A visual representation of this dreadful disease is given in a painting of her crippled hands made by the painter Henry Moore, which is on display at the Royal Society of London. Yet she continued to be active in scientific meetings and traveled widely, even if in a wheelchair, until her death in 1994.

World peace, disarmament, and international cooperation remained important issues for Hodgkin throughout her life. She had shared even as a student J. D. Bernal's left-wing political views, and later her husband's, who also sympathized with left-wing political groups. Her political views were far enough removed from the conservative, cold war mentality that was in effect in Britain and the United States that she was refused a visa to enter the United States in 1953 to attend a scientific conference organized by Linus Pauling, who himself was considered to be a political extremist by the U.S. State Department.

Hodgkin was involved with organizing the International Union of Crystallography, a group formed after World War II to encourage international cooperation and the exchange of information among crystallographers. She supported science in third world countries and, by traveling widely, helped to initiate active research programs in Africa and Asia. For example, since her husband was working at the University of Ghana, she maintained a research program there and spent part of the year in Africa. She was active both in establishing a crystallographic laboratory in North Vietnam in 1974 while the country was at war with the United

States, and in campaigning against the war in Vietnam. She made many visits to China, Vietnam, and the Soviet Union.

Hodgkin also served as a mentor to an enormous number of currently successful crystallographers, including many women who have continued in her footsteps such as Jenny Pickworth Glusker, Barbara Low, Judith Howard, Maureen Mackay, Eleanor Dodson, and Margaret Adams. J. D. Dunitz has written that

Dorothy had an unerring instinct for sensing the most significant structural problems in this field, she had the audacity to attack these problems when they seemed well-nigh insoluble, she had the perseverance to struggle onward where others would have given up, and she had the skill and imagination to solve these problems once the pieces of the puzzle began to take shape. It is for these reasons that Dorothy's contribution has been so special.[5]

Her contributions to science and world peace were truly remarkable. She died on July 30, 1994.

Notes

1. C.H. Carlisle and D. Crowfoot, "The Crystal Structure of Cholesteryl Iodide," *Proceedings of the Royal Society* A184 (1945): 64.

2. D. Crowfoot et al., "X-Ray Crystallographic Investigation of the Structure of Penicillin," in *Chemistry of Penicillin*, eds. H.T. Clarke et al. (New Brunswick, N.J.: Princeton University Press, 1949), p. 310.

3. D.C. Hodgkin et al., "Structure of Vitamin B_{12}," *Nature* 173 (1956): 64.

4. M.J. Adams et al., "Structure of Rhombohedral 2–zinc Insulin Crystals," *Nature* 231 (1969): 506.

5. J.D. Dunitz, in *Structural Studies on Molecules of Biological Interest; A Volume in Honour of Dorothy Hodgkin*, eds. G. Dodson, J.P. Glusker, and D. Sayre (Oxford: Oxford University Press, 1981), p. 59.

Bibliography

Adams, M.J., et al. "Structure of Rhombohedral 2–Zinc Insulin Crystals." *Nature* 231 (1969): 506–511.

Carlisle, C.H., and D. Crowfoot. "The Crystal Structure of Cholesteryl Iodide." *Proceedings of the Royal Society* A184 (1945): 64.

The Collected Works of Dorothy Crowfoot Hodgkin. Edited by G.G. Dodson, J.P. Glusker, S. Ramaseshan, and K. Venkatesan. New Delhi: Indian Academy of Sciences, 1995.

Crowfoot, D. "X-Ray Single-Crystal Photographs of Insulin." *Nature* 135 (1935): 591–592.

Crowfoot, D., et al. "X-Ray Crystallographic Investigation of the Structure of

Penicillin," in *Chemistry of Penicillin*. Edited by H.T. Clarke et al., pp. 310–366. New Brunswick, N.J.: Princeton University Press, 1949.

Hodgkin, D.C., et al. "Structure of Vitamin B_{12}." *Nature* 173 (1956): 64–66.

Structural Studies on Molecules of Biological Interest: A Volume in Honour of Dorothy Hodgkin. Edited by G. Dodson, J.P. Glusker, and D. Sayre. Oxford: Oxford University Press, 1981.

MIRIAM ROSSI

E. DORRIT HOFFLEIT

(1907–)

Astronomer

Birth	March 12, 1907
1928	A.B., Radcliffe College
1929–38	Research Assistant, Harvard College Observatory
1932	M.A., Radcliffe College
1938	Ph.D., Radcliffe College; Carolyn Wilby Prize, Radcliffe College
1938–43	Research Associate, Harvard College Observatory
1943–48	Mathematician, Ballistic Research Laboratories, Aberdeen Proving Ground, MD
1947	Certificate of Appreciation, U.S. War Department
1948–56	Astronomer, Harvard Observatory
1948–61	Expert, Ballistic Research Laboratories, Aberdeen Proving Ground, MD
1956–69	Research Associate, Yale University Observatory
1957–78	Director, Maria Mitchell Observatory, Nantucket, MA
1964	Graduate Society Medal, Radcliffe College
1969–75	Senior Research Astronomer and Lecturer, Yale University
1975–83	Senior Research Scientist, Yale University
1983	Alumnae Recognition Award, Radcliffe College
1983–	Consultant and volunteer researcher
1987	Asteroid Dorrit named in her honor

| 1988 | George van Biesbroeck Award, University of Arizona |
| 1993 | Annenberg Foundation Award, American Astronomical Society |

(Ellen) Dorrit Hoffleit has embodied the spirit of independent research. The daughter of German immigrants Fred and Kate Sanio Hoffleit, Dorrit was born on her father's farm in Alabama. When she was 9 years old she moved with her mother and older brother Herbert to Pennsylvania after the breakup of her parents' marriage. Dorrit recounts watching Perseid meteors with her brother as a child in Pennsylvania as an important step toward becoming an astronomer.[1] The young Dorrit fell into her brilliant older brother's shadow, facing constant comparisons from teachers who were impressed with Herbert's natural talent for languages:

One of my grade school classes I had the same teacher that my brother had had a few years previously. My mother and I were walking down the street one day and we bumped into my teacher and Mother and teacher started talking . . . and the teacher says, "Dorrit isn't as bright as her brother, is she?" whereupon my mother says, "What can you expect, she's only a girl."

Dorrit was deeply hurt by that remark, but years later her mother explained that she was referring to the teacher's intelligence.[2]

Dorrit was proud of her brother, who received a Ph.D. from Harvard in classics at the young age of 21 and subsequently became a professor at UCLA.[3] Dorrit was sent to Radcliffe College by her mother "so that her brilliant son wouldn't be ashamed of his 'dumb' sister." At Radcliffe, Dorrit became a mathematics major because only two astronomy courses were offered at the time. She had her first taste of independent research quite by accident at Radcliffe when she incorrectly conducted an astronomical transit experiment. For her, it was a valuable learning experience, but she says, "I don't think my professor appreciated the educational value of that experiment. I think I got a lot more out of the pole star than I did out of what the thing was intended for. So you see, independence wasn't appreciated even then."[4] Dorrit graduated from Radcliffe *cum laude* in 1928 and began taking graduate classes while looking for work. Through a classmate she landed a job at the Harvard College Observatory (HCO) for forty cents per hour, half of a man's salary. She turned down a higher-paying statistician job to work at the HCO as a research assistant and several times subsequently turned down other, higher-paying offers because of her growing love for the HCO and respect for its director, Harlow Shapley. Shapley encouraged independent thinking.

Dorrit completed an M.A. in astronomy at Radcliffe in 1932, "the highest degree for which I felt qualified."[5] She continued her work on variable stars during the day and pursued independent research projects at night on her own time, including a pioneering study of the light curves of meteors using the accidental photographs of meteors in the Harvard plate collection. She brought her research to Shapley, who submitted it for publication.[6] He then

> called me into his office and said, "We were wondering why you were not continuing to work for your Ph.D. Go back to your office and think it over." I had never been particularly bright, and this was the greatest expression of confidence in my abilities I had ever heard.[7]

Dorrit completed her Ph.D. in 1938 with work on the absolute (intrinsic) magnitudes of stars from their spectra.[8] She continued her work at the HCO as a research associate and then as an astronomer in diverse areas of astronomy including meteors, variable stars (including the discovery of approximately 1,270 variables), novae, and absolute magnitudes. She also began her work on the history of astronomy and popular writings, including a column in *Sky and Telescope* from 1941 to 1956.[9]

During World War II, Dorrit volunteered to work at the Aberdeen Proving Ground in Maryland on the preparation of aircraft firing tables, and she came up against a wall of discrimination. As an academic with a Ph.D. she was clearly eligible for a professional rating, but she was instead relegated to a subprofessional class by a colonel even though she was assigned professional class work. This led to a conflict that Dorrit rates as a defining experience in her career:

> After I'd been there for about a year the inspector general of the Baltimore district, where Aberdeen is, discovered there was a woman Ph.D. with the subprofessional rating, and he came around on a day when the colonel was down at Washington instead of in Aberdeen and he wanted to find out all about the story about why I was on a subprofessional [rating]. . . . So when the colonel came back the next day and heard about what had happened . . . [he] told the major to tell me that there was no room for professional women . . . that I'd have my choice—either I could transfer to the Pentagon where women were welcome . . . or the poor major was to make sure that I did nothing but subprofessional work because if I didn't do anything but subprofessional [work] then it would be all right to keep me on subprofessional [rating]. So I told the poor quaking

major that he, since the colonel wouldn't talk to me himself, he the major could go back and tell the colonel "thanks, I don't accept either alternative, that isn't what I came down here for."[10]

Dorrit eventually won her "war" with the colonel and, later, returned to Harvard. But she continued as a consultant at the Proving Ground until 1961.

Dorrit's life was drastically changed by Shapley's retirement from Harvard in the 1950s. His replacement did not value independence and, much to Dorrit's horror, began indiscriminately throwing away large sections of Harvard's unique and valuable photographic plate collection. In spite of having tenure at Harvard, Dorrit was forced to follow her conscience and "defect" to Yale, where she was placed in charge of a large ongoing project on stellar proper motions (the apparent changes in the positions of stars over time) and where, to her surprise, she was not afforded the same independence she had enjoyed at Harvard. At the same time, she was offered the directorship of Nantucket's Maria Mitchell Observatory. Owing to the financial situation of the observatory, she held a split six-month/six-month appointment between Yale and Nantucket. During Dorrit's twenty-one years on Nantucket she encouraged a new generation of astronomers through her summer variable star research program for undergraduates, with a special emphasis on female students: "I encouraged them to give talks and prepare research papers for publication and presentation. Of the one hundred girls who participated, some twenty-five now have Ph.Ds and their achievements are a joy to behold."[11]

Dorrit remained an untenured research associate and astronomer at Yale (supported entirely through grants) even after her "official" retirement in 1975. Her main contributions include the first paper on the light variability of quasars, catalogues containing the proper motions of 30,000 stars, and the third and fourth editions of the *Bright Star Catalogue* and its *Supplement*.[12] After 1983 she concentrated on historical writings, including several publications on women in astronomy published by the American Association of Variable Star Observers. Dorrit has been strongly involved with this organization for several decades, including serving as its president from 1961 to 1963.[13] She has also published a history entitled *Astronomy at Yale*.[14]

In addition to having an asteroid named after her, Dorrit has received numerous awards, including the George van Biesbroeck award from the University of Arizona for outstanding service to astronomy and the Annenberg Foundation Award from the American Astronomical Society for "service to the community in education."[15] She continues to be active in research on topics of her choice. She notes that "I have become as happy and independent as I had been in my youth at Harvard."[16]

Notes

1. Dorrit Hoffleit, personal communication with Christine Larsen, October 30, 1994.

2. Dorrit Hoffleit, interview with Christine Larsen, October 7, 1994. Much of this essay is based on this interview.

3. Dorrit Hoffleit, personal communication, October 30, 1994.

4. Dorrit Hoffleit, interview, October 7, 1994.

5. Dorrit Hoffleit, "Some Glimpses from My Career," *Mercury* 21 (January/February 1992): 16.

6. Dorrit Hoffleit, "A Study of Meteor Light Curves," *Proceedings of the National Academy of Sciences* 19 (1933): 212–221.

7. Dorrit Hoffleit, "Harvard Career Profiles," *Graduate School of Arts and Sciences* (1987): 20.

8. Part of this was published as Dorrit Hoffleit, "Spectroscopic Absolute Magnitudes of Three Hundred and Seventy Southern Stars," *Proceedings of the National Academy of Sciences* 23 (1937): 111–114.

9. For example, see Dorrit Hoffleit, *Bibliography on Meteoric Dust with Brief Abstracts* (Cambridge, Mass.: Harvard College Observatory, 1952); Dorrit Hoffleit, "Variable Stars in a Field of Sagittarius," *Astronomical Journal* 62 (1957): 120–126; Dorrit Hoffleit, *Women in the History of Variable Star Astronomy* (Cambridge, Mass.: AAVSO, 1993); Dorrit Hoffleit, "Van Houten's Nova Velorum 1940," *Astronomical Journal* 55 (1950): 149–150; Karl G. Henize, Dorrit Hoffleit, and Virginia McKibben Nail, "Magellanic Clouds. XI. Survey of the Novae," *Proceedings of the National Academy of Sciences* 40 (1954): 365–372; Dorrit Hoffleit, "The Spectra and Absolute Magnitudes of 500 A3–G2 Stars," *Annals of the Harvard College Observatory* 119, no. 1 (1950); Dorrit Hoffleit, *Some Firsts in Astronomical Photography* (Cambridge, Mass.: Harvard College Observatory, 1950); and Dorrit Hoffleit, "The Discovery and Exploitation of Spectroscopic Parallaxes," *Popular Astronomy* 58 (1950): 428–438.

10. Dorrit Hoffleit, interview, October 7, 1994.

11. Dorrit Hoffleit, "Harvard Career Profiles," p. 20.

12. Harlan J. Smith and Dorrit Hoffleit, "Light Variations in the Superluminous Radio Galaxy 3C2273," *Nature* 198 (1963): 650–651; see *Transactions of the Yale University Observatory* 28–30 (1967–1970); Dorrit Hoffleit and Carlos Jaschek, *The Bright Star Catalogue*, 4th rev. ed. (New Haven: Yale University Observatory, 1982); Dorrit Hoffleit et al., *A Supplement to the Bright Star Catalogue* (New Haven: Yale University Observatory, 1983).

13. Dorrit Hoffleit, *Maria Mitchell's Famous Students; and, Comets over Nantucket* (Cambridge, Mass.: AAVSO, 1983); *Women in the History of Variable Star Astronomy* (Cambridge, Mass.: AAVSO, 1993); *The Education of American Women Astronomers before 1960* (Cambridge, Mass.: AAVSO, 1994).

14. Dorrit Hoffleit, *Astronomy at Yale, 1701–1968* (New Haven: Connecticut Academy of Arts and Sciences, 1992).

15. Martha Hazen, "Minutes of the 82nd Meeting of the AAVSO," *Journal of the AAVSO* 22 (1993): 89.

16. Dorrit Hoffleit, "Some Glimpses from My Career," p. 18.

Bibliography

Hoffleit, Dorrit. *Astronomy at Yale, 1701–1968*. New Haven: Connecticut Academy of Arts and Sciences, 1992.

———. *The Education of American Women Astronomers before 1960*. Cambridge, Mass.: AAVSO, 1994.

———. *Maria Mitchell's Famous Students; and Comets over Nantucket*. Cambridge, Mass.: AAVSO, 1983.

———. *Some Firsts in Astronomical Photography*. Cambridge, Mass.: Harvard College Observatory, 1950.

———. "Some Glimpses from My Career." *Mercury* 21 (January/February 1992): 16–19.

———. *A Supplement to the Bright Star Catalogue*. New Haven: Yale University Observatory, 1983.

———. *Women in the History of Variable Star Astronomy*. Cambridge, Mass.: AAVSO, 1993.

Hoffleit, Dorrit, and Carlos Jaschek. *The Bright Star Catalogue*. 4th rev. ed. New Haven: Yale University Observatory, 1982.

KRISTINE M. LARSEN

HELEN SAWYER HOGG

(1905–1993)

Astronomer

Birth	August 1, 1905
1926	A.B., Mt. Holyoke College
1928	M.A., Radcliffe College
1930	Married Frank Hogg (died 1951)
1930–31	Instructor, Mt. Holyoke College
1931	Ph.D., Radcliffe College
1931–35	Research Assistant, Dominion Astrophysical Observatory
1935–41	Research Assistant, David Dunlap Observatory
1941–50	Lecturer, University of Toronto
1950	Annie Jump Cannon Prize, American Astronomical Society
1951–93	University of Toronto: Assistant Professor, 1951–55; Associate Professor, 1955–57; Professor, 1957–76; Emerita, 1976–93

Helen Sawyer Hogg. Photo courtesy of Mount Holyoke
College Archives.

1957–59	President, Royal Astronomical Society of Canada
1967	Rittenhouse Medal, Rittenhouse Astronomical Society; Service Award Medal, Royal Astronomical Society of Canada; Graduate Achievement Medal, Radcliffe College; Centennial Medal, Canada
1968	Medal of Service, Order of Canada
1971–72	President, Canadian Astronomical Society
1976	Companion, Order of Canada
1983	Klumpke-Roberts Award, Astronomical Society of the Pacific
1984	Asteroid 2917 named Sawyer Hogg
1985	Sandford Fleming Medal, Royal Canadian Institute; married F.E.L. Priestley (died 1988)

1992	Commemorative Medal for the 125th Anniversary of the Confederation of Canada
Death	January 28, 1993

Helen Sawyer Hogg, a Canadian astronomer who studied variable stars in globular clusters and promoted public interest in astronomy, began her long, productive life in the United States. Born in 1905 in Lowell, Massachusetts, Helen Battles Sawyer was the second daughter of Edward Everett Sawyer, a banker, and Carrie Sprague Sawyer, a former schoolteacher. Helen graduated from Lowell High School at the age of 15 but stayed for a fifth year of high school before going to college. After entering Mount Holyoke College in 1922, she decided to major in chemistry. Helen changed her major to astronomy during her junior year while taking an introductory course from the dynamic Dr. Anne Sewell Young. On January 24, 1925, a total eclipse of the sun was visible less than a hundred miles south of Mount Holyoke. Dr. Young and the entire student body traveled by train to view the eclipse. All her life, Helen remembered watching the spectacular eclipse and deciding to study the stars while standing knee deep in snow on a golf course in Connecticut.

Astronomy classes introduced Helen to variable stars and globular clusters. Astronomers had realized only a decade earlier that some stars regularly vary in brightness because their atmospheres are pulsating. Vast numbers of such stars exist in globular clusters—huge, symmetrical agglomerations of stars that surround the center of the Milky Way galaxy.

One year after witnessing the eclipse, Helen met **Annie Jump Cannon**, an astronomer employed at Harvard Observatory. Dr. Cannon helped Helen obtain a fellowship for graduate study at Harvard starting in the fall of 1926. The caption below Helen's senior picture in Mount Holyoke's 1926 yearbook reads: "Helen is on strike for more working hours in the day. Twenty-four hours is hardly enough." And printed below the caption is her advice: "Make your hours count."

Harvard Observatory and its recently established graduate program in astronomy were directed by Dr. Harlow Shapley. Young, intense, and energetic, he habitually worked long hours and expected others to do the same. He set Helen to work measuring the size and brightness of several globular clusters. Her work later expanded to include a study of the number of variable stars in globular clusters. She published several papers on these subjects and received her Ph.D. from Radcliffe, Harvard's sister institution, in 1931.

Harvard's first astronomy Ph.D. was awarded in 1929 to a Canadian student named Frank Hogg. Helen and Frank Hogg married a year later in 1930. Helen recalled the following year as one of the busiest times of her life. While finishing her degree, she taught astronomy at Mount Holyoke. (She also taught astronomy at Smith College in 1927.)

After their first year of marriage, Frank received job offers from Harvard Observatory and from the Dominion Astrophysical Observatory in British Columbia. Helen sensed that her life's direction depended on Frank's decision. Frank finally chose the Canadian position, with the result that Helen spent most of her remaining years in Canada. In August 1931 the Hoggs drove their Model A Ford coupe from the Atlantic to the Pacific, occasionally crossing rocky pastures between sections of finished road.

During the Great Depression the Canadian government did not allow both a husband and wife to be employed, but the Dominion Astrophysical Observatory's director, Dr. J. S. Plaskett, let Helen use the 72–inch reflecting telescope. Years later, upon learning that a talented but unmarried woman astronomer had been previously passed over for a position at Dominion Astrophysical Observatory because of the "difficulty" that nighttime observing entailed, Helen realized that her own observing privileges were due in part to the presence of a husband who also acted as chaperone.

Regardless of the reasons she was allowed to observe, Helen made good use of the opportunity and began photographing globular clusters. Her graduate research had made use of photographs that already existed in Harvard's immense collection, so taking her own photographs was a new experience.

Giving birth to a daughter in June 1932 was another new experience, but Helen soon discovered that baby Sally did not interfere much with research. On observing nights, Sally slept in a basket at the observatory. Later, Dr. Plaskett received a research grant of $200, which he gave to Helen. She used this sum to hire a live-in housekeeper for an entire year.

Helen amassed several hundred photographs in three years' time and compared them in a search for new variable stars. She found 132 new variables in five globular clusters and also determined the periods (the time stars take to complete one cycle of brightening and dimming) and light curves (plots of brightness over time) for seventeen variable stars in the globular cluster called Messier 2.

In 1934, Frank received an offer to work at the University of Toronto's soon-to-be-completed David Dunlap Observatory. He accepted the position, and in 1935 the family moved to Ontario. For a year Helen continued her research on an unpaid basis, using the David Dunlap Observatory's 74–inch telescope. In 1936 she received her first paid appointment as a research assistant. Helen's son David was born in January of the same year, and another son, James, was born in September 1937.

As Helen's children grew, so did her professional reputation. Her globular cluster photographs, which ultimately numbered in the thousands, helped her to identify many thousands of variable stars. In 1939 she published a *Catalogue of 1116 Variable Stars in Globular Clusters*, the first of three such catalogues completed during her lifetime. (A fourth was in progress when she died.)

Through her studies of variable stars in globular clusters, Helen made significant contributions to understanding the age, size, and structure of our galaxy. Certain variable stars called Cepheids allow astronomers to calculate cosmic distances due to the fact that their periods relate directly to their luminosity, or true brightness. By comparing a Cepheid's true brightness to its apparent brightness, astronomers can calculate the star's distance from other stars and the earth. This "period-luminosity relationship" was discovered in 1908 by **Henrietta Leavitt** at Harvard Observatory.

Helen served as president of the American Association of Variable Star Observers from 1939 to 1941 and as acting chairman of Mount Holyoke's astronomy department from 1940 to 1941. During these years, Canada was caught up in World War II. When Helen returned to the David Dunlap Observatory several staff members were away at war, so she took on teaching duties at the University of Toronto. Domestic help was hard to find. Helen remembered these years as a very difficult time.

A few calm years followed the war. Then, on the first day of 1951, Frank died of a heart attack. One of the many tasks Helen took on after her husband's death was the writing of a weekly astronomy column for *The Toronto Star* called "With the Stars," which she continued for thirty years. Helen also popularized astronomy by publishing a book, *The Stars Belong to Everyone*, in 1976. In this and other writings she describes how dazzlingly star-studded the sky would be if the earth were located within a globular cluster. Not limited to the printed page, Helen also promoted public interest in astronomy with an eight-show series that ran on Canadian educational television in 1970.

As a single parent and scientist with teaching, observing, research, and writing responsibilities, Helen's hours were fuller than ever. Her son David recalls that she managed her time by accurately estimating the length of projects, by making lists, and by accomplishing odd tasks like writing letters during streetcar and subway rides. Helen helped all three of her children complete college. Sally majored in English, David majored in physics (he is now a radio astronomer), and James majored in chemistry.

Helen's colleague at the University of Toronto, Dr. Christine Clement, recalls that "even on cloudy nights when she was scheduled to observe at the David Dunlap Observatory, she always watched for breaks in the clouds. . . . She never missed an opportunity." Helen advanced to assistant professor in 1951, associate professor in 1955, professor in 1957, and professor emerita in 1976.

Always in demand outside the University of Toronto, Helen presided over several Canadian astronomical and scientific organizations. As director of the (U.S.) National Science Foundation's astronomy program in 1955, Helen helped determine sites for the National Radio Astronomy Observatory and for Kitt Peak National Observatory. She also served on the board of directors of Bell Telephone Company of Canada from 1968 to 1978.

Helen received many honorary degrees and awards for her scientific and educational accomplishments. In 1950 she won the prestigious Annie Jump Cannon Prize of the American Astronomical Society. She was also quite proud of receiving the Companion of the Order of Canada in 1976. An asteroid, a telescope, and an observatory are named after her.

At age 80, Helen married Dr. F. E. L. Priestly, a colleague at the University of Toronto. Up until the last days of her life, she actively encouraged young women to pursue science. Following a heart attack, Helen died on January 28, 1993, in Richmond Hill, Ontario. Friends remember her as a great scientist and a gracious person.

Bibliography

Hogg, Helen Sawyer. "Catalogue of 1116 Variable Stars in Globular Clusters." *Publications of the David Dunlap Observatory* 1, no. 4 (1939).

———. "One Hundred and Thirty-Two New Variable Stars in Five Globular Clusters." *Publications of the Dominion Astrophysical Observatory* 7, no. 5 (1938).

———. "Second Catalogue of Variable Stars in Globular Clusters." *Publications of the David Dunlap Observatory* 2, no. 35 (1955).

———. *The Stars Belong to Everyone*. New York: Doubleday, 1976.

———. "Third Catalogue of Variable Stars in Globular Clusters." *Publications of the David Dunlap Observatory* 3, no. 6 (1973).

Shapley, H., and H. Sawyer. "Variable Stars in Globular Clusters." *Popular Astronomy* 38 (1930): 408.

Webb, Michael. *Helen Sawyer Hogg: A Lifetime of Stargazing*. Mississauga, Ontario: Copp Clark Pitman, 1991. [This book is part of the "Scientists and Inventors" series for children.]

Yost, Edna. *Women of Modern Science*. New York: Dodd Mead, 1959. (Reprint: Westport, Conn.: Greenwood Press, 1984.)

JANE OPALKO

Icie Gertrude Macy Hoobler
(1892–1984)
Biochemist

Birth	July 23, 1892
1914	A.B., English, Central College for Women, Lexington, MO
1916	B.S., chemistry, University of Chicago
1916–18	Assistant Chemist, University of Colorado, Boulder
1918	Master's degree, University of Colorado, Boulder
1920	Ph.D., physiological chemistry, Yale University
1923–30	Director, Nutrition Research Laboratories, Merrill-Palmer School and Children's Hospital of Michigan
1930–54	Director, Research Laboratory, Children's Fund of Michigan
1938	Norlin Achievement Award, University of Colorado; married B. Raymond Hoobler
1939	Borden Award, American Home Economics Association
1940	Certificate of Merit Group I, American Medical Association
1946	Garvan Medal, American Chemical Society
1952	Osborne and Mendel Award, American Institute of Nutrition
1954–74	Research Consultant, Merrill-Palmer Institute
1955	Modern Medicine Award for Distinguished Achievement, Children's Hospital of Michigan
1972	Distinguished Service Award, Michigan Public Health Association
Death	January 6, 1984

Icie Gertrude Macy Hoobler called herself a "pioneer woman scientist." She blazed the trail for women in chemistry at a time when women scientists were few and not universally accepted by their male colleagues. Her research and leadership in the areas of the nutritional biochemistry and health of children and mothers, in particular, were vital contributions to science and the well-being of future generations.

The evolution of Macy's interest in science and the welfare of children

Icie Gertrude Macy Hoobler. Photo courtesy of the Bentley
Historical Library, University of Michigan.

can be traced to her childhood. Icie Gertrude Macy grew up on her parents' farm near Gallatin, Missouri, at the turn of the century. Of her early interest in nature she wrote:

> My curiosity was constantly motivated as we roamed freely among the pastures, woods and streams. I longed to understand how plants and animals came into being, then grew and matured and finally became food for the family and the birds and animals on the farm.[1]

During a childhood trip to Arkansas with her grandmother, Macy went on a horseback trip into the nearby mountains and encountered some "sick, desperately poor and deprived little children."[2] Macy was saddened by the children's plight and her compassion was aroused. Her interest in the welfare of children would endure and become a primary motivation for her research on the nutritional requirements of children.

Macy credited her success as a scientist to "my love of learning, determination, hard work, perseverance and the stimulation, encouragement and guidance of parents, teachers and counselors."[3] She had three particularly inspirational teachers, including two women science teachers. Macy received a master's degree from the University of Colorado in 1918 and entered a Ph.D. program in physiological chemistry at Yale University the same year.

Finding a place to live near the Yale campus was not easy for women graduate students. Macy reported that "most of the landlords and landladies preferred men roomers and rejected women because they felt they were a bother, especially in the use of the bathroom where they washed their clothing and hung them up to dry, cluttering the bath for others."[4] At that time, Yale accepted a few women into graduate school but "made no provision for the quality of their living conditions."[5] Macy became a member of the Graduate Women's Club that worked to improve living arrangements for graduate women. Eventually the university was persuaded to provide an on-campus house for the women.

While at Yale, Macy attended a lecture on the dairy and milk industry in which the professor, Lafayette B. Mendel, pointed out how little human nutritional research was being done. He showed the class a picture of a malnourished infant, emphasizing that health research was vital for the future of humankind and that women could make important contributions in this area. Macy later said that Mendel looked directly at her as he spoke. She wrote that she "got his message and left the lecture hall inspired and determined that the health of mothers, infants, and children was to be my first priority in research in the future."[6]

When Macy finished her studies at Yale, she went to work as an assistant chemist at the Western Pennsylvania Hospital in Pittsburgh. At the hospital, Macy struggled with sexual discrimination of various kinds. The hospital had restrooms only for men, and Macy had to use a public restroom in a building about a half-block away. Consequently, she limited herself to as few trips to the restroom as possible. Within a few months she developed acute nephritis (inflammation of the kidney), and her doctors recommended "a year's leave of absence with complete rest."[7] In addition, Macy was not allowed to eat in the doctors' dining room, since all the other doctors were men; nor could she eat in the nurses' dining room for "bureaucratic reasons."[8] She was assigned to eat with the hospital employees, but her schedule was such that she could not eat lunch until the dining room was closed.

When Macy discussed her complaints with the chief of the laboratory, he said that she would "soon get used to them."[9] Two weeks later, when her situation had not improved, she submitted her letter of resignation.

The president of the board of trustees came to see her the next day to discuss her decision to resign. He also asked her why she had not attended the annual staff banquet the night before. Macy told him that she had not known about the banquet. The president asked the chief of the laboratory if he had invited her, and he replied that "being a woman, I did not think that Dr. Macy would want to be with all those men!"[10] The president chided the chief by saying, "Dr. Macy was appointed a member of this hospital staff, the same as you were, and she is to be given the respect and privileges that all staff members enjoy."[11] After the president's visit, Macy's work conditions were adjusted to her satisfaction and she was promised that she would be "accepted as an equal with my peers of professional rank."[12]

Over the course of her career, Macy endured several "humorous situations of discrimination" in men's clubs, where she was invited to attend a scientific meeting and was the only woman present.[13] In one much-publicized case, she was invited to speak at a meeting at the Chicago Club. Her host had not realized that "Icie" was a woman's name. When she arrived and her gender was clear, her entry was barred until Macy's husband negotiated with the manager and the board of trustees took a vote to allow her to speak. Although Macy faced obstacles based on her gender during her career, she did not let them hinder her.

> In my educational training and in the practice of a woman in the science profession, I found that discrimination of women in science was an accepted fact but I was prepared to cope with it for to me it seemed ridiculous, humorous at times, and on occasion depressing healthwise. I had prepared myself over the years to accept the facts of life but challenged those where health and well-being were concerned.[14]

Macy required a year of complete rest to recuperate from her nephritis and was invited to spend that year teaching a course in food chemistry at the University of California at Berkeley. She ended up teaching other courses as well. She also was asked by the president of the university to serve as an inspector of institutions that were to become part of the State of California educational system. While she was at Berkeley she was offered, and accepted, a position as director of the Nutrition Research Project of the Merrill-Palmer School for Motherhood and Child Development and the Children's Hospital of Michigan in Detroit.

In the fall of 1923, at 31 years of age, Macy started her position at the Merrill-Palmer School. The Nutrition Research Project, founded that same year, was established "as a cooperative community project directed towards improving and extending scientific and technological aspects of safe and responsible childbearing and child rearing and to accumulate new and improved knowledge concerning the needs for safe mother-hood."[15] In 1931, for financial reasons, Macy's research laboratory became the research division of the Children's Fund of Michigan and Macy remained the director. In all, Macy devoted 31 years to directing the research of her laboratory, which was very productive. Staff associated with the laboratory published over 300 journal articles and several books. In the 1930s, Macy became an authority on nutrition and child development. The research she directed included studies on the metabolism of women during the reproductive cycle; the composition and secretion of human milk; nutrition and chemical growth in childhood; the evaluation of food purchasing and preparation; nutritional status in children's institutions in Michigan; and the chemistry of the red blood cell. Macy was known as much for her strong administrative ability and inspiring leadership as she was for her individual research accomplishments. Colonel Dewey, the president of the American Chemical Society, once proclaimed that Macy had "developed one of the outstanding research programs of America." According to Dewey, "her work has been characterized by thorough planning, skillful execution, critical interpretation of data, and accurate reporting."[16] From 1954 to 1974, Macy continued to serve the Merrill-Palmer Institute as a research consultant.

Macy believed in the importance of participation in professional societies. She served on numerous committees and boards, and in 1931 she was selected as the first woman chair of a local section of the American Chemical Society. She received 22 citations, honors, and awards, including the Norlin Award, the Garvan Award, the Borden Award, and the Osborne and Mendel Award. She also was involved in a number of community projects. For instance, in the 1960s she was a member of the Board of Control of Grand Valley State College, which was then in its infancy.

When Macy was 46 years old she married Dr. Raymond Hoobler, twenty years her senior, a physician and former director of Children's Hospital. During their marriage Hoobler was respectful of Macy's professional obligations and urged her to keep her own name:

> My husband was determined not to reduce the scale of my professional life, but to enhance it. He treated me in a caring and loving way and sought to provide all the amenities for my dual role of homemaker and professional scientific worker.[17]

Dr. Hoobler died a short five years later in 1943. Almost forty years later, after leading a rich and rewarding professional and personal life, Macy returned to her birthplace in Gallatin, Missouri, in 1982 and passed away two years later. Icie Gertrude Macy Hoobler was a remarkable person whose intelligence, curiosity, compassion, perseverance, and sense of humor allowed her to not only succeed but thrive in her profession and her life as a whole.

Notes

1. Icie Gertrude Macy Hoobler, *Boundless Horizons: Portrait of a Pioneer Woman Scientist* (Smithtown, N.Y.: Exposition Press, 1982), p. 6.
2. Ibid., p. 32.
3. Ibid., p. 146.
4. Ibid., p. 52.
5. Ibid., p. 54.
6. Ibid.
7. Ibid., p. 72.
8. Ibid.
9. Ibid., p. 73.
10. Ibid.
11. Ibid.
12. Ibid.
13. Ibid., p. 161.
14. Ibid., p. 141.
15. Ibid., p. 81.
16. Ibid., p. 185.
17. Ibid., p. 112.

Bibliography

Hoobler, I.G.M., with H.H. Williams and A.G. Williams. *Boundless Horizons: Portrait of a Woman Scientist*. Smithtown, N.Y.: Exposition Press, 1982.

Kopperl, S.J. "Icie Macy Hoobler: Pioneer Woman Biochemist." *Journal of Chemical Education* 65, no. 2 (1988): 97–98.

Macy, I.G. *Nutrition and Chemical Growth in Childhood*, Vol. 1 (1942): *Evaluation*; Vol. 2 (1946): *Original Data*; Vol. 3 (1951): *Calculated Data*. Springfield, Ill.: Charles C. Thomas.

Macy, I.G., and H.H. Williams. *Hidden Hunger*. Lancaster, Pa.: Jaques Cattell Press, 1945.

Macy, I.G., and H.J. Kelly. *Chemical Anthropology*. Chicago: University of Chicago Press, 1957.

Macy, I.G., H.J. Kelly, and R.E. Sloan. "The Composition of Milks." Publ. No. 254. Washington, D.C.: National Research Council—National Academy of Sciences, 1953.

Toverud, K.U., G. Stearns, and I.G. Macy. "Maternal Nutrition and Child Health." *Bulletin of the National Research Council*, No. 123. Washington, D.C.: National Research Council—National Academy of Sciences, 1950.

Williams, H.H. "Icie Gertrude Macy Hoobler (1892–1984): A Biographical Sketch." *Journal of Nutrition* 114 (1984): 1353–1362.

NANCY F. STIMSON

ALLENE ROSALIND JEANES

(1906–1995)

Chemist

Birth	July 19, 1906
1928	A.B., highest honors, Baylor University
1929	M.A., organic chemistry, University of California, Berkeley; Teacher, advanced mathematics, Laredo High School, Laredo, TX
1930–35	Head, Dept. of Science, Athens College, Athens, AL
1936–37	Instructor in Chemistry, University of Illinois, Urbana
1937–38	Chemical Foundation Fellow, University of Illinois
1938	Ph.D., organic chemistry, University of Illinois
1938–40	Corn Industries Foundation Fellow, National Institutes of Health
1940–41	Research Associate, National Bureau of Standards
1941–76	Northern Regional Research Laboratory, U.S. Department of Agriculture, Peoria, IL
1953	Distinguished Service Award, U.S. Department of Agriculture
1956	Distinguished Service Award to NRRL Dextran Team, U.S. Department of Agriculture
1956	Garvan Medal, American Chemical Society
1962	Federal Woman's Award, U.S. Civil Service Commission
1962–66	Subcommittee on Plasma and Plasma Substitutes of the Medical Division, National Academy of Science, National Research Council
1968	Outstanding Alumna Award, Baylor University; Superior Service Award to Biopolymer Research Team, U.S. Department of Agriculture
Death	December 11, 1995

Allene Jeanes's life of research at the U.S. Department of Agriculture involved finding a blood plasma substitute as well as new uses for cereal starches that have commercial application to the food industry. The

Allene Rosalind Jeanes. Photo courtesy of Mary P. Kerrey, Agriculture Laboratory, Peoria, Illinois.

holder of several patents, she was widely recognized for her pioneering work.

Allene Rosalind Jeanes was born to Largus Elonzo and Viola Herring Jeanes in 1906. She and her older brother, Perry, grew up among the 20,000 residents of Waco, Texas. Their father was a switchman, and later yardmaster, for the Cotton Belt Route of the St. Louis Southwestern Railroad. Allene graduated "with honors" from Waco High School in 1924. She continued studies with Dr. Wilbur G. Gooch at Baylor University, served as lab assistant, and graduated *summa cum laude* with departmental honors.[1]

At the University of California at Berkeley, her research on "Esters of Bromosuccinic Acid" was sponsored by C. W. Porter. Allene graduated in 1929 and found temporary employment as a high school mathematics teacher in Laredo. Within six months she obtained the position of science department head at Athens College, Alabama, where she taught biology, chemistry, and physics from 1930 to 1935.[2]

Having decided to return to graduate school, Allene went north to Illinois, where she joined the laboratory of Roger Adams. She majored in organic chemistry with a minor in biochemistry. Her work with Roger Adams focused on the addition of sodium to phenanthrene, a crystalline aromatic hydrocarbon, and she earned her first patent with Dr. Adams in 1941 on phenanthrenedicarboxylic anhydrides. (Anhydrides are compounds obtained from others by removing the elements of water.) At the time she completed her Ph.D., she had a strong desire to do pharmaceutical research. However, opportunities for women were very limited in chemistry at this time, because of both discrimination and the economic depression.[3]

Jeanes was offered an opportunity to work in Claude Hudson's laboratory at the National Institutes of Health on a Corn Industry Research Foundation fellowship. She worked with Hudson on periodate oxidation of starches, a technique she applied to dextran in one of her early papers in 1950. (Dextran is a general term used to identify any of numerous polysaccharides that yield only glucose on hydrolysis, particularly high molecular weight compounds that are obtained by the fermentation of sugar.) Her work with Hudson was published in 1955. She moved from the National Institutes of Health to what was then the National Bureau of Standards (now the National Institute of Standards and Technology), where she continued her work on carbohydrates while working with Horace Isbell. Her work with Dr. Isbell on chlorite oxidation of aldoses, a certain kind of sugars, would also have later application to her work with polysaccharides, the more complex carbohydrates.

In 1941, reflecting the country's interest in developing agricultural products, the U.S. Department of Agriculture opened the Northern Regional Research Laboratory (NRRL) in Peoria, Illinois, a center for corn production in the United States. Jeanes moved back to Illinois shortly after the opening of the laboratory and worked there until her retirement. Dr. G. E. Gilbert, who hired her, employed both men and women and African Americans at a time when some of his associates flatly refused to do so.[4] Jeanes remained in Illinois after her retirement and continued to explore her professional interests. In addition to chemistry, she had a strong interest in classical music and ballet and was active in amateur groups in Peoria.[5]

Jeanes's early work on the structure of starch and other polysaccharides led her to look at dextrans, which are microbially produced from sucrose and are based on the same linkage that is found in the isomaltose residue in starch. In 1950 she published her first bibliography on dextrans, which was 13 pages long. In 1978 her bibliography on dextrans, which excluded clinical aspects, was 368 pages long. Much of her early work reflected her own careful experimental technique, for although she was a Ph.D. chemist, as a woman she was not given a large group to supervise.

In 1950 there was a need for an acceptable blood-plasma substitute because of the Korean conflict. Dextran had been used for this purpose in Sweden, and it was decided to mount a crash program at the Northern Regional Laboratory taking advantage of Dr. Jeanes's work. According to Benjamin Alexander, who asked to work with her as a technician at that time, there was a reluctance among some of the technicians to be supervised by a woman. He said that she taught him a great deal, both experimentally and theoretically.[6]

As a result of her work on dextrans, Jeanes was honored with the Distinguished Service Award, the highest award given by the Department of Agriculture. She was the first woman in the Agriculture Department to be so honored. She was also recognized for her leadership in the pioneering chemical research on dextrans with the presentation of the Garvan Medal of the American Chemical Society. At that time it was pointed out that she had made significant contributions to knowledge of the structure of starch and other polysaccharides and that she had developed assay methods for carbohydrates and their derivatives and degradation products. A former associate praised her: "By her own accomplishments in research, by the training of her assistants, and above all, by the high standards of her work and ideals, she has influenced those who know and have worked with her."[7]

Although Jeanes and her research group characterized and isolated dextrans from more than one hundred bacterial strains, the NRRL B-512 strain, isolated originally from root beer by Dr. Robert G. Benedict, was one of the most vigorous. Jeanes developed a simple, rapid method for structure determination applying periodate oxidation techniques. The method provided chemical proof that dextran structures were directly related to the strain of the microorganism used to produce the dextran. Jeanes and her colleagues at NRRL received several patents for their work, as well as the Distinguished Service Award of the U.S. Department of Agriculture in 1956.

The Department of Agriculture was interested in exploring other types of microbial polysaccharides that could be produced by microorganisms from cereal grains. A biopolymer project was established with the objective of establishing new outlets for cereal starches; finding new types of water-soluble polysaccharides that have thickening, stabilizing, and suspending action when in solution; and replacing imported gums with domestic products. In 1962 the industrial production of xanthan was initiated from *Xanthomonas campestris* NRRL B1459. In 1968 the Biopolymer Research Team was honored by the USDA Superior Service Award for "outstanding creativity and highly significant microbiological, chemical, and engineering team research resulting in the discovery, development, industrial acceptance, and commercialization of a new industrial gum (xanthan) of cereal grain origin."[8] Jeanes continued her studies on the extracellular microbial polysaccharides, leading to a series of papers on their properties, applications, use to the food industry, and other fundamental and practical applications.

Throughout her career and prior to the advent of computers and word processors, Jeanes published seminal reviews of the literature on dextrans. Her final paper published in 1986, well after her official retirement from NRRL, is a review on the "Immunochemical and Related Interactions with Dextrans Reviewed in Terms of Improved Structural Information." In this paper, published in *Molecular Immunology*, she considers the work from her laboratory and others that had been published over the past thirty years on dextrans. She indicates that the proper interpretation of past and future studies necessitates pointing out previous inadequacies of dextran structural data. She indicates that new techniques can provide a better picture of the structure of many of the dextrans produced by various strains of bacteria, which can lead to a better understanding of the immunochemical interactions.

Dr. Morey Slodki, who worked with her in her later years, has noted that Dr. Jeanes had a terrific analytical mind and was almost always right. He felt that very few people have the insight and foresight about their work that Jeanes had. He believes it was through her leadership that the dextran industry and microbial gum industry developed in the United States.[9]

Notes

1. Private communication between William P. Jeanes and Nina Matheny Roscher, 1995.

2. Ibid.

3. P.A. Sandford, "Allene R. Jeanes," *Carbohydrate Research* 66 (1978): 3.

4. Private communication between William P. Jeanes and Nina Matheny Roscher, 1995.

5. "Allene R. Jeanes," *Chemical and Engineering News* 34 (1956): 1984.

6. Private communication between Benjamin Alexander and Nina Matheny Roscher, 1995.

7. "Allene R. Jeanes," *Chemical and Engineering News* 34 (1956): 1984.

8. Sandford, "Allene R. Jeanes," p. 4.

9. Private communication between Morey Slodki and Nina Matheny Roscher, 1995.

Bibliography

"Allene R. Jeanes." *Chemical and Engineering News* 34 (1956): 1984.

American Men and Women of Science. 12th ed. *New York*: Jaques Cattell Press, R.R. Bowker, 1972.

Jeanes, Allene. "Digestibility of Food Polysaccharides by Man." *ACS Symposium Series: Physiological Effects of Food Carbohydrates* 15 (1974): 336–347.
———. "Extracellular Microbial Polysaccharide-Polyelectrolytes: Properties and Applications." *Polyelectrolytes* (1976): 207–225.
———. "Immunochemical and Related Interactions with Dextrans as Reviewed in Terms of Improved Structural Information." *Molecular Immunology* 23 (1986): 999–1028.
Sandford, P.A. "Allene R. Jeanes." *Carbohydrate Research* 66 (1995): 3–5.

NINA MATHENY ROSCHER

IRÈNE JOLIOT-CURIE
(1897–1956)
Physicist

Birth	September 12, 1897
1914	Baccalauréat degree, Collège Sévigné, Paris
1918	Assistant to Marie Curie, Radium Institute, Paris; French Medal of Recognition
1920	Licentiate in Physics and Mathematics, University of Paris
1921–35	Researcher, Curie Laboratory, Paris
1927	Married Frédéric Joliot on October 9
1932	Mateucci Medal, Italian Society for Sciences
1933	Henri Wilde Prize, France
1934	Marquet Prize, Academy of Sciences, Paris
1935	Nobel Prize in Chemistry; Chevalier Légion d'Honneur, France
1935–36	Head of Research, Caisse Nationale de la Recherche Scientifique
1936	Undersecretary of State for Scientific Research, France
1937–56	Researcher, Radium Institute, Paris (appointed head, 1946)
1940	Bernard Gold Medal, Columbia University, NY
1946–50	Member, Commisariat à l'Énergie Atomique
Death	March 17, 1956

The French physicist Irène Joliot-Curie is known for her pioneering work in radioactivity. Her fame derives principally from the discoveries she made with her husband, Frédéric Joliot. They discovered artificial radioactivity, for which they shared the Nobel Prize in chemistry in 1935. Dr. Curie's subsequent investigations of radioactive elements ensured her position among the great modern scientists.

Irène was the older daughter of **Marie Sklodowska Curie** and Pierre Curie, the discoverers of radium. She was born in Paris in 1897, delivered by Eugène Curie, her paternal grandfather. While Madame Curie was in the laboratory, Irène was in the loving care of her grandfather, who had come to live with the family in 1898 after his wife died. Eugène Curie also helped to care for Irène's sister, Eve, when she was born in 1904.

At the age of 48, in 1906, Pierre Curie was hit by a horse-drawn wagon and killed. After Pierre's death, Eugène Curie's presence in their home became even more vital. He befriended Irène, who was very much like the son he had lost. Eugène introduced Irène to natural history and botany and communicated to her his enthusiasm for the works of Victor Hugo. When he expressed a desire to live in the country, the family moved to a suburb of Paris.

Irène and Eve benefited from walks in the open country. However, because of the separation from their mother, it became necessary for them to have governesses. Marie chose Polish women to fill this role. In this way Irène and Eve learned their mother's native tongue.

Irène was a strange young creature: tall, awkward, gray-eyed, with short-cropped hair. She inherited the shyness and the intellectual abilities of her parents. She was not quick, but she had great reasoning abilities. Marie Curie taught her children to be affectionate but always restrained, and never to raise their voices in anger or joy. Irène was devoted to her mother but developed a strong bond of affection with Eugène Curie. By the time she was 12, the old doctor had imprinted his democratic and social ideals on her. It was to her grandfather, a convinced freethinker, that Irène owed her atheism and her attachment to liberal socialism.

Marie Curie took charge of Irène's education early on. She believed that it was unhealthy for children to be shut up in school for long hours. As a result, Irène did not attend school until the age of 12. Every day began with an hour's work, intellectual or manual. When she finished her daily task, she was sent outside. She mastered gardening, gymnastics, cooking, and sewing. The sisters spent their vacations either in the mountains or at the seashore, where they learned to canoe and swim and took long walks and bicycle rides.

When it was time for more formal schooling, Irène studied for two years at a teaching cooperative established by several of Marie's colleagues, including physicists Paul Langevin and Jean Perrin. Each took charge of the teaching of a particular subject. The school was scientifically overbalanced, but Irène thrived on the diet of mathematics, physics, and chemistry. Following this preparation, she was able to enter a higher class in the Collège Sévigné and had no difficulty passing her bachelor's examination at the age of 17. She received her baccalauréat just before the outbreak of World War I and then studied physics and mathematics at the Sorbonne until 1920.

During World War I, Irène learned nursing and radiology. She assisted her mother in setting up the apparatus for the radiography of the wounded and helped to maneuver them in front of the equipment. There was considerable resistance among older, male physicians to the use of X-rays as a means of diagnosis. Both women experienced significant resentment to their presence. Nonetheless, at the age of 18 Irène had sole responsibility for installing the radiographic equipment in a hospital a few miles from the front in Flanders. Irène later recalled, "My mother had no more doubts about me than she doubted herself."[1] She received a medal in recognition of her work. During this time, she was frequently exposed to X-rays. The protective measures that were used—a few small metal screens, cloth gloves, and avoiding the direct beam of the X-rays when possible—were quite inadequate.

Irène never hesitated about her vocation: she would become a physicist. In 1918 she began to work as her mother's assistant at the Radium Institute of the Sorbonne. Her first important investigation concerned the fluctuations in the alpha rays ejected from polonium. In 1925 she presented this work in her doctoral thesis, dedicated "To Madame Curie from her daughter and pupil." For this offspring of famous parents, a thousand people turned out to hear her first public contribution to science, and it was reported in the *New York Times*.

Marie Curie hired a young physicist each year as her laboratory assistant, giving him a chance at a career in science. Her choice for 1925, Frédéric Joliot, was handsome and outgoing, highly recommended for his quick mind. "Fred" soon formed a bond with Madame Curie's other assistant, Irène. Despite outward differences, the two assistants found that they complemented each other, sharing a love of science, outdoor sports, and radical politics. Within two years they married.

By 1929, Fred and Irène had each published some useful pieces of research, but neither had made a mark as an exceptional scientist. When they began to work in collaboration, they chose to work with polonium, a radioactive metallic element. Polonium is highly toxic but has the useful property of emitting abundant high-energy alpha particles.

Fred and Irène had a major breakthrough in 1934. They found that when polonium was placed next to aluminum, radioactive emissions could be detected from the aluminum. To their surprise, even after the polonium was removed, the aluminum's radioactive discharge continued. Fred and Irène immediately realized that they had stumbled on an entirely new reaction. The aluminum nuclei were swallowing up alpha particles. The new element produced was an unstable isotope of phosphorus not found in nature. In 1935 they received the Nobel Prize in chemistry for their discovery of "artificial radioactivity."

Irène's fame opened new opportunities to her. In 1936, Léon Blum, the head of the Popular Front government, created the position of Undersecretary of State for Scientific Research. The job was offered to Irène Joliot-Curie. Her prestige as a Curie helped to ensure government support of scientific research. It was one of the first high government posts to be offered to a woman. After four months, however, Jean Perrin replaced her by prearrangement, so that she could return to her scientific work.

Following this short tenure in government, Irène was elected professor at the Sorbonne in 1937. She continued to work at the Radium Institute, analyzing the complex phenomena that resulted from bombarding uranium with neutrons. Collaborating with Pavle Savich, Irène performed the original research that motivated Otto Hahn to devise an experiment that showed it was possible for a neutron to divide a uranium atom into two atoms of comparable mass, that is, atomic fission. Her work had instigated the discovery for which Hahn won the Noble Prize in chemistry in 1944.

When the Germans invaded France in 1940, Irène decided to remain in her laboratory in Paris. However, with the signing of the armistice, the Joliot-Curies moved to a small town near Bordeaux to avoid Nazi interference with their work. In 1944, a few months before the liberation of Paris, Irène and her children were smuggled into Switzerland by the Resistance, who feared that she might suffer reprisals for the underground activities of her husband.

Returning to Paris after World War II, Irène was named director of the Radium Institute. From 1946 to 1950, she was also one of the directors of the Commisariat à l'Énergie Atomique. Fred was admitted to the Academy of Sciences in 1943; Irène was not. She applied every few years, and after each rejection she publicized the Academy's continued discrimination against women. Irène divided her efforts in the last year of her life between setting up the Radium Institute's new laboratories at Orsay and working for women's pacifist movements. Her experiences during World War I had given her a lifelong hatred of war. She died at the age of 58, a victim (like her mother) of acute leukemia. The disease was undoubtedly a consequence of the radiation to which she had been exposed.

Irène Joliot-Curie maintained throughout her life a simple, direct, and honest manner. She loved nature and enjoyed rowing, sailing, and swimming during vacations in Brittany. Her favorite writers were Victor Hugo and Rudyard Kipling. Irène was also fond of taking long walks in the mountains, where she often went because of her twenty-year battle with tuberculosis. Irène found much joy in motherhood and, despite the hours spent in the laboratory, devoted much time to her two children, Hélène and Pierre. Both became brilliant researchers: Hélène, like her mother and grandmother, in nuclear physics, and Pierre, in biophysics.

Note

1. Sharon Bertsch McGrayne, "Irène Joliot-Curie: Radiochemist," in *Nobel Prize Women in Science: Their Lives, Struggles, and Momentous Discoveries* (New York: Carol Publishing Group, 1993), p. 117.

Bibliography

Chadwick, James. "Obituary, Mme. Irène Joliet-Curie." *Nature* 177 (1956): 964–965.
Curie, Eve. *Madame Curie.* Translated by Vincent Sheean. Garden City, N.Y.: Doubleday, Doran, 1937.
Curie, Marie. *Pierre Curie.* Translated by Charlotte Kellogg and Vernon Kellogg. New York: Macmillan, 1923.
Joliot, F., and I. Curie. "Mass of the Neutron." *Nature* 132 (1934): 721.
———. "New Evidence for the Neutron." *Nature* 130 (1932): 57.
McGrayne, Sharon Bertsch. "Irène Joliet-Curie: Radiochemist," in *Nobel Prize Women in Science: Their Lives, Struggles, and Momentous Discoveries.* New York: Carol Publishing Group, 1993.
Perrin, Francis. "Irène Joliet-Curie," in *Dictionary of Scientific Biography.* Edited by Charles Coulston Gillespie. New York: Charles Scribner's Sons, 1970.
Reid, Robert. *Marie Curie.* New York: E.P. Dutton, 1974.
Weart, Spencer R. *Scientists in Power.* Cambridge, Mass.: Harvard University Press, 1979.

CLARA A. CALLAHAN

MADELEINE M. JOULLIÉ
(1927–   )
Chemist

Birth	March 29, 1927
1949	B.Sc., chemistry, Simmons College
1950	M.Sc., chemistry, University of Pennsylvania
1953	Ph.D., chemistry, University of Pennsylvania

Madeleine M. Joullié. Photo courtesy of Madeleine M. Joullié.

1953–	University of Pennsylvania, Dept. of Chemistry: Instructor (1953–57); Research Associate (1957–59); Assistant Professor (1959–68); Associate Professor (1968–74); Professor (1974–)
1959	Married Richard Prange
1972	Philadelphia Section Award, American Chemical Society
1978	Garvan Medal, American Chemical Society
1984	American Cyanamid Faculty Award, University of Pennsylvania
1985	Scroll Award, American Institute of Chemists
1991	Lindback Award for Distinguished Teaching, University of Pennsylvania; Philadelphia Section Award, Association of Women in Science
1994	Philadelphia Organic Chemists Club Award; Henry Hill Award, American Chemical Society

Madeleine Joullié is an award-winning chemist renowned both for the elegance of her research that has had direct application to medicine and for her inspirational teaching.

Madeleine M. Joullié was born in Paris, France. Her father was very adventurous and traveled all over the world, landing in Brazil. He sent for the family when she was young, so she grew up in Rio de Janeiro. Women in Brazil were very protected—she never even went to downtown Rio by herself. Her father thought she was being brought up too narrowly and felt she should go to the United States for her education.[1]

Madeleine's father had a Dutch partner who recommended Simmons College. She found the transition to Boston to be very difficult; the winter weather alone was a significant change. Moreover, in Brazil she had grown up in a very limited circle and schools did not have social functions. She lived in the International House at Simmons, but she did not fit in. She did not go to parties or other activities. Madeleine wanted only to take science courses, so she took all the chemistry courses that Simmons offered. She also took, beyond the required courses, as many languages as her advisor would permit. Her roommate was very interested in art and she spent a lot of time in the Boston museums. The Boston Public Library had a good set of *Chemical Abstracts*, so she often spent her weekends there studying.

At that time most of the students at Simmons did not go on to graduate school, so Madeleine did not find the faculty very helpful as she was trying to decide what to do after college. She met someone at the Massachusetts Institute of Technology who told her the University of Pennsylvania was a good place. She went to visit and liked it. There she was the only woman in her class of about fifty, but she found that everyone was generally friendly both at the University and in the city of Philadelphia. She earned her M.Sc. and a Ph.D. there.

Madeleine stayed on to teach and taught undergraduate organic lectures and laboratories. She taught five days a week; in those early days, the professor actually taught the laboratory. After several years of teaching, she became a research associate and was able to develop her research career. She had no graduate students for the first five years, doing all her work with the undergraduate students. Initially only the women graduate students, limited in number, wanted to work with her. At that time no starting money or assistance to develop a research program was made available. Nowadays the university provides money for postdoctoral fellows to help a new faculty member get her or his research efforts started. Joullié still teaches the sophomore organic course at the University of Pennsylvania, but she now supervises the graduate students who teach the laboratories.

Joullié met her husband over a hot dog in Houston Hall at the University of Pennsylvania. As they talked and got to know each other, they found they had many common interests. Having been married since 1959, they represent an early example of a commuting couple. Her husband is a physics professor at the University of Maryland and has a townhouse near the train station in Washington, D.C. Madeleine herself has a home in Philadelphia. As a theoretical physicist, it is easier for him to travel back and forth. However, bicycling is a shared interest and they have explored much of Washington on the bike trails that run throughout the city.

In 1976, Joullié's research group synthesized tilorone, an important interferon inducer. (Interferons are low molecular weight proteins that inhibit infection by a wide range of viruses.) Tilorone was found to be a broad-spectrum, orally active, antiviral agent. In order to make the more complex molecules that are useful in medicine, Joullié's research group studied some reactions of simple compounds using ketene and sulfur dioxide. Ketene is a small, simple compound that is very reactive. By working with the ketene family of compounds the researchers were able to develop new addition products that would react further with a large variety of substrates. Not only were they able to make many new compounds, but they were better able to understand the chemistry of the simple molecules.

Joullié has always been willing to explore new fields. For example, in the carbohydrate field she examined little-known rearrangements that could convert carbon-oxygen bonds into carbon-carbon bonds. She synthesized furanomycin, an antibiotic isolated from natural sources. She produced the first stereoisomeric synthesis of this compound from a sugar precursor and also unambiguously established the structure, which had been erroneously assigned by previous investigators. She also developed methodology to transform other commercially available sugars (such as D-ribonolactone and vitamin C) into versatile intermediates to be used in natural product synthesis.

Joullié's most recent and notable achievements are in the area of cyclopeptide alkaloids and cyclodepsipeptides. The synthesis of such highly strained molecules is one of the most difficult tasks in organic synthesis. In addition to completing the first total synthesis of a 14–membered dihydro derivative, she pioneered the use of a "four-component condensation" for synthesizing cyclopeptide alkaloids and prolyl peptides. More recently, using organic synthesis, she has revised the stereochemistry of the didemnins well before the X-ray structure of didemnin B was reported. (Didemnins are a new class of depsipeptides isolated from a Caribbean marine chordate animal of the family *Didemnidae*, a species of the genus Trididemnum.) She has carried out the total synthesis of didemnins A, B, C, and several analogs, whose activities are equal or superior to those of natural products. The elegant strategy involved in these syntheses, combined with the development of new technology, illustrates a superb blend of target-oriented synthesis with the discovery of new synthetic methods. The reproducible synthetic strategies that were used served to produce gram quantities of several didemnins and didemnin analogs, which are affording more useful drugs.

In 1972, Joullié received the American Chemical Society Philadelphia Section Award. She was the first woman to receive this award. In 1978, she received the Garvan Medal of the ACS "in recognition of her many contributions in heterocyclic and medicinal chemistry, exemplifying her broad approach, experimental finesse and originality; of her deep and abiding interest in the training and development of her research advisees; and of her devoted and inspirational teaching of graduate and undergraduate students."[2] She is considered to be an exceptional teacher; it was pointed out that she personally teaches more than 250 students each semester, conducting a number of recitations as well as supervising a group of graduate students.[3]

In 1994, Joullié received the Henry Hill Award of the American Chemical Society Division of Professional Relations. The award honors individuals who, in the judgment of their peers, have demonstrated outstanding dedication to promoting fair treatment and the well-being of chemists. It was noted that "Joullié has a distinguished record of substantive and sustained service to professional relations principles by promoting equity and fairness in science and academe and the professional well-being of scientists." Joullié worked against discrimination as the first faculty affirmative action officer at the University of Pennsylvania from 1975 to 1980.[4]

At the time she received the Hill Award, Dr. Joullié spoke on "Chemistry: Is It a Profession?" She discussed the early struggles to have chemistry recognized as a profession. She was quoted in the *ACS Division of Professional Relations Bulletin*: "The role of scientists in advancing scientific knowledge and technology is not understood by our leaders, policy makers and most of the American population. . . . The majority of the general public either believes that science can solve all problems at the drop of a hat or rejects all contributions made by science and technology." She was concerned that we need to "publicize the contribution of chemistry to society. . . . Teach the public to better understand the scientific method so research can be better utilized by society. . . . Develop programs to retrain and reorient individuals and find ways to increase our ability to attract the best and the brightest into our profession."[5]

Joullié is a member of many honorary and professional societies. She has authored three books, numerous review articles, and over 185 publications.

Notes

1. Unless otherwise noted, quotations in this essay are from conversations between Madeleine Joullié and Nina M. Roscher and from her personal biography.

2. 1978 Awards Program, 175th National Meeting, American Chemical Society, Anaheim, California.

3. "Garvan Medal," *Chemical and Engineering News* (August 29, 1977): 50–51.

4. "Madeleine Joullié Wins Henry Hill Award," *Chemical and Engineering News* (September 5, 1994): 45.

5. "Henry Hill Award 1994," *American Chemical Society Professional Relations Bulletin* (October 1994): 4.

Bibliography

"Garvan Medal." *Chemical and Engineering News* (August 29, 1977): 50–51.
"Henry Hill Award 1994." *American Chemical Society Professional Relations Bulletin* (October 1994): 4.
"Madeleine Joullié Wins Henry Hill Award." *Chemical and Engineering News* (September 5, 1994): 45.

NINA MATHENY ROSCHER

ISABELLA L. KARLE
(1921–   )
Crystallographer

Birth	December 2, 1921
1941	B.S., University of Michigan
1942	M.S., University of Michigan; married Jerome Karle
1944	Ph.D., University of Michigan
1946–	U.S. Naval Research Laboratory
1964	Navy Superior Civilian Service Award
1968	Society of Women Engineers Achievement Award
1969	Hillebrand Award, American Chemical Society
1973	Federal Woman's Award, U.S. Civil Service Commission
1976	Garvan Medal; D.Sc., University of Michigan
1979	D.Sc., Wayne State University

1980	Robert Dexter Conrad Award
1984	Pioneer Award, American Institute of Chemists LHD, Georgetown University
1986	D.Sc., University of Maryland
1987	Secretary of the Navy Distinguished Achievement in Science Award
1988	Gregori Aminoff Prize, Royal Swedish Academy of Science; Rear Admiral William S. Parsons Award, Navy League of United States
1989	Michigan Women's Hall of Fame
1990	Bijroet Medal, University of Utrecht, Netherlands
1992	Vincent du Vigneaud Award
1993	Bower Award, Franklin Institute

In 1993, Dr. Isabella L. Karle became the first woman to receive the Bower Award and Prize for Achievement in Science, which included a gold medal and $250,000. The Franklin Institute selects the annual recipient of this prestigious award from a slate of international nominees. For Karle, the award was a capstone to a career spanning nearly fifty years with the U.S. Naval Research Laboratory in Washington, D.C., where she is a senior scientist specializing in crystallography.[1] Crystallographers, Karle explains, determine "which atoms are connected to which, and how they are arranged in respect to each other in the crystal lattice."[2] Karle's pioneering work makes it possible to determine this structure immediately through electron and X-ray diffraction. The "symbolic addition procedure," as her famous method is called, revolutionized the process of crystal structure analysis and has applications in diverse areas of science.[3]

Once the crystal structure of a molecule is known, it becomes possible to create a synthetic in the laboratory. For example, Karle's current work involves *designed* peptides, which take advantage of beneficial properties while minimizing harmful ones, thereby making it possible to alleviate the negative side effects of therapeutic drugs.[4] In other cases, a synthetic might be cheaper to produce than the naturally occurring substance. The Navy, for instance, uses an enormous amount of wood to build ships and docks; and it needs wood that is resistant to destructive creatures. A scarce Panamanian wood meets this requirement: a substance in it interferes with the calcium of worms and termites, so they will not eat it. Karle and her colleagues analyzed this insect-repelling substance and successfully created a synthetic to treat the wood used by the Navy.[5]

The Bower Award has given Karle a degree of notoriety today, but there were "many trips downward" before reaching a "final high," she said.[6] She was born Isabella Helen Lugoski in Detroit in 1921 to Polish immigrant parents. Her father was a house painter and her mother a seamstress.[7] The family had only a meager income, but the children were not deprived. Having had little formal education themselves, Isabella's parents gave constant academic encouragement to her and her younger brother. They were also able to take the family on vacations throughout the Depression years because they were thrifty and managed their money well. Since their neighborhood was populated by Polish and other immigrants, Isabella did not learn English until she started school, although her mother taught her to read and write in Polish and do basic arithmetic before first grade. Isabella was interested in her schoolwork and found it easy. Her parents expected that she might become a schoolteacher or lawyer. It was not until she went to high school that Isabella got her first exposure to science. By chance she chose a chemistry class to fulfill college preparatory requirements, and she found it fascinating.[8] "It was different from everything else I had studied up to that point," she said. Today, looking back on those years, Karle describes her attraction to chemistry using the words of her young granddaughter, who also wants to be a scientist: "I want to know how things work," she declared.[9] Although Isabella's high school teacher told her that chemistry was "not a proper field for girls," she had made up her mind to study it in college.[10]

Her first semester at Wayne State University was critical. An assistant professor of chemistry, Dr. Joseph Jasper, talked to her about going to graduate school, which was a new concept for Isabella. "He was influential," she said, and he continued to provide encouragement and advice through correspondence when, after one semester, she transferred to the University of Michigan.[11] Michigan had awarded her a four-year undergraduate scholarship, and she also obtained an assistantship in the chemistry department there. After graduating with highest honors in 1941, however, she found the road blocked to further study: graduate teaching assistantships in the chemistry department were awarded only to male students. Fortunately, the American Association of University Women granted her a fellowship that allowed her to start her graduate work, and the following year she received a Horace H. Rackham fellowship.[12]

During graduate school Isabella met fellow student Jerome Karle, whom she married in 1942. Early on, the Karles developed a collaborative working relationship: Jerome contributed most of the theory, and Isabella concentrated on the practical aspects of application.[13] They were both interested in the work of Dr. Lawrence Brockway, an associate professor at the University of Michigan who was conducting experiments on electron diffraction by gaseous molecules.[14] Brockway was passing electron beams through gas molecules and producing photographs of the diffracted, concentric rings that resulted. The diameters and relative intensities of these rings could be measured and analyzed mathematically, making it possible to calculate the distances between atoms and the angles in the gas molecules.[15] The Karles would later speculate about diffraction analysis of *solid* matter in the form of crystals.[16] They finished their Ph.D.s in 1944 within a few months of each other. The Karles worked briefly on the Manhattan Project in Chicago after graduation. When that work was finished, they started looking for research jobs where they could work together. Isabella said, "Although academia appealed to us, there were no opportunities for women in institutions with reasonable research facilities."[17] The U.S. Naval Research Laboratory (NRL) offered them employment as physicists beginning in 1946, and they have been there ever since.

Karle said that when she and her husband arrived at NRL, they "were immediately thrust into an atmosphere of exciting research and surrounded by exceptional people."[18] For a while the Karles continued their research into the molecular structure of gases, but at the same time they began to think about the molecular arrangements of atoms in crystals.[19] Although the structure of some crystals had been solved using what is known as the *heavy-atom* technique, for the large number of compounds with nearly equally weighted atoms—and this includes most organic crystals—the technique failed.[20] Moreover, photographs of X-rays diffracted by crystals produced images that were different from the images of gases. Instead of concentric rings with measurable diameters and intensities, a crystal showed only spots, the intensities of which could be measured, but not the *phases*. "A key problem existed," explained Karle: "half of the necessary information seemed to be lost in the diffraction procedure." The phase had to be known before the more difficult crystal structures could be solved. Isabella's husband, Jerome, and a colleague, Herbert Hauptman, solved the problem of obtaining the phases theoretically, and for this work they shared the 1985 Nobel Prize in chemistry.[21] Isabella Karle translated their mathematical results into practical procedures. In an article published in 1966 in *Acta Crystallographica*, the Karles described the theory and procedures for determining the phases of X-rays diffracted by crystals. If the phases are known, the crystal structure can be determined immediately. The article had an enormous impact on the world of crystallography and quickly became a classic.[22]

Isabella Karle was the one to put the method to work solving crystal structures. One of her first major successes was to establish the structure of the venom of a South American frog. Natives used the frog skin secretions as arrow-tip poison. Since these toxins block specific nerve impulses, they are invaluable in the medical study of nerve transmission. A synthetic form of the venom was made for the National Institutes of Health.[23] Over the years, Karle herself has done hundreds of structures in various areas of science. Demand for her to teach the method to others led Karle to conduct advanced-study workshops worldwide. The workshops, she said, "were instrumental in expanding the number of crystal structures solved around the world, from a handful each year to much more than 10,000 a year at present."[24] The speed and facility with which crystal structures could be solved had increased exponentially as a result of her work. The author of nearly 300 scientific papers, Karle revolutionized crystal-structure analysis.[25]

Karle is concerned about political pressures for quick results in scientific research. "Legislators elected for two years," she said, "like to see some progress," but "basic research doesn't work this way."[26] At the reception honoring her for receiving the Bower Award, Karle stated the importance of funding for basic research. She said: "Often it takes 10 or 50 or even more years for the knowledge derived from basic research to be applied to high technology, but the pool containing basic information must be constantly replenished, otherwise it will run dry."[27]

Today, at age 73, Karle is as busy as ever and has no plans to retire. She said, "I feel fine and the lab is happy with what I do. There are new problems to work on." She attributes her vitality and health to physical activity: ice skating, swimming, hiking, and doing yard work.[28] She has three daughters, Louise (Hanson), Jean M., and Madeleine (Towney). "Personally," she said, "I've had a satisfying and fascinating career."[29]

Notes

1. Tom Kenworthy and John Schwartz, "Navy Scientists Honored," *Washington Post* (November 18, 1993): A21.

2. Iris Noble, *Contemporary Women Scientists of America* (New York: Julian Messner, 1979), p. 107.

3. "Remarks at the Bower Award Reception for Isabella Karle," *Naval Research Reviews* 46, no. 2 (1994): 3.

4. Dr. Isabella L. Karle, telephone conversation with Janet Owens, July 11, 1995.

5. Noble, *Contemporary Women Scientists*, pp. 108–109.

6. Karle, telephone conversation.

7. Noble, *Contemporary Women Scientists*, p. 109.

8. Isabella L. Karle, "Crystallographer," *Annals of the New York Academy of Sciences* 208 (March 15, 1973): 11.

9. Karle, telephone conversation.

10. Karle, "Crystallographer," p. 11.

11. Karle, telephone conversation.

12. Karle, "Crystallographer," p. 11.

13. Ibid., p. 14.

14. Noble, *Contemporary Women Scientists*, p. 111.

15. Maureen M. Julian, "Profiles in Chemistry: Isabella L. Karle and New Mathematical Breakthrough in Crystallography," *Journal of Chemical Education* 63 (January 1986): 66.

16. Noble, *Contemporary Women Scientists*, p. 113.
17. Karle, "Crystallographer," p. 11.
18. "Remarks at the Bower Award Reception," p. 6.
19. Noble, *Contemporary Women Scientists*, p. 112.
20. Julian, "Profiles in Chemistry," p. 67.
21. "Remarks at the Bower Award Reception," p. 6.
22. Julian, "Profiles in Chemistry," p. 67.
23. Noble, *Contemporary Women Scientists*, p. 116.
24. "Remarks at the Bower Award Reception," pp. 6–7.
25. Julian, "Profiles in Chemistry," p. 67.
26. Karle, telephone conversation.
27. "Remarks at the Bower Award Reception," p. 8.
28. Karle, telephone conversation.
29. "Remarks at the Bower Award Reception," p. 7.

Bibliography

American Men and Women of Science, 1995–96. 19th ed. New Providence, N.J.: R.R. Bowker, 1994.

Julian, Maureen M. "Isabella L. Karle and a New Mathematical Breakthrough in Crystallography." *Journal of Chemical Education* 63 (January 1986): 66–67.

Karle, Isabella L. "Crystallographer." *Annals of the New York Academy of Sciences* 208 (March 15, 1973): 11–14.

Karle, Isabella L., and J. Karle. "An Application of the Symbolic Addition Method to the Structure of L-Arginine Dihydrate." *Acta Crystallographica* 17, pt. 7 (July 1964): 835–841.

Karle, Isabella L., et al. "Crystal Structure of [Leu^1]zervamicin, a Membrane Ion-Channel Peptide: Implications for Gating Mechanisms." *Proceedings of the National Academy of Sciences of the United States of America* 88, no. 12 (June 1991): 5307–5311. Biochemistry.

Karle, Isabella Lugoski, and Jerome Karle. "Internal Motion and Molecular Structure Studies by Electron Diffraction." *Journal of Chemical Physics* 17, no. 11 (November 1949): 1052–1058.

Karle, J., and I.L. Karle. "The Symbolic Addition Procedure for Phase Determination for Centrosymmetric and Noncentrosymmetric Crystals." *Acta Crystallographica* 21, pt. 6 (December 1966): 849–859.

Noble, Iris. "Isabella Karle, Crystallographer," in *Contemporary Women Scientists of America.* New York: Julian Messner, 1979.

JANET OWENS

JOYCE JACOBSON KAUFMAN

(1929–   )

Chemist

Birth	June 21, 1929
1949	B.S., Johns Hopkins University
1959	M.A., Johns Hopkins University
1960	Ph.D., Johns Hopkins University; Visiting Scientist, Quantum Chemistry Group, University of Uppsala
1960–62	Staff Scientist, Martin Company, Research Institute for Advanced Studies
1963	*D.E.S. Tres Honorable*, La Sorbonne, Paris
1963–69	Head, Quantum Chemistry Group, Research Institute for Advanced Studies
1964	Martin Company Gold Medal for Outstanding Scientific Accomplishments (also received in 1965 and 1966)
1969	Une Dame Chevalier, France, Chapitre Centre National de la Recherche Scientifique
1969–	Associate Professor, Dept. of Anesthesiology, Johns Hopkins School of Medicine; and Principal Research Scientist, Dept. of Chemistry, Johns Hopkins University
1970–	Consultant, National Institutes of Health
1974	Garvan Medal, American Chemical Society; Outstanding Chemist, Maryland Section, American Chemical Society; honored as one of ten Outstanding Women in the State of Maryland
1977	Associate Professor, Dept. of Surgery, Johns Hopkins School of Medicine; Scientific Advisory Committee to the Pentagon, Deputy Director of Research and Engineering, Department of Defense
1981	Elected to American Society for Pharmacology and Experimental Therapeutics

Joyce Kaufman is distinguished in a wide variety of fields: chemistry, physics, biomedicine, and supercomputers. Her research interests include theoretical quantum chemistry, a branch of physical chemistry that attempts to explain chemical phenomena by means of the laws of quan-

Joyce Jacobson Kaufman. Photo courtesy of Walter S. Koski, Department of Chemistry, Johns Hopkins University.

tum mechanics. She studies the chemical physics of energetic compounds such as explosives, rocket fuels, and oxidizers and the applications of these subjects to biomedical research in pharmacology, drug design, toxicology, and molecular modeling—the use of computers for the simulation of chemical processes. Kaufman designed a computer model for quantum chemical predictions in the areas of toxicity, toxicology, and pharmacology. Her work resulted in a clearer understanding of the structure of tranquilizers, narcotics, and carcinogens. She has also published in nuclear chemistry (the study of the atomic nucleus) and radiochemistry (the study of radioactive substances).

Joyce was born in 1929 in New York City to Robert and Sara (Seldin) Jacobson. Both parents were Jewish, and her father was a salesman. When her parents separated in 1935, Jacobson and her mother moved to Baltimore, Maryland, to live with her maternal grandparents, Jacob and Mary Seldin. In December 1940 her mother married Abraham Deutch, a successful businessman who owned and operated the American Cornice and Roofing Company. Deutch raised Jacobson as his own daughter, and although her mother died in 1964 she remained close to her stepfather.

Jacobson was a bright and precocious child who began reading before she was 2 years old. By the age of 6 she had read all the children's books in the local library. A librarian recommended that she read about lives of famous people. Biographies and autobiographies of scientists became her favorite reading matter. The biography of **Marie Curie** had a profound influence on her, and by the age of 8 she had decided that she wanted to be a chemist.

Jacobson was an outstanding student. In 1937 she was selected to attend a summer program at Johns Hopkins University (JHU) for children who were gifted in science and mathematics. Later she was chosen to attend the Robert E. Lee High School 49, where exceptional students were permitted to cover three years of study in two years. Despite carrying a heavier load than other students, Jacobson graduated seventh in a class of almost three hundred students. Her family was supportive of her interest in science. In fact, relatives on both sides of the family had specialized in mathematics and science in college.

Jacobson attended Johns Hopkins University and received her B.S. degree "with honor" in 1949. She had wonderful professors there who had a great influence on her career in science. In particular, Clark Bricker, her freshman chemistry professor, allowed her and two other freshmen to do research with him in quantitative analysis; and Walter S. Koski taught her physical chemistry.

In December 1948, Jacobson married Stanley Kaufman, an engineer. They had one daughter, Jan Caryl, who was born on June 24, 1955. Three months after Jan was born, Joyce Kaufman returned to work. A long-time employee of the family provided child care for Jan until she was 12 years old. Family was important to Kaufman, and she was frequently accompanied by her daughter or her mother when she traveled. Jan grew up to be a rabbi, distinguished in her own right.

Kaufman worked as a technical librarian in 1949–1950 at the Army Chemical Center in Baltimore, where she established a scientific indexing system for the Center's technical reports. In 1950 she became a research chemist at the Center, developing new methods of analysis and micro-analysis. In 1952, Professor Koski invited her to work with him at JHU on a research contract. She conducted research in kinetics of isotope exchange reactions involving the boron hydrides and the chemical physics of related compounds. In the summer of 1960 she attended the Summer Institute of the Quantum Chemistry Group at the University of Uppsala, Sweden. This was an important step for her career: it was the beginning of her efforts in the application of quantum mechanics to chemistry, biology, and medicine.

Professor Koski encouraged Kaufman to get her Ph.D. She earned both her master's degree (1959) and her Ph.D. (1960) in physical chemistry. She also received a D.E.S. "Tres Honorable" in theoretical physics from the Sorbonne in 1963. After receiving her Ph.D., Kaufman was invited to join the staff at the Martin Company's Research Institute for Advanced Studies to do theoretical research on the application of quantum mechanics to problems in chemistry. She worked there until 1969, when she joined JHU with a joint appointment in the Department of Chemistry and the Department of Anesthesiology and Clinical Care Medicine at the School of Medicine. Kaufman and Koski currently have a joint group in experimental physical chemistry/chemical physics and theoretical quantum chemistry.

Kaufman's most significant contribution concerns the application of quantum chemistry to calculations involving large molecules. At a time when most scientists did not think it would be possible or productive to apply this research approach to large molecules, she persisted and made significant contributions to structural properties of molecules, drug design, and predictive toxicology. She used her calculations to predict a new series of neuroeleptics, drugs that are useful in the treatment of mental disorders, especially psychoses. At the time, there was no work of this type going on in the large pharmaceutical laboratories. Kaufman was invited to speak about her methods at many of the large pharmaceutical companies. As a result of her research, these companies now routinely have quantum chemistry groups supporting their efforts in drug design. The experimental determination of the toxicity of a chemical involves intensive, expensive, and time-consuming animal studies. The theoretical studies advocated by Kaufman have the potential to reduce the expense, time, and number of animals used. In 1974, Kaufman was awarded the Garvan Medal from the American Chemical Society, an award granted annually to the outstanding woman chemist in the United States.

Kaufman has contributed much to the training of students. At Johns Hopkins School of Medicine, she has worked with interns and residents in the area of theoretical quantum chemical and experimental physico-chemical studies of central nervous system drugs, psychotropic drugs, and general and spinal anesthetics. At the JHU Department of Chemistry, she works with several postdoctoral fellows in her lab as well as senior scientists and a wide range of visiting scientists. She also has about ten undergraduates doing research under her supervision, primarily pre-medical students. Kaufman has an unusual rapport with students and inspires them to do their best work.

In addition to her academic and research activities, Kaufman has served in many administrative positions. She was a senior councillor for the American Chemical Society's Division of Physical Chemistry and has been active on numerous committees of the Society. She has also served in an editorial advisory capacity for John Wiley Interscience Publishers as advisor in theoretical chemistry (1965–1980) and was editor of the Benchmark Book Series in Physical Chemistry (1977–1988). She has been on the editorial boards of *Molecular Pharmacology* (1970–1988) and the *International Journal of Quantum Chemistry* (1967–1985). She has been a member of the editorial advisory board of the *Journal of Computational Chemistry* since 1984 and the *Journal of Explosives* since 1981. Kaufman has been a consultant to the National Institutes of Health since 1970 in the area of quantum chemical calculations of molecules of pharmacological and biological significance, and in the field of computers and supercomputers. In 1989 she was appointed a member of the National Science Foundation Evaluation Committee for NSFNET, the network that links the NSF supercomputers to each other and to the world. Joyce Kaufman is widely published and has presented numerous invited lectures at international conferences, research institutes, and academies of science throughout the world. She has achieved a high level of success in her career and with her family.

Bibliography

American Men and Women of Science. 19th ed. New Providence, N.J.: R.R. Bowker, 1994.

Kaufman, J. "Strategy for Computer Generated Theoretical and Quantum Chemical Prediction of Toxicity and Toxicology (and Pharmacology in General)." *International Journal of Quantum Chemistry* (1981): QBS8, 419–439.

———. "Theoretical Approaches to Pharmacology." *International Journal of Quantum Chemistry* (1978): QBS4, 375–412.

Kaufman, J., and W.S. Koski. "Physicochemical, Quantum Chemical and Other Theoretical Studies of the Mechanism of Action of CNS Agents: Anesthetics, Narcotics and Narcotic Antagonists and Psychotropic Drugs." *International Journal of Quantum Chemistry* (1975): QBS2, 35–57.

Kaufman, J., W.S. Koski, and D. Peat. "A Systems and Control Theory Approach to Dynamic Neurotransmitter Balance in Narcotic Addiction and Narcotic Antagonism." *Life Sciences* 17 (1975): 83–84.

Koski, Walter S. "Joyce Jacobson Kaufman," in *Women in Chemistry and Physics: A Biobibliographic Sourcebook.* Edited by L.S. Grinstein, R.K. Rose, and M.H. Rafailovich, pp. 302–313. Westport, Conn.: Greenwood Press, 1993.

McGraw-Hill Modern Scientists and Engineers. New York: McGraw-Hill, 1980.

Roscher, Nina M. "Women Chemists." *Chemtech* 6 (1976): 738–743.

WALTER S. KOSKI and MARY ANN MCFARLAND

VERA E. KISTIAKOWSKY

(1928–)

Physicist

Birth	September 9, 1928
1948	A.B., chemistry, *magna cum laude*, Mt. Holyoke College; Phi Beta Kappa
1952	Ph.D., chemistry, University of California, Berkeley
1957–59	Instructor in Physics, Columbia University
1959–62	Assistant Professor of Physics, Brandeis University
1963–69	Staff Member, Laboratory for Nuclear Science, Massachusetts Institute of Technology (MIT)
1969–72	Senior Research Scientist, Dept. of Physics, MIT
1972	Centennial Alumnae Award, Mt. Holyoke College
1972–	Professor of Physics, MIT
1983–84	Phi Beta Kappa Visiting Scholar
1983–	Director, Council for a Liveable World
1984	Fellow, American Association for the Advancement of Science
1990–92	Sigma XI Visiting Lecturer

Vera Kistiakowsky, Ph.D., is a professor of physics at the Massachusetts Institute of Technology (MIT) in Cambridge, Massachusetts. When she was young, her father advised her to plan a career that would allow her to support herself and not to depend on someone else to support her. This was very unusual advice for a father to give a girl who had been born in 1928 in Princeton, New Jersey. But Vera's lifetime of accomplishments make it clear that this advice made a significant impact on her as research physicist, activist, and mother. She also experienced several major events that stirred her to take a stand rather than just be an observer. These included the return of World War II veterans to the labor market, the advent of affirmative action programs, the women's movement, and the arms race. Her experiences growing up in a very supportive environment, especially her close relationship with her father, prepared her well for her masterful balance of a career in physics research, a conscience that stirred her to action, and motherhood.[1]

Vera did not live an ordinary life for a middle-class girl during her formative years. She was an only child who was immersed in the scientific community. An interest in science was natural for her. Luckily she did not notice that most scientists were men, nor did anyone bring that to her attention. Her late father, George Kistiakowsky, was a well-known and respected physical chemist and a professor at Harvard University. He also served as the presidential science advisor to Dwight D. Eisenhower. Her father took her along on hiking and skiing trips with his graduate students and postdoctoral students. (She thinks that it was probably on one of these trips that he gave her the career advice.) She did very well in school, skipping grades and excelling in math. However, she remembers the science classes in high school as being "dreadful" owing to the way they were taught. She completed high school at age 15 and had just turned 16 when she started college.

Entering college as a pre-med student at the age of 15 (her birthday was September 9) would have been quite a challenge for an ordinary coed. But not for Vera. She did not have to contend with male/female issues because she attended Mount Holyoke, an all-girls college (as it was referred to at that time). She was considered different from the other students for another reason. In 1944 there was a great psychological push to get women out of the labor market and back into the home to make way for the returning World War II veterans. Therefore, most of Vera's classmates were looking forward to getting married and having degrees that would make them better wives and mothers. Out of her class of about 350, only about 30 of the girls had their sights set on furthering their education.

Vera soon realized that pre-med was not for her; instead she was attracted to the chemistry department because of its atmosphere and research environment. At that time, women who had degrees in chemistry and other sciences got jobs in women's colleges. Mount Holyoke had a very impressive and lively chemistry department, with a number of women on the staff who also did research. The physics department was just the opposite: it had poor professors and was not a thriving concern. Much of the research that Vera did in undergraduate and graduate school could have been in either the chemistry or physics department. For example, for her undergraduate thesis she built a Geiger counter to look for radioactivity.

When Vera was ready for graduate school after graduating *magna cum laude* from Mount Holyoke, she decided to major in nuclear chemistry. There were only three universities in the country that offered this major, and the University of California at Berkeley was considered the best. It was the home university of Glen Seaborg, who won a Nobel Prize for discovering transuranic elements. Vera did her Ph.D. thesis there on nuclear energy states of promethium, an element that has no isotope that is stable against radioactive decay. It does not occur in nature and for this reason was particularly interesting to her. At other universities, her major would have been considered nuclear physics.

In stark contrast to Vera's undergraduate experience, there were no women professors in her graduate school at Berkeley and no women students. Of the three or four who entered with her, a couple earned only master's degrees and the others earned no advanced degrees. By this time Vera was accustomed to being different; even though her classes were very large, she did not have any problems being the only woman. Her professors knew and respected her father, so she was treated very well. (Vera does not know if the same would have been true if she were just another woman student.) Her fellow students also treated her well. In the International House where she lived, for example, there were separate areas for the men and women and common areas that both groups could share. The men made a point of holding their study group meetings in the common area so that she could join them to do the problem sets and participate in discussions. During her stay at Berkeley, Vera took herself very seriously; she attributes that to the tendency of the faculty and students to take her seriously also.

After receiving her Ph.D. in chemistry from Berkeley in 1952, Vera discovered that finding a job was quite a challenge. Initially her career options were limited because her husband, whom she had married in June 1951, would not be receiving his Ph.D. degree until 1954. So she stayed in Berkeley and took a job at the Naval Radiological Defense Laboratory. Her assignment was to set up a spectroscopy lab for studying radioactive decay. However, when a battleship returned with a carboy (a large bottle encased in a box for holding corrosive liquids) containing the two radioactive isotopes from the explosion of the first hydrogen bomb prototype, everyone had to work on finding out what was in the water. When Vera complained that she had not been hired for that, the response was that she was in the Navy (even though she was a civilian) and had to do it. Vera stayed there until the end of the year, when she received a fellowship from the American Association of University Women to do nuclear physics research.

While her husband was finishing his degree, Vera realized that it would be easier for him to get a job. She had sent out about 100 letters to colleges and universities in the United States and received only one response, a letter from a Jesuit in the physics department at Boston College. He apologetically informed her that Boston College was an all-male school and, therefore, he could not offer her a job. He at least gave her the courtesy of a response. Vera's husband, on the other hand, was offered a job as an instructor at Columbia University. She was then offered a research associateship at Columbia. It turned out that the most distinguished woman physicist in the United States was at Columbia: the internationally known **Chien-Shiung Wu**. Although she was a rather reserved person, she was encouraging and supportive to Vera. Because of her stature, no one could say that women could not be physicists.

When Vera's husband left Columbia to take a new job in Boston, she was expecting their first child. Vera left Columbia too and got a job as assistant professor at Brandeis University. She was then doing nuclear physics, which was clearly distinct from chemistry. While at Brandeis, Vera had her second child. Then, for a variety of reasons, she left Brandeis and took a job as staff scientist at MIT. By this time she had switched from nuclear physics to particle physics. She was following her interests; "that's where the interesting stuff was going on." In 1969 she became a senior research scientist at MIT. Around the same time, Jerome Wiesner became president of MIT. He said that since the number of women undergraduates had increased dramatically, the department should consider women seriously for faculty positions. (This statement by Wiesner predated the Affirmative Action program.) Thus Vera was made a full professor in 1972. She worked in particle physics until 1987 when she switched to astrophysics, in which she continued her research until she retired.

Apparently, being at MIT under the leadership of a forward-thinking president such as Wiesner awakened the activist in Vera. Concurrent with her physics research, which resulted in 86 technical publications during that period, she was very active on women's issues. In 1969 she and two friends, Vera Pless, a mathematician, and Elisabeth Baranger, a theoretical nuclear physicist, organized a women in science workshop for the National Organization of Women's (NOW) conference. The workshop attracted 60 women scientists from all over the greater Boston area and as far away as Providence, Rhode Island. According to Vera, "it was good to see so many other women scientists." There was so much interest that Vera and her two friends founded the Boston-area group called WISE (Women In Science and Engineering), which continued until the Boston-area chapter of AWIS (Association for Women in Science) was established in 1980. Also in 1971, Vera obtained the signatures of twenty women physicists on a letter requesting the American Physical Society to set up an ad hoc committee to study the status of women physicists. Vera chaired the committee that was subsequently established, which assembled a roster of women physicists and conducted a detailed and comprehensive study of their situation. She also conducted other studies, served on other committees, and gave public lectures on the topic through 1992. She was equally active within MIT, serving as the affirmative action officer (1973–1976); as chair of the physics department's Equal Opportunity Committee (1973–1976 and 1992–1994); as a member of a variety of committees that dealt with minority and women's issues; and as co-organizer of the MIT Women Faculty Network (1991–1993).

By the end of the 1970s, Vera was very busy making a positive impact in the area of women's issues. Then, in 1979, President Carter announced the reactivation of the draft. Before then Vera had been aware of the arms race but not actively involved with the issues. Her father had been effectively involved, and because of her time constraints, she had chosen to focus on the women's issues, her physics research, and motherhood. But now the arms race was hitting too close to home. Her son just missed being of draft age by six months. Moreover, Vera was concerned that the arms race had been so severe and was escalating so fast that she decided to focus her energies on it.

When President Reagan later announced the Strategic Defense Initiative (SDI), also referred to as Star Wars, Vera became an arms race activist. By reading extensively and talking with her father, she prepared herself in order to speak knowledgeably about the weapons and defense issues. In 1983 she organized the collection of faculty signatures on a pledge not to request or accept funding from the SDI. This effort caught on and quickly went from campus to campus. It was very effective in helping to stop the increase in funding of SDI. In 1984 she collected the signatures of twenty faculty members on a request to set up a committee to study the impact of military funding on MIT's research and education. She served on the committee that was formed and the presidential committee that succeeded it. In 1988 she started the *MIT Faculty Newsletter* as a vehicle for public expression of faculty opinion and served on its editorial board. Vera presented the results of her studies on the escalation of the arms race mostly in public lectures. She gave 52 such lectures on 78 occasions from 1981 to 1991. She also served on the board of directors for the Council for a Liveable World; chaired the Boston Regional United Campuses to Prevent Nuclear War (1986–1987); and participated in and/or led other related organizations. For all her hard work in this arena, she was named an outstanding woman military expert by the Center for Defense Information in 1986.

Vera has had a very rewarding life that successfully combined women's issues and arms race activities while nurturing the development of the next generation (her children). Her reward has been a personal sense of satisfaction and much recognition: Phi Beta Kappa; Centennial Alumnae Award and Honorary Sc.D., Mount Holyoke College; Phi Beta Kappa visiting scholar; fellow of the American Association for the Advancement of Science; and Sigma XI visiting lecturer. In addition, she has a son and daughter who are scientists.

Note

1. This essay is based on a November 18, 1995, telephone interview with Vera Kistiakowsky by Valerie L. Thomas, and on Dr. Kistiakowsky's curriculum vitae.

Bibliography

Kistiakowsky, Vera. "Keep the Pentagon Out of Civilian Economy." *Bulletin of Atomic Scientists* 45 (April 1989): 5.

———. "Military Funding of University Research." *Annals of the American Academy of Political and Social Science* 502 (March 1989): 141.

———. "Women in Engineering, Medicine and Science." A report prepared for the Conference on Women in Science and Technology. Washington, D.C.: National Research Council, June 1973.

———. "Women in Physics: Unnecessary, Injurious, and Out of Place?" in *History of Physics*. Edited by Spencer R. Weart and Melba Phillips. New York: American Institute of Physics, 1985.

Kistiakowsky, Vera, and D.J. Helfand. "Observations of [SIII] Emission from Galactic Radio Sources: The Setection of Distant Planetary Nebulae and a Search for Supernova Remnant Emission." *Journal of Astronomy* 105 (1993): 2199.

Kistiakowsky, Vera, and M.E. Tidball. "Baccalaureate Origins of American Scientists and Scholars." *Science* 193 (1976): 646.

VALERIE L. THOMAS

HENRIETTA SWAN LEAVITT
(1868–1921)

Astronomer

Birth	July 4, 1868
1885–88	Attended Oberlin College
1892	A.B., Radcliffe College (then called the Society for the Collegiate Instruction for Women)
1892–95	Traveled in the United States and Europe; lost most of her hearing sometime during this period
1895	Began work as a volunteer research assistant at the Harvard College Observatory
1902	Appointed to the permanent staff of the Harvard College Observatory by Edward Pickering
1908	Published discovery of the period-luminosity relation of the Cepheid variable stars

1912	Published findings on the magnitudes of the stars in the North Polar sequence
1913	International Committee on Photographic Magnitudes adopted Leavitt's system for determining star brightness
Death	December 21, 1921

Bias against women working in a "male" profession made it difficult for women to get a job or be trained in astronomy at the turn of the twentieth century. Henrietta Leavitt had to deal with this fact in her life. Additionally, she had to overcome the loss of her own hearing to achieve success. Yet in spite of open bias and physical disability, Henrietta Leavitt became one of the most important astronomers of her time. Her work with the period-luminosity relationship of stars helped other astronomers determine interstellar distance and the size of the universe.

Henrietta Leavitt was born in Lancaster, Massachusetts, in 1868. Her family moved to Cleveland, Ohio, early in her life. She attended Oberlin College from 1885 to 1888, where she studied music, and then went to the Society for the Collegiate Instruction of Women (now called Radcliffe College) in 1888. She graduated in 1892 with an A.B. degree. During her senior year at Radcliffe, Leavitt took a course in astronomy. This stirred her interest in the subject, and after graduation she did some graduate work in astronomy.

Disaster struck Leavitt after she graduated from college. After having traveled in Europe and the United States for several years, she became seriously ill and subsequently was struck almost completely deaf. Despite this, she continued to pursue her interest in astronomy. Leavitt volunteered as a research assistant at the Harvard College Observatory in 1895 and was appointed a permanent staff member in 1902 at the wage of thirty cents an hour.

Although Leavitt started as a research assistant, she soon became chief of the photographic photometry department. She was given little theoretical work to do. Instead, she was assigned the difficult work of making sure that the telescopes always had the proper corrections as well as doing the hard work of research using photographic plates. Leavitt still managed, however, to make several important contributions to astronomy.

The first problem Leavitt worked on was determining the magnitude of a star from a photographic image. (Magnitude in this case meant the brightness of the star.) This was not a simple task. Photographs did not record the exact brightness of a star, and different telescopes gave different results for the brightness of a star. Leavitt came up with a standard for determining star brightness by examining the north polar sequence. This sequence allowed star brightness to be determined by comparing one star with another. The scientific community immediately realized the importance of this work. In 1913, the International Committee on Photographic Magnitudes adopted her work as the standard for determining star brightness. Her research remained the standard for determining star brightness until about 1940.

Leavitt's most important contribution was the discovery of the period-luminosity relationship of stars. This came about directly from her research on variable stars. (In fact, at the time of her death there were 2,400 known variable stars, half of which had been discovered by Leavitt.) This study of variable stars led her directly to the period-luminosity relationship of stars.

Variable stars can be determined by the regularity in which they go through their light cycles. Leavitt studied the Cepheid variable stars in the Small Magellanic Cloud. These stars were known to all be about the same distance from Earth. This was important because if they were all approximately the same distance away, then one that appeared brighter actually *was* brighter. Leavitt wrote that "since the variables are probably at nearly the same distance from the earth their periods are apparently associated with their actual emission of light as determined by their mass, density and surface brightness."[1]

Since the stars were the same distance away, Leavitt was able to move beyond the apparent brightness of stars and examine their luminosity. Writing about the stars in the Small Magellanic Cloud, Leavitt noted "that in Table VI [sixteen of the variables from the Small Magellanic Cloud] the brighter variables have longer periods. It is also noticeable that those having the longest periods appear to be as regular in their variations as those which pass through their changes in a day or two."[2]

Leavitt stated the period-luminosity relationship quite simply: "A straight line can readily be drawn among each of the two series of points corresponding to maxima and minima, thus showing that there is a simple relation between the brightness of the variables and their periods."[3] This discovery permitted the determination of distances in space. If two variable stars have equal periods but one seems brighter than the other, their luminosity is probably equal and the difference between them in brightness is a result of distance. Leavitt's determination of the period-luminosity relationship in the Cepheids of the Small Magellanic Cloud gave a measuring stick for determining the distances of other stars in other parts of the galaxy.

Before Leavitt discovered the period-luminosity relationship of stars, astronomers could only determine cosmic distances up to one hundred light years. After Leavitt's discovery, distances could be determined up to ten million light years. This was extremely significant. With this information, astronomer Einar Hertsprung determined the distances of stars from Earth. Harlow Shapley used the information to determine the size of the galaxy.

Leavitt was elected an honorary member of the Associated Variable Star Observers. She was a member of the American Astronomical and Astrophysical Society as well as the American Association for the Advancement of Science. Additionally, she was a member of Phi Beta Kappa and the American Association of University Women.

Leavitt died in 1921 at the age of 53 as a result of cancer. Her death was immediately noted by the scientific community. The following year her colleague at the Harvard College Observatory, Solon Bailey, wrote an article lamenting her death in *Popular Astronomy*. Her death at an age when she should have had many more years of productive research ahead of her was a tragedy. She almost certainly would have made more astronomical discoveries. Her obituary in the *Boston Transcript* had the headline "Noted Woman Astronomer": the fact that she was a brilliant astronomer and a woman was so unusual that the newspaper immediately drew attention to it.

Henrietta Leavitt made several important astronomical discoveries in her lifetime. She determined the magnitude of stars, and her work in this area was adopted as the standard for her time. She formulated the period-luminosity relationship that allowed for measuring interstellar distances, which led others to determine the size of the galaxy. More important, she helped open up the study of astronomy to women. She was one of the many talented scientists who helped blaze a path for other women to follow.

Notes

1. Quoted in G. Kass-Simon and Patricia Farnes, *Women of Science: Righting the Record* (Bloomington: Indiana University Press, 1990), p. 104.
2. Quoted in Helen Buss Mitchell, "Henrietta Swan Leavitt and the Cepheid Variables," *Physics Teacher* 14, no. 3 (1976): 164.
3. Ibid.

Bibliography

Bailey, Martha J. "Leavitt, Henrietta Swan," in *American Women in Science: A Biographical Dictionary*. Santa Barbara, Calif.: ABC-CLIO, 1994.
Bailey, Solon I. "Henrietta Swan Leavitt." *Popular Astronomy* 30, no. 4 (1922): 196–199.
———. *The History and Work of the Harvard College Observatory*. New York: McGraw-Hill, 1931.
Dugan, Raymond S. "Leavitt, Henrietta Swan," in *Dictionary of American Biography*. Edited by Dumas Malone. New York: Charles Scribner's Sons, 1961.
Gingerich, Owen. "Leavitt, Henrietta Swan," in *Dictionary of Scientific Biography*. Edited by Marshall De Bruhl. New York: Charles Scribner's Sons, 1980.

Hoffleit, Dorrit. "Leavitt, Henrietta Swan," in *Notable American Women, 1607–1950*. Edited by Edward James. Cambridge, Mass.: Belknap Press of Harvard University Press, 1971.

Kass-Simon, G., and Patricia Farnes. *Women of Science: Righting the Record*. Bloomington: Indiana University Press, 1990.

Leavitt, Henrietta S. "The North Polar Sequence." *Annals of the Astronomical Observatory of Harvard College* 71, no. 3 (1971).

———. "1,777 Variables in the Magellanic Clouds." *Annals of the Astronomical Observatory of Harvard College* 60, no. 4 (1980): 87–108.

Mitchell, Helen Buss. "Henrietta Swan Leavitt and the Cepheid Variables." *Physics Teacher* 14, no. 3 (1976): 162–167.

[Obituary]. "Noted Woman Astronomer." *Boston Transcript* (December 13, 1921): sect. 2, p. 7.

Siegel, Patricia Joan, and Kay Thomas Finley. *Women in the Scientific Search: An American Bio-Bibliography, 1724–1979*. Metuchen, N.J.: Scarecrow Press, 1985.

Thaler, Sebastian. "Henrietta Leavitt," in *Notable Twentieth-Century Scientists*. Edited by Emily McMurray. Detroit: Gale Research, 1995.

MICHAEL LORENZEN

LEONA WOODS MARSHALL LIBBY
(1919–1986)
Physicist, Molecular Spectroscopist

Birth	August 9, 1919
1938	B.S., University of Chicago
1942–44	Chicago Metallurgical Laboratory, Chicago, Manhattan District
1943	Ph.D., University of Chicago
1944–46	Consulting Physicist, Hanford Engineer Works, Hanford, WA
1946–47	Fellow, Institute for Nuclear Studies, University of Chicago
1947–54	Research Associate for Nuclear Studies, University of Chicago
1954–60	Assistant Professor of Nuclear Studies, University of Chicago

Leona Woods Marshall Libby. Photo courtesy of John Marshall.

1958–60	Visiting Scientist, Brookhaven National Laboratory, NY
1960–64	Associate Professor, New York University
1964–73	Professor of Physics, University of Colorado, Boulder
1970–86	Adjunct Professor, Engineering School, University of California, Los Angeles
Death	November 10, 1986
1992	50th Anniversary Honoree, Women Pioneers in Nuclear Science

Leona Woods was the only woman and the youngest member of the Chicago Metallurgical Laboratory. This secret war project group consisted mainly of very bright young male physicists who worked on the first nuclear fission pile in the basement of Stagg Field at the University of Chicago. Led by Enrico Fermi, the group successfully built the Chicago Pile (CP-1), the world's first nuclear fission reactor, on December 2, 1942.

Woods's expertise as a molecular spectroscopist was essential for the construction of the boron trifluoride neutron detector that was required for the CP-1. Describing construction of the pile in her book *The Uranium People*, she retrospectively considered her own work: "After the fifteenth layer was reached, a boron trifluoride counter (one of my better creations) was put into the lattice."[1] The first piles were graphite-moderated, and her detector was used to monitor the progress of the fission chain reaction. In addition to asking Leona to join the research team, Fermi gave her the assignment of taking notes at his series of lectures to the team; as a result, Leona acquired state-of-the-art knowledge of nuclear physics.

As one of five children, Leona Woods was the second born to Wreightstill and Mary Holderness Woods on August 9, 1919. She was raised on the family farm in the (then) small town of LaGrange, Illinois, a suburb of Chicago. Her father, a lawyer, raised three daughters and two sons. After his death, Mrs. Woods ran the small farm and Leona helped. Others who joined in periodically included members of the Metallurgical Lab. Laura Fermi described Leona in her book, *Atoms in the Family*, as "a tall young girl built like an athlete, who could do a man's job and do it well." Mrs. Fermi also helped with picking apples and believed both Leona and her mother were "endowed with inexhaustible energy."[2]

Leona earned her B.S. degree in chemistry from the University of Chicago in 1938 when she was only 19 years old. She worked diligently to pass her three-day qualifying exams quickly, which was necessary before she was allowed to begin graduate research. Upon completion, Leona asked Nobelist James Franck to be her graduate advisor. Since he knew of her participation in his research seminar in the new field of quantum chemistry, he readily agreed. But he gave Leona a warning similar to the one he had received as a Jewish graduate student—that as a woman scientist she might starve to death. "I looked at his corpulence; with not the least indication of any starvation . . . and decided that I could not tolerate such a depressing idea that I might starve to death. . . . I worked instead for Robert Mulliken, who had a pleasant attitude, neither encouraging nor discouraging me."[3] Mulliken told Leona to choose her own research problem within the area of diatomic molecular spectroscopy. She completed her Ph.D. in 1943 after joining the team led by Fermi.

Leona married one of the members of the Metallurgical Lab team, John Marshall Jr., in July 1943. She described him as being "of medium height and very handsome. . . . He has a beautiful baritone voice that he still keeps in practice."[4] The Marshalls continued to work with Fermi designing nuclear reactors.

While working on the CP-2 operations at the Argonne Forest Lab, Leona became pregnant and told Fermi about it. In order to keep working in the reactor building, Leona hid her first pregnancy by wearing overalls and a denim jacket. She continued to work up until two days before her son's birth, when she was hospitalized owing to high blood pressure. A week after Peter Marshall's birth, Leona returned to work and soon thereafter joined her husband in Hanford, Washington, where they worked with plutonium production reactors. The young couple received help with child care from Leona's mother and occasionally from John Baudino, Fermi's bodyguard. The Marshall's second son, John Marshall III, was born in Chicago in 1949.

After the war Fermi headed the Institute for Nuclear Studies at the University of Chicago, and the Marshalls moved back to Chicago to work with Leona's mentor. She became a fellow at the Institute in 1947. Fermi was one of the most influential figures in Leona's life intellectually, and he encouraged her as a woman physicist. In addition to assisting her with his ideas, he encouraged her to develop her own.

Perhaps the most influential person in my life was Enrico Fermi, not only scientifically but also philosophically. He set the example of how best to deal with other people, how to anticipate change, how to put up with the ambient indignities and humiliations of the world, and how to cope with the inevitable spiritual charges of taxes and death. He managed this instruction with pomposity. As he frequently said, he was amazed when he thought how modest he was. Others didn't share his amazement. Clearly he knew the extent to which he excelled in mental ability, and often said that to think people are born equal was idiocy, that they differed by orders of magnitude in ability. Although I didn't believe him then, I do now.[5]

In 1958 Leona attended the International Astronomical Meeting held in Moscow. At the time, she was a visiting scientist at Brookhaven National Laboratory on Long Island and was separated from her husband. In an unpublished work entitled "Trip to U.S.S.R.," she wrote of trips to the ballet, opera, record shops, bookstores, and art at the Hermitage. But she had to return from this sojourn. Pursuing her own career and divorced, Leona spent a year at the Institute for Advanced Studies in Princeton working on problems in astrophysics. She accepted a paid full professorship with New York University in 1960, where she worked jointly between teaching there and conducting high energy particle research at Brookhaven National Laboratory. She developed a full analysis for the particle physics experiments with graduate students at the university. Then Frank Oppenheimer suggested she apply to the University of Colorado.

In 1963, Leona moved her analysis operation to Boulder. The program grew so rapidly that she was the largest student employer on campus.

In December 1966, Leona married Willard Libby, who had received the 1960 Nobel Prize in chemistry for his work with radiocarbon dating. Leona worked on numerous problems and issued nearly sixty publications during her time in Colorado.

Leona worked in conjunction with Libby on about ten papers. After moving to the University of California in Los Angeles in 1970, she created a research program to examine past climates by analyzing isotope data in ancient wood samples. Her second book, *Past Climates: Tree Thermometers, Commodities, and People*, was published in 1983. *The Isotope People* remains an unpublished manuscript.

Throughout the years and every place Leona traveled, she and her companions would talk of problems from many different fields. One example shared while traveling to her work at the Argonne Lab was the computation of "the cross section of an automobile for collision with other automobiles, using 50,000 crashes per year, 20 million cars, and putting them all on two-lane roads."[6] In an interview with John Marshall, her second son, he recounted continual conversations between Leona and Bill Libby in their home, at their work, and when they traveled "because Leona was having fun constructing all sorts of theoretical scenarios."[7]

Leona Libby died on November 10, 1986, at Saint John's Hospital in Santa Monica, California. She had most recently been examining spectroscopic data from quasars and found tentative evidence for the fission products of superheavy elements. At the time of her death she was an adjunct professor of environmental science and engineering at the University of California at Los Angeles and a consultant to the Los Alamos Scientific Laboratory. Her published works include her books and two hundred articles.

At the fiftieth anniversary of the CP-1 in November 1992, four women were honored as pioneers in nuclear science. Leona Woods Marshall Libby was one of those women. She was in the company of **Lise Meitner, Marie Curie**, and **Irène Joliot-Curie**.

Notes

1. Leona M. Libby, *The Uranium People* (New York: Crane, Russak, 1979), p. 119.

2. Laura Fermi, *Atoms in the Family* (Berkeley: University of California Press, 1954), p. 179.

3. Libby, *The Uranium People*, p. 30.

4. Ibid., p. 6.

5. Ibid.

6. Ibid., p. 152.

7. Private communication between John Marshall and Connie Nobles, 1995.

Bibliography

Fermi, Laura. *Atoms in the Family*. Chicago: University of Chicago Press, 1954.

Libby, Leona M. *Past Climates: Tree Thermometers, Commodities, and People*. Austin: University of Texas Press, 1983.

———. *The Uranium People*. New York: Crane, Russak, 1979.

Libby, Leona M., and L.J. Pandolfi. "Melting the Moon with Superheavy Elements," in *Superheavy Elements: Proceedings of an International Symposium* (1978): 184–190.

Libby, Leona M., S.K. Runcorn, and L.H. Levine. "Systematics of Quasi-Stellar-Object Spectra." *Astronomical Journal* 89 (1984): 311–315.

Libby, Leona M., S.K. Runcorn, and W.F. Libby. "Primeval Melting of the Moon." *Nature* 270 (1977): 676–681.

CONNIE H. NOBLES

PAULINE BEERY MACK
(1891–1974)
Chemist

Birth	December 19, 1891
1913	B.A., Missouri State University
1913–15	Science Teacher, Norborne High School
1915–18	Science Teacher, Webb City High School
1918–19	Science Teacher, Springfield High School
1919	M.A., Columbia University
1919–52	Faculty Member, Pennsylvania State University
1923	Married Warren Bryan Mack on December 27
1927	Began publication of the *Chemistry Leaflet*
1932	Ph.D., Pennsylvania State College
1940–52	Director, Ellen H. Richards Institute
1945–48	National President, Iota Sigma Pi
1949	Distinguished Daughters of Pennsylvania Medal
1950	Garvan Medal, American Chemical Society
1952	D.Sc., Moravian College for Women, Western College for Women

Pauline Beery Mack. Photo courtesy of the Moravian College Archives.

1952–62	Dean and Director of Research, College of Household Arts and Sciences, Texas Woman's University
1962–73	Director, Research Institute, Texas Woman's University
1970	Astronauts Silver Snoopy Award
Death	October 22, 1974

Pauline Beery Mack was born in 1891, a time when women were expected to conform to conventional standards. Yet during her lifetime she was able to make significant contributions to the world of science and technology. She was renowned for her teaching and research in chemistry, a field not traditionally receptive to women. In addition, she dedicated herself to stimulating interest in science as a career choice for young people, especially women.[1]

She was born Pauline Gracia Beery in Norborne, Missouri, where her parents, John and Dora (Woodford), operated the general store. She was a precocious child who spoke several words before she was 9 months old. Before she was 5 years old she was working in the store and had her own special department with thread, ribbons, and trimmings. Her mother was a strict disciplinarian, and Pauline always understood that much was expected of her. She led her class throughout her public schooling and also demonstrated talent as a singer. At Norborne High School she followed the classical course of that time, which included four years of Latin, three years of German, and four years of mathematics.[2] She also played forward on the school's basketball team, which was ranked first in the state.

Pauline went to Missouri State University in 1910, accompanied by her mother, with the intention of pursuing Latin studies. However, the head of the Latin department proved to be an elderly spinster not sympathetic to young women and lacking in inspiration. Pauline much preferred the marvelous course of freshman chemistry given by Dr. Herman Schlundt. So she chose that subject as her major with minors in physiology, mathematics, and English.

Mother and daughter lived in a small apartment near the university, which became a social center for Pauline and her friends, male and female. Pauline maintained her interests in music and basketball and worked as a student assistant in the physiology laboratory. She received honors in chemistry and mathematics and became a member of Phi Beta Kappa, the nation's most prestigious honor society. Her bachelor's degree was earned in three years because of a provision granting extra credit for high grades.

Her first job after graduation in 1913 was as an assistant principal and science teacher at Norborne High School, where she faced senior class students who had been freshmen during her own senior year. Pauline possessed great poise for a young woman and did not let this potentially awkward situation interfere with her responsibilities. After two years she became head of the science department of the Webb City (Missouri) High School. While there, she also coached girls' basketball and conducted the school orchestra. She then spent one year as head of the physical science department of the Springfield (Missouri) High School. During these years Pauline spent her summer vacations continuing her studies at Columbia University and was awarded the M.A. degree in 1919.

Singing remained a vital part of her life during this time. Her voice had a wide range, but she sang best at coloratura soprano. She was selected to take part in a recital at Carnegie Hall, and at one point she received an offer from the Metropolitan Opera Company to join the chorus. She briefly considered the offer but ultimately declined.[3]

Finding a research and teaching position in physical chemistry was difficult for a woman of her times. She decided to accept a position at Pennsylvania State College teaching freshman chemistry and household chemistry to students of home economics. Her schedule was rigorous. She taught thirty to thirty-five hours a week and conducted independent research studies in food, clothing, and housing with occasional help from graduate students or associates. Pauline tutored many students on her own time, never accepting fees. Her dynamic teaching style won her a reputation as a colorful and entertaining character. During her thirty years at Penn State, more than 12,000 undergraduate students attended her lectures.

Her research projects in the early years received funding as a result of her efforts in personally approaching numerous sources for assistance. The Pennsylvania Department of Welfare lent its support for work on textile durability testing methods, which led to a formal fellowship from Pennsylvania's Departments of Health, Public Instruction, Military Affairs, and Welfare. In 1932 she was given the direction of a research fellowship of the Pennsylvania Association of Dyers and Cleaners after having been associated with them informally for five years. A short time later, a similar fellowship of the Pennsylvania Laundryowners' Association was put under her charge. Both of these afforded opportunity for advanced study in textile chemistry and textile maintenance.

Pauline Beery married Dr. Warren Bryan Mack, professor and head of the Department of Horticulture at Penn State, on December 27, 1923. They had met four years earlier while she was chaperoning a picnic for undergraduates at the college. She did not like housework or cooking and felt she had not been trained for them. Her professional career, for which she had spent a great deal of money and time, was not something she would give up. Dr. Mack willingly agreed.

Together they started the Old Main Art Shop in State College, a gift and art store. After two years of marriage, Pauline induced her parents to move nearer to them and they took over the running of the store. They also helped produce the *Chemistry Leaflet*, a weekly magazine that Pauline began publishing in 1927. This publication was designed to appeal to beginning students of chemistry in high school and college, showing them how chemistry can be applied to everyday living. She had originally financed the project herself, but within a few years subscriptions were strong enough to cover the costs of production—but not to compensate her for her efforts. The publication was later published by the American Chemical Society as its official periodical and eventually became *Chemistry*, a magazine available from Science Service.

In 1932, Pauline received her Ph.D. degree from Pennsylvania State College with a major in biological chemistry and a minor in physics. Gradually she was able to turn over most of her undergraduate teaching to associates and concentrate on furthering her research interests. A series of studies begun in 1935, known as the Pennsylvania Mass Studies in Human Nutrition, was a comprehensive research program that had a continuous fifteen-year history. The Ellen H. Richards Institute was set up as a department of the College in 1940, with Dr. Pauline Mack as director, in order to continue these unique studies.

Her work led to the development of a measuring technique to determine the mineral content of bones using X-rays. This provided a way to analyze the calcium content in the living human skeleton and made it possible to evaluate an individual's calcium needs for proper growth and function. In recognition of this work, Mack received the Francis P. Garvan Gold Medal in 1950. (This honor is presented by the American Chemical Society annually to one American woman chemist for outstanding achievement.)

After thirty years at Penn State, Mack decided to accept a position as the dean of the College of Household Arts and Sciences at Texas State College for Women, now Texas Woman's University (TWU). Her husband was looking forward to his retirement in Texas, but he died in 1952 shortly before they were to move. Mack went to Texas alone to start a new life.

In 1962, Mack decided to give up her administrative duties as dean to concentrate on research. Under her leadership the TWU Research Institute conducted studies in food, nutrition, clothing textiles, and textile technology. She was most taken with the work she did with NASA on both the Gemini and Apollo programs. Here she had the opportunity to study calcium loss from bones during the periods of inactivity experienced by astronauts during extended space missions. The results of her work were used in planning NASA's Skylab program and earned her the Silver Snoopy Award in 1970. This award was created by the astronauts for recognition of people who had done exemplary work for the space program. Mack had developed personal friendships with several astronauts including Jim Lovell, Frank Borman, and Wally Schirra; she was the only woman to receive the award.

Until the age of 79, Mack maintained a full-time work schedule. Ill health forced her into semi-retirement, but she was still able to spend a few hours a day on the work she so enjoyed. She died in a Texas nursing home on October 22, 1974, at the age of 83, leaving no known survivors.

Pauline Beery Mack excelled at many things during her life and was nationally known for her work. She was a woman of energy and charm who loved teaching and, especially, chemistry. It was important to her that chemistry be seen as applicable to everyday life and as a career path within anyone's reach.

Notes

1. Much of the material for this essay comes from a photocopy of Warren B. Mack, "Biographical Material about Pauline Beery Mack, Director of the Helen

H. Richards Institute, Professor of Textile Chemistry," Penn State Room, Pennsylvania State University, University Park.

   2. See *Current Biography*, 1950 (New York: H.W. Wilson, 1950), p. 373.

   3. See Nancy Folkenroth, "She Took Science by Storm," *Town and Gown* 26 (1991): 59.

Bibliography

American Men and Women of Science. 13th ed. New York and London: R.R. Bowker, 1978.

American Women 1935–1940: A Composite Biographical Dictionary. Detroit: Gale Research, 1981.

Current Biography, 1950. New York: H.W. Wilson, 1950.

Folkenroth, Nancy. "She Took Science by Storm." *Town and Gown* 26 (1991): 58–61.

"For Better Living." *Pathfinders* 53 (1946): 39.

"Garvan Medal to Pauline Mack." *Chemical and Engineering News* 28 (1950): 1032.

Mack, Pauline Gracia Beery. *Chemistry, Applied to Home and Community*. Philadelphia: J.B. Lippincott, 1923.

———. *Stuff: The Story of Materials in the Service of Man*. New York: D. Appleton, 1923.

Rickard, Dorothy. "Gold Medalist in Chemistry Research." *Independent Woman* 29 (1950): 201.

World Who's Who in Science. 1st ed. Chicago: A.N. Marquis, 1968.

ELEANOR L. LOMAX

MARGARET ELIZA MALTBY
(1860–1944)
Physicist

Birth	December 10, 1860
1882	B.A., Oberlin College
1883–87	Taught in Wellington and Massilon, OH, high schools
1889–93	Instructor in Physics, Wellesley College
1891	B.S., Massachusetts Institute of Technology; M.A., Oberlin College
1895	Ph.D., Göttingen University
1896	Head, Dept. of Physics, Wellesley College

1897–98	Instructor in Physics, Lake Erie College, Painesville, OH
1898–99	Research Assistant, Physikalisch-Technische Reichsanstalt, Charlottenburg, Germany
1899–1900	Clark University in Worcester, MA
1900–03	Instructor in Chemistry, Barnard College
1903	"Starred" in *American Men of Science*
1913–24	Chair, Fellowship Committee, American Association of University Women
1913–31	Chair, Dept. of Physics, Barnard College
Death	May 3, 1944

Margaret Maltby was born in 1860 on a farm in Bristolville, Ohio, to Edmund and Lydia Maltby. Her parents allowed her sisters, ages 15 and 11 at the time, to name the baby Minnie. Maltby disliked the name so much that she legally changed it to Margaret at the earliest opportunity. But no matter her first name, Margaret went on to become a famous and pioneering physicist of firsts.

The education generally available to women in the late nineteenth century did not prepare them for college. Maltby's high school education included mathematics but no science. She enrolled in the preparatory department of Oberlin College in Ohio and became a freshman in 1878. Her transcripts showed she took courses in several sciences, including civil engineering. The only mathematics courses she took were geometry, algebra, and trigonometry. In 1882 she graduated with a B.A. degree.

Having a strong interest and talent in art, she studied for a year at the Art Students' League in New York City. In 1883 she returned to Ohio and taught high school for four years, but her interest in physics led her to enroll as a special student at the Massachusetts Institute of Technology (MIT) in 1887. Women often had to be "special students" because they were not allowed to enroll in coeducational universities or earn graduate degrees. Maltby received a B.S. degree in 1891, becoming one of the first women to receive such a degree from MIT. Recognizing her additional study, Oberlin awarded her an M.A. the same year. She had been a physics instructor at Wellesley College since 1889, and during her last year there (1893) she began two years of graduate studies at MIT.

To encourage her graduate studies, the Association of Collegiate Alumnae (later known as the American Association of University Women) awarded Maltby a European Fellowship in 1893 to travel to Göttingen University in Germany. In 1895 she became the first American woman to receive a Ph.D. from Göttingen and the first woman to receive a Ph.D. in physics from any German university. She stayed for another year of postdoctoral work and then returned to Wellesley College, where she became head of the physics department for one year. Maltby then taught at Lake Erie College in Painesville, Ohio, for the 1897–1898 academic year.

Impressed by her work, Friedrich Kohlrausch, president of the Physikalisch-Technische Reichsanstalt, invited Maltby to return to Germany in 1898 as a research assistant. Their research involved measuring conductivities of aqueous solutions of alkali chlorides and nitrates. Their methods took into consideration all sources of experimental error and became standards for conductivity measurements.

Maltby returned to the United States in 1899. She taught at Clark University and worked with A. G. Webster for a year in theoretical physics. A year later she became a chemistry instructor at Barnard College; in 1903 she transferred to the physics department as adjunct professor. Maltby became an associate professor and department chair at Barnard in 1913, where she remained until her retirement in 1931. Shortly before her retirement, she taught the first course in the physics of music.

Her peers recognized her stature as a scientist when they chose to list her in the first edition of *American Men of Science* (1906). They also honored her with a starred entry, meaning they considered her one of America's most important scientists. Only one other woman physicist, **Katharine Blodgett**, received a star in the first seven editions of the publication.[1] Maltby was also elected a fellow of the American Physical Society.

Maltby was very active in the American Association of University Women (AAUW), especially in providing opportunities for women to obtain graduate degrees. She chaired the Fellowship Awards Committee from 1913 to 1924 and helped define the policies and standards that shaped the awards. In 1929 she published a *History of the Fellowships Awarded by the American Association of University Women, 1888–1929*, a compilation of brief biographies of AAUW fellowship recipients. In honor of her contributions to the association, the AAUW established the Margaret E. Maltby Fellowship in 1926.

After her retirement Maltby spent time attending the opera and philharmonic concerts, traveling, reading, and visiting friends. She also spent a great deal of time with her adopted nephew's family. During the last few years of her life she was incapacitated by arthritis; she died on May 3, 1944, at the age of 83.

Maltby broke down many barriers by obtaining graduate degrees in a male-dominated field. Recognizing the struggles of women to gain an education or to enter a scientific field, she spent the later years of her career expanding opportunities for women to pursue higher education. In 1927 she said:

Up to this time, many women capable of outstandingly original work have been forced to direct their ideas and their energy to the attaining and the keeping of any kind of opening in the field of science, to winning for themselves and others the right to do their special kind of work. But as I have said this situation is gradually being bettered and for this reason I believe we shall have more important and more original work from our young women scientists.[2]

Notes

1. Margaret W. Rossiter, *Women Scientists in America* (Baltimore and London: Johns Hopkins University Press, 1984), p. 293.
2. Helen Ferris and Virginia Moore, *Girls Who Did: Stories of Real Girls and Their Careers* (New York: E. P. Dutton, 1927), p. 225.

Bibliography

Bailey, Martha J. *American Women in Science: A Biographical Dictionary*. Denver: ABC-CLIO, 1994.

Barr, E. Scott. "Anniversaries in 1960 of Interest to Physicists." *American Journal of Physics* 28 (May 1960): 462–475.

Cross, Charles R., and Margaret E. Maltby. "On the Least Number of Vibrations Necessary to Determine Pitch." *Proceedings of the American Academy of Arts and Sciences* N.S. 19 (1892): 222–235.

Harrison, Shirley W. "Margaret Eliza Maltby (1860–1944)," in *Women in Chemistry and Physics: A Biobibliographic Sourcebook*. Edited by Louise S. Grinstein, Rose K. Rose, and Miriam H. Rafailovich. Westport, Conn.: Greenwood Press, 1993.

History of the Fellowships Awarded by the Association of University Women 1888–1929. Washington, D.C.: American Association of University Women, 1929.

Notable American Women 1607–1950: A Biographical Dictionary. Cambridge, Mass.: Belknap Press of Harvard University Press, 1971.

TERESA BERRY

INES MANDL

(1917–)

Biochemist

Birth	April 19, 1917
1936	Married Hans Alexander Mandl
1944	Diploma in chemistry, National University of Ireland
1945–49	Assistant, Dept. of Interchemical Corporation and Chemistry, New York University
1947	M.S., Polytechnic Institute of Brooklyn
1949	Ph.D., Polytechnic Institute of Brooklyn
1949–55	Research Associate, Columbia University College of Physicians and Surgeons
1971	Distinguished Alumnus Award, Polytechnic Institute of Brooklyn
1973	Associate Professor, Columbia University College of Physicians and Surgeons
1974	Carl Neuberg Medal, American Society of European Chemists and Pharmacists
1976	Professor, Columbia University College of Physicians and Surgeons
1982	Garvan Medal, American Chemical Society
1983	D.Sc., University of Bordeaux, France
1992	Austrian Honor Cross First Class; Golden Honor Emblem of the City of Vienna

A leader in the research on enzymes and elastic tissue, Ines Mandl was probably the best-known female biochemist of her generation in the United States. She approached biomedical problems from the perspective of chemistry, which resulted in tremendous advances in the understanding of pulmonary emphysema. Mandl initiated new directions for biochemical research and very early recognized the importance of interdisciplinary cooperation to combat disease. A member of the American Academy of Arts and Sciences and the American Chemical Society and a fellow of the New York Academy of Sciences, she participated in many international symposia and had gained a worldwide reputation when she retired in 1986.

Ines Hochmuth Mandl was born in Vienna, Austria, in 1917 as the only child of Ernst and Ida (Bassan) Hochmuth. Her father, a prominent industrialist, belonged to the well-to-do, cultured, and cosmopolitan Jewish bourgeoisie that formed a distinctive part of Viennese society before 1938. Ines attended a state-supported primary school until age 14. No further public schooling was available for girls at that point, but private secondary schools offered an additional curriculum of two to three years. Ines graduated from such an academy and followed the conventional pattern for young ladies of her class by marrying; she wed Hans Alexander Mandl in 1936.

When the National Socialists came to power in Austria in 1938, the Mandls emigrated to England. After the start of World War II they moved to Ireland and Ines enrolled at the National University of Ireland at Cork. She chose biochemistry as her specialty because she saw it as a very versatile field and felt it would permit her to work anywhere. She graduated in 1944 with a diploma in chemistry. In the meantime, her parents had escaped from Austria to the United States, and in 1945 she and her husband followed them.

Through an organization that found employment for Jewish immigrants, Ines heard about a position at International Chemical Corporation with Carl Neuberg, one of the foremost biochemists at that time. He had taught at the University of Berlin and had founded the journal *Biochemische Zeitschrift*, in which he had created the term "biochemistry" for this newly emerging specialty. After the National Socialists dismissed all Jewish professors from German universities, he emigrated first to Palestine and then to the United States. He became a professor at New York University and also served as a consultant for the Interchemical Corporation.

After working for Apex Chemical Company during the summer of 1945, Ines joined Carl Neuberg as his assistant at Interchemical as well as at New York University. Having recognized her intellectual potential and her talent for research, Neuberg suggested she take advanced courses. She enrolled at the Polytechnic Institute of Brooklyn for evening courses and in 1947 earned a Master of Science degree with a thesis entitled "Transosazonation." She also published her research results jointly with Carl Neuberg in a journal article. They had already collaborated on an article on the identification of fructose, and they subsequently published a whole series of investigations of sugar derivatives.

In 1949, Ines earned her doctorate from the Polytechnic Institute of Brooklyn, the first woman to do so at that institution. In the same year she was offered a position as research associate in the Department of Surgery at the Columbia University College of Physicians and Surgeons. Although she no longer worked for Carl Neuberg, they kept up some of their joint research. Their experiments with the solubilization of normally insoluble natural matter culminated in their development of a patented procedure for the solution of certain amino acids by means of enzymes.

Mandl then investigated collagenase, a group of about twenty enzymes capable of breaking down collagen into a soluble form. She was the first researcher to isolate and purify one of those enzymes—the collagenase of the bacterium *Clostridium histolyticum*, which is now produced commercially for use in the treatment of third-degree burns, bedsores, and herniated disks. In 1959 she started as director of the obstetrics/gynecology laboratories of Delafield Hospital, which is affiliated with Columbia University. She also kept her teaching appointment as assistant professor of biochemistry at the College of Physicians and Surgeons. Her studies of bacterial enzymes now concentrated on the role of those enzymes in the growth of tumors in the female reproductive system.

At Delafield Hospital, Mandl changed the focus of her research somewhat. She became interested in the enzymic processes involved in the breakdown of the elastic tissues in the human body. When she examined the causes of pulmonary emphysema and respiratory distress syndrome in newborn babies, she identified the function of pulmonary elastin. She was then able to show the disease process of emphysema, in which the deterioration of elastin destroyed lung tissue. Elastin, a structural protein, is the main component of the elastic fibers of the connective tissues (i.e., tendons and ligaments as well as arteries and bronchi). Ines Mandl had found her life's work in the study of elastin and related collagen fibers and the enzymic reactions involved in elastin's breakdown, especially as they occur in lung tissue.

Increasingly Mandl's research transcended the distinction between biochemical and medical fields, and she gained international renown as a pioneer in that area. She developed chemical solutions to biomedical problems and often initiated new directions for research. When she investigated the effects of smoking on lung function, she showed conclusively that smoking damaged the elastin of lung tissue. Her discoveries also led to a possible therapy for preventing a particular type of pulmonary emphysema caused by a missing enzyme inhibitor.

In 1970, Mandl set up the first interdisciplinary symposium on collagenase at Columbia University and edited the proceedings. With her research associates, she published over 140 articles. In 1972 she founded the journal *Connective Tissue Research*, serving as its editor-in-chief until her retirement. Despite these achievements, she was not promoted to associate professor until 1973 and to full professor until 1976.

Women scientists were often slighted by the professional organizations in their fields. They were for a long time dissuaded from participation in the meetings, they were not elected as officers, and they rarely received prizes as recognition for outstanding research. This particular form of discrimination was still fairly prevalent when Ines Mandl was honored in 1977 with the Carl Neuberg Medal by the American Society of European Chemists and Pharmacists. This prize, named after Mandl's erstwhile mentor and established in 1947, is awarded annually to "an outstanding personality credited with special contributions to science."

Until 1980, the Garvan Medal had been the only prize the American Chemical Society (ACS) awarded to female chemists. Women were not encouraged to compete with men for other ACS awards. (Only in 1967 did the ACS give a general award to a woman.) Francis P. Garvan (1875–1938), a patent lawyer and the director of Chemical Foundation, had endowed the prize in 1935 to honor the achievements of women in the field. In 1982 the selection committee awarded the prize to Ines Mandl for her outstanding contributions to research in biochemistry.

Bibliography

American Men and Women of Science: A Biographical Directory of Today's Leaders in Physical, Biological and Related Sciences. 19th ed. New Providence, N.J.: R.R. Bowker, 1995.

Collagenase: First Interdisciplinary Symposium. Edited by Ines Mandl. New York: Gordon and Breach, 1972.

"Garvan Medal." *Chemical and Engineering News* 60 (September 13, 1982): 54–55.

Herzenberg, Caroline L. *Women Scientists from Antiquity to the Present: An Index.* West Cornwall, Conn.: Locust Hill Press, 1986.

Hochberg, Edward. "Ines Hochmuth Mandl (1917–)," in *Women in Chemistry and Physics: A Biobibliographic Sourcebook.* Edited by Louise S. Grinstein, Rose K. Rose, and Miriam H. Rafailovich. Westport, Conn.: Greenwood Press, 1993.

Mandl, Ines. "Collagenase." *Science* 169 (1970): 1234–1238.

Mandl, Ines, J.O. Cantor, M. Osman, et al. "Elastin Biosynthesis." *Connective Tissue Research* 15, no. 1–2 (1986): 9–12.

Woman's Who's Who of America. Edited by John William Leonard. New York: American Commonwealth, 1914. (Reprint: Detroit, Gale Research, 1976).

World Who's Who in Science: A Biographical Dictionary of Notable Scientists from Antiquity to the Present. Edited by Allen G. Debus. Chicago: Marquis Who's Who, 1968.

IRMGARD WOLFE

ANTONIA MAURY
(1866–1952)
Astronomer

Birth	March 21, 1866
1887	B.S., Vassar College
1888–96	Assistant, Harvard College Observatory (intermittent)
1891–94	Science Teacher, Gilman School, Cambridge, MA
1896–1918	Teacher and Lecturer, including physics and chemistry, Castle School, Tarrytown, NY
1918–35	Assistant, Harvard College Observatory
1935–38	Custodian, Draper Park Observatory Museum
1943	Annie J. Cannon Prize, American Astronomical Society
Death	January 8, 1952

Antonia Maury revolutionized the system used to classify stars and participated in the discovery of a number of binary stars.

Antonia Caetana de Paiva Pereira Maury, daughter of the Reverend Mytton Maury and Virginia Draper Maury, was born in Cold Spring, New York, in 1866. She was the granddaughter of Dr. John William Draper and the niece of Dr. Henry Draper, prominent physicians as well as noted amateur astronomers specializing in astrophotography. In fact, it was her uncle's work on stellar spectra that eventually led to Antonia's career at the Harvard College Observatory. Antonia is said to have helped her famous uncle in his lab when she was only 4 years old; and her father had her reading Virgil in the original Latin at age 9.[1]

Antonia was a student of famed astronomer and teacher **Maria Mitchell** at Vassar; she graduated in 1887 with honors in astronomy, physics, and math. In 1888 her father inquired at the Harvard College Observatory (HCO) about employment for Antonia as a computer. Edward Pickering, the HCO director, was famous for hiring women.[2] He asked Antonia's aunt, Anna Draper, about Antonia's qualifications and willingness to work in such a tedious position. Mrs. Draper, "whose attitude toward her niece by marriage remained somewhat ambivalent throughout ... doubted that Antonia had been consulted at all [by her father] or that, having been prepared to teach chemistry and physics, she would have the slightest interest in such dull routine as computing."[3] However,

Antonia enthusiastically accepted the position. She began working at the HCO that same year and continued to do so on and off until 1935.

Antonia joined the HCO during the Henry Draper Memorial project, a monumental stellar classification project funded by Anna Draper in the name of her late husband, who had been the first to photograph a stellar spectrum showing distinct spectral lines. The first analysis was done beginning in 1886, chiefly by **Williamina Paton Fleming**. With the advent of better telescopes and spectroscopy equipment, Pickering decided to re-examine stellar spectra in much greater detail, even while the first Draper catalogue was still being prepared. He divided the project into two groups, Northern and Southern stars. The Northern stars were tackled by Antonia Maury. But personality conflicts with Pickering hampered her work. According to **Dorrit Hoffleit**, who knew Maury at the HCO, "she was one of the most original thinkers of all the women Pickering employed; but instead of encouraging her attempts at interpreting observations, he was only irritated by her independence and departure from assigned and expected routine."[4] However, Antonia's intellect and education made her especially well suited for the task of re-evaluating the new and greatly improved stellar spectra photographs.

Maury reordered some of Fleming's original spectral classes to represent a sequence in temperature and decided that they were inadequate to describe the complexity of the spectral lines she saw. She noted that "it also appeared that a single series was inadequate to represent the peculiarities which presented themselves in certain cases, and that it would be more satisfactory to assume the existence of collateral series."[5] Maury therefore added a second "dimension" to the system, a letter that described the appearance of the spectral lines: "a" for wide and well defined; "b" for hazy but relatively wide and of same intensity as "a"; and "c" for spectra in which the H lines (contours in which the electromagnetic field strength is constant) and "Orion lines" (now known to be due to helium) were narrow and sharply defined, whereas the calcium lines were more intense. She also had a class "ac" for stars having characteristics of both "a" and "c."[6]

Maury left the HCO for a teaching post at the Gilman School in Cambridge, Massachusetts, in 1891, her work unfinished. However, she returned in 1893 after receiving hints that her work might have to be finished by someone else.[7] Maury worked on the catalogue for a year and a half but suffered from burnout and left for Europe. When she returned in December 1895, she wrote to Pickering asking to complete any loose ends in the catalogue and aid in its publication. It finally appeared in print in 1897. In it Maury emphasized the importance of the "c-characteristic," which she firmly believed represented a fundamental property of the stars.[8] Thereafter Maury left the HCO and lectured on astronomy to public and professional audiences. She accepted private pupils and occasional teaching positions, including the Castle School in Tarrytown, New York. In 1908 she returned to the HCO as a part-time research associate.

After Pickering's death in 1919, Maury continued to work on and off at the HCO under the auspices of the new director, Harlow Shapley. Her work at this point was mainly on spectroscopic binaries, stars whose duplicity is known though irregularities in their spectra. Maury had aided in the discovery of the first such star, Mizar in the Big Dipper, in 1889 and had discovered the second, Beta Aurigae, that same year. She was the first person to find the orbital periods of these double stars and also the first person to compute their orbits.[9] Colonel John Herschel called Maury's work on spectroscopic binaries "one of the most notable advances in physical astronomy ever made."[10] She was appointed Pickering fellow for 1919–1920 to aid in her spectroscopic work. After her official retirement from Harvard in 1935 she spent three years as custodian of the Draper Park Observatory Museum in Hastings-on-Hudson, where she lived until her final illness. She continued to visit Harvard yearly to check on observations of her final project, the enigmatic double Beta Lyrae. In 1943 the American Astronomical Society awarded Maury the Cannon Prize for her work on stellar spectra, and Dr. W. W. Morgan, who dedicated a version of his famous atlas of stellar spectra to Maury, called her "the single greatest mind that has ever engaged itself in the field of the morphology of stellar spectra."[11] In fact, a lunar crater has been named in her honor.

The fate of Maury's spectral classification system should be discussed at length. Pickering believed it to be too cumbersome and never utilized it in any further HCO publications. However, other astronomers realized the importance of Maury's "second dimension." The Danish astronomer Ejnar Hertzsprung published a paper in 1905 that tackled the problem of determining the absolute magnitudes or true luminosities of stars. Stars with high proper motions (visible motions relative to other stars) were known to be close to earth and, thus, do not have to be intrinsically bright to be easily visible; whereas stars with small proper motions had to be truly luminous to be visible from large distances. Hertzsprung discovered that the red stars appeared to be of two types: nearby (and hence intrinsically dim) stars or dwarfs, and distant (and hence intrinsically bright) stars or giants. Interestingly, these "red giant" stars turned out to be the very same ones that Maury had labeled with her "c-characteristic." Hertzsprung wrote to Pickering describing the importance of Maury's work and questioning why it was not utilized in all HCO catalogues:

In my opinion the separation of Antonia C. Maury of the c- and ac- stars is the most important advancement in stellar classification since the trials by Vogel and Secchi. . . . To neglect the c-properties in classifying stellar spectra, I think, is nearly the same thing as if the zoologist, who has detected the deciding differences between a whale and a fish, would continue classifying them together.[12]

Pickering replied that the spectra were not good enough to distinguish characteristics as finely as Maury had claimed. In the words of Dorrit Hoffleit, "it wasn't that he [Pickering] couldn't see them [the c-properties] but that he was miffed that she was the one that discovered something."[13] During the same period, Henry Norris Russell of Princeton was independently moving toward the same conclusion, that two types of red stars exist. Russell was the first to make an actual diagram plotting absolute magnitude versus spectral class. The diagonal progression of hot-bright to cool-dim stars he called dwarfs (now called main sequence), and the others he called giants. The resulting diagram is known as the Hertzsprung-Russell diagram and is the backbone of stellar astrophysics.

Antonia Maury had wide cultural interests besides astronomy, including philosophy and ornithology. She was also an outspoken conservationist of historic sites and "could talk about every subject imaginable."[14] Although she was amazed by the immensity of the universe, she believed "the human brain is greater yet, because it can comprehend it all."[15]

Notes

1. Dorrit Hoffleit, *Maria Mitchell's Famous Students; and, Comets over Nantucket* (Cambridge, Mass.: AAVSO, 1983).

2. For more information on the hiring of women at the HCO, see Dorrit Hoffleit, *The Education of American Women Astronomers before 1960* (Cambridge, Mass.: AAVSO, 1994); Bessie Zaban Jones and Lyle Gifford Boyd, *The Harvard College Observatory: The First Four Directorships, 1839–1919* (Cambridge, Mass.: Belknap Press of Harvard University Press, 1971); Pamela E. Mack, "Straying from Their Orbits: Women in Astronomy in America," in *Women of Science: Righting the Record*, eds. G. Kass-Simon and Patricia Farnes (Bloomington: Indiana University Press, 1990); Margaret W. Rossiter, *Women Scientists in America: Struggles and Strategies to 1940* (Baltimore: Johns Hopkins University Press, 1982).

3. Jones and Boyd, *The Harvard College Observatory*, p. 396.

4. Dorrit Hoffleit, *Women in the History of Variable Star Astronomy* (Cambridge, Mass.: AAVSO, 1993), p. 3.

5. Antonia C. Maury, "Spectra of Bright Stars," *Annals of the Harvard College Observatory* 28, pt. 1 (1897).

6. For an in-depth discussion of Maury's classification system, see Jones and Boyd, *The Harvard College Observatory*; Solon I. Bailey, *The History and Work of Harvard Observatory, 1839 to 1927* (New York: McGraw-Hill, 1931); Dorrit Hoffleit, "The Evolution of the Henry Draper Memorial," *Vistas in Astronomy* 34 (1991): 107–162.

7. Maury's correspondence with Pickering at this time appears in Jones and Boyd, *The Harvard College Observatory*.

8. Antonia C. Maury, "Spectra of Bright Stars."

9. For a discussion on this aspect of Maury's work, see Hoffleit, *Maria Mitchell's Famous Students*; and Bailey, *The History and Work of Harvard Observatory*.

10. Jones and Boyd, *The Harvard College Observatory*, p. 244.

11. Hoffleit, *Maria Mitchell's Famous Students*, p. 6.

12. Jones and Boyd, *The Harvard College Observatory*, p. 240.

13. Dorrit Hoffleit, interview with Christine Larsen, October 7, 1994.

14. Ibid. Dorrit Hoffleit has long been a champion of Maury and speaks very fondly of her as a person and brilliant astronomer.

15. Hoffleit, *Maria Mitchell's Famous Students*, p. 7.

Bibliography

Bailey, Solon I. *The History and Work of Harvard Observatory, 1839 to 1927.* New York: McGraw-Hill, 1931.

Hoffleit, Dorrit. "Antonia C. Maury." *Sky and Telescope* 11, no. 5 (1952): 106.

———. "The Discovery and Exploitation of Spectroscopic Parallaxes." *Popular Astronomy* 58 (1950): 428–438, 483–501.

———. *The Education of American Women Astronomers before 1960.* Cambridge, Mass.: AAVSO, 1994.

———. *Maria Mitchell's Famous Students; and Comets over Nantucket.* Cambridge, Mass.: AAVSO, 1983.

Jones, Bessie Zaban, and Lyle Gifford Boyd. *The Harvard College Observatory.* Cambridge, Mass.: Harvard University Press, 1971.

Mack, Pamela E. "Straying from Their Orbits: Women in Astronomy in America," in *Women of Science: Righting the Record.* Edited by G. Kass-Simon and Patricia Farnes. Bloomington: Indiana University Press, 1990.

Maury, Antonia C. "Spectra of Bright Stars." *Annals of the Harvard College Observatory* 28, pt. 1 (1897).

———. "The Spectral Changes of Beta Lyrae." *Annals of the Harvard College Observatory* 84, no. 8 (1933).

KRISTINE M. LARSEN

GRACE MEDES
(1886–1967)
Chemist

Birth	November 9, 1886
1904	B.A., University of Kansas
1913	M.A., University of Kansas

1916	Ph.D., zoology, Bryn Mawr
1916–19	Instructor in Zoology, Vassar College
1919–22	Assistant Professor of Physiology, Vassar College
1922–24	Associate Professor of Physiology, Wellesley College
1924–25	Fellow, University of Minnesota Medical School
1925–32	Assistant Professor of Clinical Chemistry, University of Minnesota Medical School
1932–52	Department Head, Metabolic Chemistry, Lankenau Hospital Research Institute (later merged with Institute for Cancer Research)
1954–60	Senior Member of the Institute for Cancer Research
1955	Garvan Medal, American Chemical Society; Fellow, American Academy of Arts and Sciences; Distinguished Service Citation, University of Kansas
1960	Distinguished Service Citation, Bryn Mawr
1960–67	Emerita Chairman, Dept. of Metabolic Chemistry, Institute for Cancer Research
Death	December 31, 1967

Grace Medes was a pioneering researcher whose studies of metabolism not only led to her naming a rare disease but also put her in the forefront of cancer research.

Grace Medes was born to William and Kate (Hagney) Medes in 1886 in Keokuk, Iowa. Her educational background was primarily in zoology, and she received her B.A. and M.A. from the University of Kansas in 1904 and 1913, respectively. Her doctoral degree in zoology, physiology, and chemistry was awarded to her in 1916 from Bryn Mawr College. Upon graduation, Medes became an instructor of zoology at Vassar College and was later appointed assistant professor of physiology in 1919. In 1922 she joined the faculty at Wellesley College as an associate professor. In 1924 she moved to the University of Minnesota Medical School as a member of the faculty in the Department of Physiological Chemistry, teaching clinical chemistry and becoming an assistant professor there in 1925.

Medes's years in clinical physiological chemical research at the University of Minnesota were very productive. It was during her tenure there that she made the most important discovery of her career, the human metabolic disorder known as tyrosinosis, which she named. Originally believed to be a rare condition, there had been only one reported case until 1955. Tyrosinosis is a rare and possibly inherited disease in which the body does not form tyrosine transaminase, which is important in the creation of certain hormones and melanin. It is detected through large amounts of tyrosine and p-hydrophenylpyruvic acid in the urine. In infants, tyrosinosis can slow development and lead to liver failure. In children and adults, it can lead to chronic liver and renal problems as well as a condition similar to rickets, including the softening of bones and other deformities. Later studies, particularly in Norway, have shown the condition not to be as rare as believed, but Medes's work in the area was so successful that it has never been challenged.[1]

For her work she was awarded the Garvan Medal from the American Chemical Society in 1955, an annual award that recognizes the outstanding achievements of a woman chemist. She was praised for her efforts by the chair of the Women's Service Committee of the American Chemical Society, Gladys Emerson, who had received the award herself in 1952.[2] Medes's studies helped other researchers in their interpretation of the abnormal metabolism of phenylalnine, one of twenty amino acids essential in the human diet, in some patients.[3] She was honored further for her research with a symposium on tyrosinosis that was held in Oslo, Norway, on June 2–3, 1965.[4]

Medes joined the Lankenau Hospital Research Institute in Philadelphia in 1932 as a member of the cancer research staff, moving from academia to private industry. The director of the Institute was Dr. Stanley P. Reichmann, a pathologist. His interest in normal and malignant growths led him to gather chemists, biochemists, physiologists, zoologists, and other researchers in the field to study "the relationships of different sulfur groups in compounds to simulation and retardation of cell proliferation."[5] Medes joined this group and began her research in cysteine metabolism, the pathway in which cysteine, a non-essential amino acid found in many proteins in the body, produces sulphur, which is important in various body functions. Medes continued to work at the Institute, which was merged with the Institute for Cancer Research in 1950, until her retirement in 1956. During her tenure there, Medes published a number of important papers. Starting in the 1940s, with the availability of carbon isotopes and the possibility of isotope tracer studies, which use radioactive materials to study metabolism, she participated in some of the earliest studies of fatty acid metabolism in cancer cells and helped to clarify the intermediate steps in metabolism.[6] Her studies on the participation of acetyl groups in acetoacetate synthesis were published and led the way to the discovery of acetyl coenzyme A, an essential agent in many biological reactions and in the creation of certain acids. In 1952, the same year she won the Garvan Medal, the University of Kansas bestowed on her its distinguished service award, as did Bryn Mawr College in 1960.[7] Medes also received the title of senior member of the Institute for Cancer Research in 1954, just two years before her retirement.[8]

Medes was known among her colleagues at the Institute for Cancer Research for her patient and careful research as well as for using herself as a guinea pig for her studies on sulfur metabolism, often at the risk of her own health. In addition, she was highly resourceful. When she started at the Institute she had neither adequate staffing nor equipment nor funding, with the exception of a $500 grant received in 1939. Only a few years later, Medes had a staff of a dozen assistants and grants totaling $32,000.[9] She published her findings—over 100 articles—in many prominent scientific journals and three books up until only a few months before her death. She also collaborated with other researchers on a number of research projects, including her studies of fatty acid metabolism.

After her retirement in 1956, Medes resumed her research on tyrosinosis at the Fels Research Institute, Temple University, where she was a visiting scientist. Her studies there led to additional publications in the field. Medes continued her work at Fels, even though she was in declining health and suffering from the results of a serious car accident. She died on December 31, 1967, in Philadelphia.

To her friends, Medes was "a woman of simple tastes with a ready smile, whose work and hobby were fused in an interest in biochemistry. She did, none-the-less, take a bit of time for cabinet-making, gardening, and camping."[10] She was also a dedicated sister, caring for a brother in Philadelphia until his death. The niece of her good friend, with whom Medes spent many holidays, quoted her as saying that she "liked to get to bed early and couldn't wait for the next day to come so she could get on with her work."[11]

Notes

1. Paris Svoronos, "Grace Medes (1886–1967)," in *Women in Chemistry and Physics: A Biobibliographic Sourcebook*, eds. Louise S. Grinstein, Rose K. Rose, and Miriam Rafailovich (Westport, Conn.: Greenwood Press, 1993), p. 388.

2. Ibid.

3. "Grace Medes. Garvan Medal," *Chemical and Engineering News* 33 (April 11, 1955): 1515.

4. *Symposium on Tyrosinosis: In Honor of Grace Medes, June 2–3, 1965* (Oslo: Universitetsforlaget, 1966).

5. Ethel Echternach Bishop, "Grace Medes: 1886–1967," in *American Chemists and Chemical Engineers*, ed. Wyndham D. Miles (Washington, D.C.: American Chemical Society, 1976), p. 330.

6. Patricia J. Siegel and K. Thomas Finley, *Women in the Scientific Search: An American Biobibliography* (Metuchen, N.J.: Scarecrow Press, 1985), p. 130.

7. Svoronos, "Grace Medes (1886–1967)," p. 388.

8. Ibid., p. 387.

9. Ibid., pp. 388–389.

10. Bishop, "Grace Medes: 1886–1967," p. 331.

11. Ibid.

Bibliography

American Women: The Standard Biographical Dictionary of Notable Women. Edited by Durward Howes. Teaneck, N.J.: Zephyrus Press, 1974.

Bishop, Ethel Echternach. "Grace Medes: 1886–1967," in *American Chemists and Chemical Engineers*. Edited by Wyndham D. Miles. Washington, D.C.: American Chemical Society, 1976.

"Deaths—Grace Medes." *Chemical and Engineering News* 46 (March 4, 1968): 68.

"Grace Medes. Garvan Medal." *Chemical and Engineering News* 33 (April 11, 1955): 1515.

Medes, Grace. "Cancer Isotopes in the Study of Cancer." *Clinics* 4 (1945): 128–134.

———. "Fat Metabolism." *Annual Review of Biochemistry* 19 (1950): 215–234.

———. "A New Error of Tyrosine Metabolism: Tyrosinosis." *Biochemical Journal* 26 (1932): 917–940.

———. "Production of Ketone Bodies." *Proceedings of the American Diabetes Association* 7 (1948): 85–93.

Medes, Grace, and J.F. McClendon. *Physical Chemistry in Biology and Chemistry*. Philadelphia: Saunders, 1925.

Medes, Grace, and Stanley P. Reimann. *Normal Growth and Cancer*. Philadelphia: Lippincott, 1963.

Siegel, Patricia J., and K. Thomas Finley. *Women in the Scientific Search: An American Bio-Bibliography*. Metuchen, N.J.: Scarecrow Press, 1985.

Svoronos, Paris. "Grace Medes (1886–1967)," in *Women in Chemistry and Physics: A Biobibliographic Sourcebook*. Edited by Louise S. Grinstein, Rose K. Rose, and Miriam Rafailovich. Westport, Conn.: Greenwood Press, 1993.

Women's Book of World Records and Achievements. Edited by Lois Decker O'Neill. Garden City, N.Y.: Anchor Press/Doubleday, 1979.

STEFANIE BUCK

LISE MEITNER
(1878–1968)
Physicist

Birth	November 7, 1878
1906	Ph.D., physics, University of Vienna
1912–15	Assistant, Institute of Theoretical Physics, University of Berlin

1918–38	Head, Dept. of Physics, Kaiser Wilhelm Institute for Chemistry, Berlin-Dahlem
1922	Lecturer, University of Berlin
1924	Leibnitz Medal, Berlin Academy of Sciences
1925	Lieban Prize, Vienna Academy of Sciences
1926	Extraordinary Professor, University of Berlin
1928	Ellen Richards Prize, jointly with Remart Lucas
1938–47	Researcher, Nobel Institute
1946	Visiting Professor, Catholic University, Washington, DC
1947	Prize in Sciences, City of Vienna
1947–60	Researcher, Royal Swedish Institute of Technology
1949	Planck Medal, German Physical Society, jointly with Otto Hahn
1954	Otto Hahn Prize, West Germany
1957	Ordre pour le Mérite, Civilian Class, West Germany
1962	Schlozer Medal, University of Göttingen
1965	Enrico Fermi Prize, U.S. Atomic Energy Commission, jointly with Otto Hahn and Fritz Strassmann
Death	October 27, 1968

Lise Meitner was one of the early pioneers in radioactivity. Although she was Austrian, she did most of her research in Berlin. In fact, Albert Einstein nicknamed her "the German **Marie Curie**." She discovered protactinium with radiochemist Otto Hahn, and she played a major role in the discovery of atomic fission, which made possible atomic power as well as atomic weapons.

Lise was the third of eight children of Hedwig Skovran, a pianist, and Philipp Meitner, a lawyer. She was born in Vienna in 1878. It was a talented family: two sisters and a brother received university degrees, and one of the older sisters became a concert pianist. Both parents were from Jewish families, but Lise was raised as a Protestant.

Lise developed an interest in physics at a young age, but her father insisted that she spend three years studying French so that she could support herself as a teacher should the need arise. The Viennese system did not allow girls to enter high school. So, after passing the French examination, Lise studied privately for the *Matura*, the university entrance examination. She completed eight years of schoolwork in two years and entered the University of Vienna in 1901, two months before her twenty-third birthday. She was the second woman to receive a doctorate in science from the university; her dissertation, presented in 1906, was on heat conduction in non-homogeneous materials.

After graduation, Lise was introduced to the new subject of radioactivity by Stefan Meyer. She designed one of the first experiments to demonstrate that alpha rays are slightly deflected in passing through matter. Lise's father provided her with a modest allowance to go to Berlin to study quantum physics with Max Planck. She stayed for thirty-one years.

Besides attending lectures, Lise was eager to continue with her experiments but had difficulty finding a laboratory. Then she met Otto Hahn, who was looking for a physicist to help in his work on the chemistry of radioactivity. Hahn was teaching at the Chemical Institute under Emil Fischer. Although Fischer later became one of Meitner's strongest supporters, at that time he did not allow women in his laboratory. Hahn and Meitner adapted a carpenter's workshop in the basement of the chemistry building and set to work. From 1907 to 1909 Lise performed radiation experiments in the cellar, careful never to be seen upstairs. She was even prohibited from using the bathroom in the chemistry building, so she used facilities at a nearby hotel. She would sometimes hide under the seats of the amphitheater to hear a lecture.

Meitner and Hahn were opposites in many ways. Lise was slender, 5 feet tall, gentle, reserved, and shy; Otto was tall and outgoing. Hahn, the chemist, was intuitive and primarily interested in the discovery of new radioactive elements; Meitner, the physicist, was a critical, systematic thinker, more concerned about understanding radiation. They developed a complementary relationship that served them well for many years.

After five years, Meitner and Hahn moved to the Kaiser-Wilhelm Institut für Chemie, a new facility that had opened in Berlin-Dahlem in 1912. That year, Max Planck also gave Meitner a prestigious assistantship with a small stipend. Yet her income was so low that she often had only black bread to eat and coffee to drink. When she was offered a professorship by the University of Prague in 1914, the Kaiser-Wilhelm Institut finally decided to pay her a salary.

World War I interrupted her work. Hahn was immediately called to military service. At first Lise remained in Berlin and worked on her own projects; but once it became clear that the war was not going to end quickly, she returned to Vienna to volunteer as a roentgenographic, or X-ray, nurse with the Austrian army. On her leaves, she went back to Berlin to measure radioactive substances. Hahn's leaves sometimes coincided, so they continued their collaboration. However, Lise did most of the work alone. The search for the precursor of actinium was successful. In 1918, Meitner and Hahn named the new element protactinium.

Later that year, Meitner was appointed head of the physics department of the Kaiser-Wilhelm Institut. Although most German academics disapproved of the Weimar Republic, the position of German women greatly improved under it. In 1922, Meitner was allowed to lecture at the University of Berlin for the first time. She was appointed "extraordinary professor" in 1926 at the age of 48, Germany's first woman professor of physics.

Lise developed a close circle of friends, mostly physicists. Music played an important role in her friendships. The Institut had a choir. Lise and her friends met for regular musical evenings. Einstein played his violin and Meitner taught her friends Brahms songs. In the laboratory, Lise's research focused on clarifying the relationships between the beta spectra and gamma rays emitted by radioactive material. She was at the height of her career. Each year she was among the people discussed for a Nobel Prize.

By the early 1930s nuclear physics had advanced dramatically, and Meitner was fascinated by the emerging information. After twelve years apart, she asked Hahn to team up with her again. She needed an expert radiochemist to identify the new heavy isotopes. To help identify their minute samples, Hahn brought in Fritz Strassmann, an analytical chemist. Two other women completed the group: an American chemist, Dr. Clara Leiber, and their technician, a German, Irmgard Bohne. Meitner was the intellectual leader of the team.

At this time, however, Meitner was forced to interrupt her research and leave Germany. Nazi racial laws were making it increasingly difficult for her to work. At first, Meitner was safe because as an Austrian she was not subject to Germany's anti-Semitic laws and the Kaiser-Wilhelm Institut was controlled by powerful industrialists who were able to protect her. However, Meitner's situation became critical following the invasion of Austria. She was banned from teaching, attending conferences, and publishing her research. When Hahn gave a talk outside the Institut on his work with Meitner, he could not mention her name. Her reputation was being obliterated. In 1938, Dirk Coster at the University of Groningen arranged for Meitner to enter Holland despite her lack of papers. She took a train there on the pretext that she was going on a week's vacation. She left behind all her belongings and her scientific papers.

Meitner remained in Holland for a short time and then went to Denmark, where she was the guest of Niels Bohr. She subsequently accepted an invitation from Manne Siegbahn to work in the new Nobel Institute in Stockholm. Meitner was 60 years old when she went to Sweden. Nevertheless, she acquired a good command of the language and built up a small research group. Her thoughts on her career never changed: "Life need not be easy, provided only that it is not empty."[1]

Meitner made her most famous contribution to science shortly after arriving in Stockholm. Mail between Stockholm and Berlin could be delivered overnight, so Hahn and Meitner continued their collaboration. They corresponded several times a week and once met secretly in Copenhagen. Hahn and Strassmann turned to her when they found that the decay products from the neutron bombardment of uranium were isotopes of barium, and they asked her to propose an explanation. No known scientific process explained their findings. Meitner discussed their findings with her nephew, the physicist Otto Frisch. When they proposed that the neutrons had split the uranium, the data all fit.

Meitner immediately sent her explanation to Hahn. Although she and Frisch also reported their interpretation in a paper that described this nuclear "fission," their paper was published shortly after the one by Hahn and Strassmann. Hahn had not acknowledged Meitner for her contribution. Otto Hahn alone won the Nobel Prize in chemistry in 1944 for the discovery of fission of heavy nuclei.

But Meitner's contribution was recognized in some circles. She was invited to join the team working on the development of the atomic bomb, but she refused and hoped until the very end that the project would prove impossible. She was voted Woman of the Year by the Associated Press; and she received honorary degrees from the University of Rochester, Rutgers University, the University of Stockholm, the University of Berlin, Adelphi College, and Smith College. In 1946, Meitner was invited to spend half a year in Washington, D.C., as a visiting professor at Catholic University. In 1947 she retired from the Nobel Institute and went to work in the laboratory that the Swedish Atomic Energy Commission had established for her at the Royal Institute of Technology.

Lise Meitner was a shy, modest person who cared little for fame or money. She enjoyed unraveling the mysteries of nuclear physics because it was fun to do so. She continued her research until she was 81 years old. After spending twenty-two years in Sweden, Meitner retired to Cambridge, England, in 1960 to be near her family. She shared the 1966 Enrico Fermi Prize of the Atomic Energy Commission with Hahn and Strassmann. Meitner continued to travel, lecture, and attend concerts until near the end of her life. She died a few days before her ninetieth birthday on October 27, 1968. German physicists named a newly created element, meitnerium, in her honor in 1992.

Note

1. Sharon Bertsch McGrayne, "Lise Meitner: Nuclear Physicist," in *Nobel Prize Women in Science: Their Lives, Struggles, and Momentous Discoveries* (New York: Carol Publishing Group, 1993), p. 40.

Bibliography

Frisch, Otto. R. "Lise Meitner." *Biographical Memoirs of Fellows of the Royal Society of London* 16 (1970): 405–420.
———. "Lise Meitner," in *Dictionary of Scientific Biography*. Edited by Charles Coulston Gillespie. New York: Charles Scribner's Sons, 1970.

————. "Lise Meitner, Nuclear Pioneer." *New Scientist* (November 9, 1978): 426–428.

Hahn, Otto. *Otto Hahn: My Life*. Translated by Ernst Kaiser and Eithne Wilkins. New York: Herder and Herder, 1970.

McGrayne, Sharon Bertsch. "Lise Meitner: Nuclear Physicist," in *Nobel Prize Women in Science: Their Lives, Struggles, and Momentous Discoveries*. New York: Carol Publishing Group, 1993.

Meitner, Lise. "Looking Back." *Bulletin of the Atomic Scientists* (November 1964): 2.

————. "Resonance Energy of the TH Capture Process." *Physical Review* 60 (1941): 58.

Meitner, Lise, and O.R. Frisch. "Disintegration of Uranium by Neutrons: A New Type of Nuclear Reaction." *Nature* 143 (1939): 276.

Nachmansohn, David. *German-Jewish Pioneers in Science, 1900–1933*. New York: Springer-Verlag, 1979.

Sime, Ruth Lewin. "Belated Recognition: Lise Meitner's Role in the Discovery of Fission." *Journal of Radioanalysis and Nuclear Chemistry* 142 (1990): 13–26.

————. "The Discovery of Protactinium." *Journal of Chemical Education* 63 (1986): 653–657.

————. "Lise Meitner and the Discovery of Fission." *Journal of Chemical Education* 66 (1989): 373–376.

Weart, Spencer R. *Scientists in Power*. Cambridge, Mass.: Harvard University Press, 1979.

CLARA A. CALLAHAN

MIRIAM THE ALCHEMIST
(1st or 2nd century A.D.)
Alchemist

Also known as Mary, Maria, and Miriam the Prophetess or Jewess, Miriam the Alchemist was born in Alexandria, Egypt, in the first or second century A.D. Writing under the name of Miriam the Prophetess, sister of Moses, Miriam the Alchemist should not be confused with the biblical Miriam. Fragments of Miriam's writings, including the *Maria Practica*, exist in collections of ancient alchemy such as the writings of the third- or fourth-century alchemist Zosimos, who repeatedly quoted Miriam's descriptions and instructions.

In the first century B.C., Alexandria was an intellectual center. Although a scientific decline had set in by Miriam's time, alchemy flourished. In their attempt to understand the secret of life, alchemists intertwined the rational and the mystical. While they conducted experiments to test their theories and invented apparatus in their laboratories, they nevertheless regarded metals as male and female and believed that laboratory products were the result of sexual generation.

Miriam the Alchemist's major inventions and improvements included the three-armed still or *tribikos*, the *kerotakis*, and the water bath. Even though these inventions reflected the alchemical beliefs that metals were living organisms, which over time transmuted into gold, and that alchemists could accelerate the process by subjecting them to high temperatures, their use in modern science and by contemporary households are legion.

The *tribikos* was an apparatus for distillation, a process of heating and cooling that imitated processes in nature. From Zosimos comes Miriam's description of this earthenware vessel containing the matter to be distilled, a copper or bronze still-head for condensing the vapor, three delivery spouts fitted into the still-head, and the receivers. A gutter or rim on the inside of the still-head collected the distillate and carried it to the delivery spouts. Cold sponges cooled the still-head and receivers. Miriam advised that the thickness of the metal for the delivery spout tubing be "little more than that of a frying pan for cakes."[1] Flour paste was recommended for sealing the joints.

The *kerotakis* was an apparatus named for the triangular palette used by artists to keep their mixtures of wax and pigment hot. Miriam adapted the *kerotakis* to her study of the effect of vapors of arsenic, mercury, and sulphur on metals. Suspended from a hemispherical cover in a closed apparatus, the *kerotakis* held metal fragments. When the solution near the bottom of the apparatus boiled, the resulting vapors acted on the metal before condensing on the cover and flowing downward to the solution. It was hoped that this continuous "reflux" would yield the desired reaction: a black sulphide called Mary's Black—the first stage of transmutation.

The water bath, or "Marie's bath" (*bain-marie*) as the device came to be called in the fourteenth century, is similar to the present-day double boiler. A container holding a substance to be heated or melted is placed over boiling water. The water, not the fire, becomes the direct heat source. Water heated a substance more slowly and evenly and was easier for the alchemist to control.

Despite the mystical elements of alchemy, Miriam's contributions yielded practical laboratory equipment with contemporary applications. From sponge-cooled condensers to the double boiler, her work has enjoyed an enviable longevity ranging from ancient and medieval chemists to modern chemists and cooks.

Note

1. F. Sherwood Taylor, "Origins of Greek Alchemy," *Ambix* (May 1937): 42.

Bibliography

Alic, Margaret. *Hypatia's Heritage.* Boston: Beacon Press, 1986.
Doberer, K. K. *The Goldmakers: 10,000 Years of Alchemy.* London: Nicholson & Watson, 1948.
Holmyard, E. J. *Alchemy.* New York: Penguin, 1957.
Lindsay, Jack. *The Origins of Alchemy in Graeco-Roman Egypt.* New York: Barnes and Noble, 1970.
Ogilvie, Marilyn Bailey. *Women in Science: Antiquity through the Nineteenth Century.* Cambridge, Mass.: MIT Press, 1986.
Read, John. *Prelude to Chemistry.* London: G. Bell and Sons, 1936.
Taylor, F. Sherwood. *The Alchemists, Founders of Modern Chemistry.* New York: Schuman, 1949.
———. "Origins of Greek Alchemy." *Ambix* (May 1937): 39–42.
Warren, Rebecca Lowe, and Mary H. Thompson. *The Scientists within You: Experiments and Biographies of Distinguished Women in Science.* Vol. 1. Eugene, Ore.: ACI Publishing, 1994.

MARY HAWORTH THOMPSON

MARIA MITCHELL
(1818–1889)
Astronomer

Birth	1818
1836–47	Librarian, Nantucket Atheneum
1847	Discovered Comet Mitchell
1849–68	Computer, *American Ephemeris and Nautical Almanac*
1850	First woman elected to American Academy of Arts and Sciences
1865–68	Professor of Astronomy and Director of Observatory, Vassar College
Death	1889

Maria Mitchell, the first woman astronomer in the United States, was born on August 1, 1818, on Nantucket Island. She was the third child of

William and Lydia Coleman Mitchell, who were Quakers and who eventually had ten children. The Quaker religion stressed education; sober, sensible lives; and the rejection of frivolous pleasure. Although Maria later gave up the Quaker religion, its early influence in her life was strong.

Her mother had worked in libraries because she loved to read. Her father, an easygoing man, was a schoolteacher and amateur astronomer. Both Maria and her father loved all of nature, but especially the night sky. She later remarked that "in Nantucket people quite generally are in the habit of observing the heavens, and a sextant will be found in almost every house. The landscape is flat and somewhat monotonous, and the field of the heavens has greater attractions there than in places which offer more variety of view."[1] Nantucket was a whaling community, and William Mitchell was consulted by captains of whalers about the weather and positions at sea. He adjusted their chronometers, and Maria learned as a child to rate them when her father was absent. She learned to use a sextant as well as a simple reflecting telescope and a Dollard telescope, with which she and her father viewed the eclipse of 1831. Her father encouraged her love of mathematics, and she attended her father's school. Later she went to Cyrus Pierce's School for Young Ladies. At age 16, Maria left school but continued to study mathematics and astronomy on her own. When she was 17 she opened her own school. She was an experimental teacher and followed this approach later in her career.

When Maria was 18 her father began making astronomical observations for the U.S. Coast Survey. A year later in 1836, William Mitchell became cashier at the Pacific Bank. A small observatory was established on the roof of this building, and the Coast Survey loaned a 4–inch equatorial telescope. Maria and her father made thousands of observations here. They calculated the altitudes of stars for the determination of time and latitude, and moon culminations and occultations for longitude. They searched for nebulae and double stars. Nantucket is a fairly cloudy place, but it suffers little interference from outdoor lights because of its location. William was acquainted with astronomers on the mainland with whom he corresponded. They visited, loaned instruments, and made good scientific contacts for Maria.

In 1836 Maria was offered the post of librarian at the Nantucket Atheneum, where she was to work for twenty years. The library was open to the public only in the afternoons and on Saturday evenings, so she had time to pursue her own studies as well. She studied the celestial mechanics of Gauss and Laplace. Maria was very independent, partly as a result of her upbringing in the Nantucket community. The men were often away for long periods on whaling voyages, so women took a strong and active role in the town. Maria felt about herself that she "was born of only ordinary capacity, but of extraordinary persistency."[2]

On October 1, 1847, Maria observed a new comet with the Dollard telescope and calculated its orbit. Her father sent word of her discovery to his friend William Bond, director of Harvard College Observatory, who confirmed it and notified the president of Harvard. The King of Denmark had announced in 1931 that a gold medal would be awarded to the first discoverer of a comet by telescope. This comet was also seen by an observer in Rome and one in England shortly after Maria saw it. After a number of letters back and forth, the award was given to Mitchell and she became famous as an astronomer in both Europe and the United States.

As a result of her discovery of the comet, in 1848 Mitchell became the first—and for a long time, the only—woman ever elected to the American Academy of Arts and Sciences. Her name was proposed by Louis Agassiz, a well-known American naturalist and geologist who identified the Ice Age. Mitchell also became a member of the newly formed American Association for the Advancement of Science in 1850 and formed a life-long friendship with Joseph Henry, director of the Smithsonian Institution. The previous year she was hired as a computer for the U.S. Coast Survey. She computed tables of the positions of planets for the *American Ephemeris and Nautical Almanac*, which gives daily information needed for ocean navigation.

In 1857, Mitchell obtained a position as chaperon for the daughter of a wealthy Chicago banker, which enabled her to travel to Europe—something she had long been anxious to do. There she met leading astronomers, including Sir John and **Caroline Herschel** and **Mary Somerville**, and visited their observatories. In addition, she met with explorer and naturalist Alexander von Humboldt, philosopher of science William Whewell, and geologist Adam Sedgewick. Her meetings with the Herschels and Somerville began to focus her attention on the role of women in science. When she returned to the United States, a group of women led by Elizabeth Peabody gave her a 5–inch Alvan Clark telescope.

In the early 1860s Mathew Vassar decided to establish a women's college that would rival the best schools for men, and he offered the post of professor of astronomy to Mitchell. She would teach as well as be in charge of an observatory that he wished to build. At this time Maria lived in Lynn, Massachusetts, where she and her father had moved after her mother's death to be near her sister. She was hesitant about Vassar's offer in part because she had never received a college education, but she accepted the position after some persuasion.

Maria continued her unconventional methods of teaching at Vassar. She did not use grades or report absences, but she gained a reputation as one of Vassar's greatest teachers. She emphasized small classes, individual attention, simple technology, and mathematical learning. She opened her advanced class with the comment "we are women studying together" and took students with her on trips to make astronomical observations.[3] They also worked with her in the Vassar observatory. She viewed science as "a way of thinking" and hoped science education would make women's minds operate in a more systematic and rational manner. She wanted to help them escape the "narrowness typical of their lives," to learn to observe carefully and question authority.[4] Mitchell said:

> women more than men, are bound by tradition and authority. What the father, the brother, the doctor and the minister have said has been received undoubtingly. Until women throw off this reverence for authority they will not develop. When they do this, when they come to truth through their investigations, when doubt leads them to discovery, the truth which they get will be theirs, and their minds will work on and on unfettered.[5]

Mitchell continued her own research at Vassar, pioneering in the daily photography of sunspots, reporting on solar eclipses, and noting changes in the surfaces of Jupiter, Saturn, and their satellites. She studied the nebulae, her observations having suggested that they were variable, and watched them for changes. Observing double stars, she speculated that one might revolve around the other. She also noted the color variations among stars, postulating that the chemical composition of the star was important as well as its distance from Earth. She inspired her students to share her research, and Vassar subsequently graduated a number of leading women scientists and astronomers. Twenty-five of them later appeared in *Who's Who in America*.

Mitchell made a second trip to Europe in 1873. One of the places she visited was the Russian observatory at Pulkova. This time she not only spoke with astronomers but also explored women's higher educational opportunities. Increasingly she was becoming concerned with recruiting women into science and encouraging public awareness of this need. When she returned from Europe she joined a national women's movement, signing a call for a meeting of leading women involved in educational reform. This group formed the Association for the Advancement of Women (AAW). At the Association's First Women's Congress in New York in 1873, Mitchell read a paper entitled "The Higher Education of Women," which was based on her observations of higher education for women in England, especially at Girton College, where she felt women had asserted themselves more than in the United States. Mitchell wanted society to change its view that the female brain was too weak for mathematics and science. She formed a Committee on Science to study women's progress in science, which compiled statistics and collected other data. Mitchell was elected president of the AAW at its second annual meeting. However, she did not stay in this role long because she felt her primary role was as a scientist.

Mitchell's published work includes seven items listed in the *Royal Society Catalog* and three articles, observational in nature, published in *Silliman's Journal*, also known as the *American Journal of Science and Arts*. In addition, she wrote three popular articles for *Hours at Home, Century*, and the *Atlantic*. During her years at Vassar she edited the astronomical column of *Scientific American*. She believed strongly in the value of imagination in science, saying, "It is not all mathematics, nor all logic but is somewhat beauty and poetry."[6] She contemplated the relationship between poetry and astronomy, writing an article on the subject about Milton. She remarked that "Paradise Lost" reflected "through a poet's lens but with considerable learning, the state of astronomical knowledge in his time."[7]

After teaching at Vassar for 23 years, Mitchell's health began to fail. She retired in December 1888 and returned to Lynn, where she died in 1889. Mitchell did not feel that she was a theoretician, but rather a teacher and observer. She saw a conflict in trying to do both, remarking that "the scientist should be free to pursue his investigations. He cannot be a scientist and a schoolmaster."[8] She felt she had only ordinary talent but "extraordinary persistence."[9] Her other important contribution, of course, was as a leader in the women's movement, focusing on the encouragement and recruiting of women into science. Mitchell received three honorary degrees: from Hanover College in Indiana, Columbia University, and Rutgers Female College. A crater on the moon was named after her and a society established in her honor—the Maria Mitchell Association of Nantucket, which maintains the Maria Mitchell Observatory.

Notes

1. Eve Merriam, "Maria Mitchell: Extracts from Her Diary," in *Growing Up Female in America* (New York: Dell, 1971), p. 73.

2. Helen Wright, *Sweeper in the Sky: The Life of Maria Mitchell* (New York: Macmillan, 1949), p. 25.

3. Phebe Mitchell Kendall, *Maria Mitchell: Life, Letters and Journals* (Boston: Lee and Shepard, 1896), p. 31.

4. Sally Gregory Kohlstedt, "Maria Mitchell and the Advancement of Women in Science," in *Uneasy Careers and Intimate Lives, Women in Science, 1789–1979*, eds. Pnina G. Abir-am and Dorinda Outram (New Brunswick, N.J.: Rutgers University Press, 1987), pp. 130–131.

5. Kendall, *Maria Mitchell*, p. 187.

6. Ibid., p. 186.

7. Maria Mitchell, "The Astronomical Science of Milton as Shown in *Paradise Lost*," *Poet-Lore* 6 (June 1894): 323.

8. Marilyn Bailey Ogilvie, *Women in Science* (Cambridge, Mass.: MIT Press, 1986), p. 138.

9. Helen Wright, "Maria Mitchell," in *Notable American Women 1607–1950*, Vol. 2, ed. Edward T. James (Cambridge, Mass.: Belknap Press, 1971), p. 555.

Bibliography

Bailey, Martha J. *American Women in Science*. Denver, Colo: ABC-CLIO, 1994.

Belserene, Emilia Pisani. "Maria Mitchell: Nineteenth Century Astronomer." *Astronomy Quarterly* 5 (1986): 133–150.

Hoffleit, Dorrit. "Maria Mitchell," in *Dictionary of Scientific Biography*, Vol. 9. Edited by Charles Coulston Gillispie, pp. 421–422. New York: Charles Scribner's Sons, 1974.

Keller, Dorothy J. "Maria Mitchell, An Early Woman Academician." Ph.D. dissertation, University of Rochester, 1974.

Kendall, Phebe Mitchell. *Maria Mitchell: Life, Letters and Journals*. Boston: Lee and Shepard, 1896.

Kidwell, Peggy Aldrich. "Three Women of American Astronomy." *American Scientist* 78 (May/June 1990): 244–251.

Kohlstedt, Sally Gregory. "Maria Mitchell and the Advancement of Women in Science," in *Uneasy Careers and Intimate Lives: Women in Science, 1789–1979*. Edited by Pnina G. Abir-Am and Dorinda Outram, pp. 129–146. New Brunswick, N.J.: Rutgers University Press, 1987.

Merriam, Eve. "Maria Mitchell: Extracts from Her Diary," in *Growing Up Female in America*, pp. 83–101. New York: Dell, 1971.

Mitchell, Maria. "The Astronomical Science of Milton as Shown in *Paradise Lost*." *Poet-Lore* 6 (June 1894): 313–323.

———. "Notes on the Satellites of Saturn." *American Journal of Science Arts* 17 (1879): 430.

Ogilvie, Marilyn Bailey. *Women in Science*. Cambridge, Mass.: MIT Press, 1986.

Opalko, Jane. "Maria Mitchell's Haunting Legacy." *Sky and Telescope* 83 (May 1992): 505–507.

Wright, Helen. "Maria Mitchell," in *Notable American Women 1607–1950*, Vol. 2. Edited by Edward T. James, pp. 554–556. Cambridge, Mass.: Belknap Press, 1971.

———. *Sweeper in the Sky: The Life of Maria Mitchell*. New York: Macmillan, 1949.

JOAN G. PACKER

AGNES FAY MORGAN

(1884–1968)

Chemist

Birth	May 4, 1884
1904	B.S., chemistry, University of Chicago
1905	M.S., chemistry, University of Chicago
1905–07	Instructor in Chemistry, Hardin College
1908	Married Arthur I. Morgan
1910–12	Instructor in Chemistry, University of Washington
1911	Established Iota Sigma Pi, honor society for women chemists
1914	Ph.D., organic chemistry, University of Chicago
1915–19	Assistant Professor of Nutrition, University of California, Berkeley
1919–23	Associate Professor, University of California, Berkeley
1923	Professor of Household Science, University of California, Berkeley
1940–42	Chair, California State Nutrition Committee
1942–45	U.S. Office of Scientific Research and Development
1946–50	Member, Committee of Nine
1949	Garvan Medal, American Chemical Society
1951	Borden Award, American Institute of Nutrition
1963	Named one of the ten outstanding women in the Bay Area by the *San Francisco Examiner*
Death	July 20, 1968

Agnes Fay Morgan was born in Peoria, Illinois, in 1884. Her father, Patrick John Fay (a laborer and builder), and her mother, Mary Dooley (second wife of Patrick), had emigrated to the United States from Galway, Ireland. The Fays had four children. Agnes, their third child and the only one in the family to attend college, went on to become a noted teacher and pioneer in vitamin research.

Because Agnes achieved excellent marks in high school, a local donor presented her with a full college scholarship. She started at Vassar but soon transferred to the University of Chicago, where she studied under

Agnes Fay Morgan. Photo courtesy of The Bancroft Library.

Professor Stieglitz, a famous analytical chemist. Morgan earned her bachelor's, master's, and doctoral degrees in organic chemistry from the university in 1904, 1905, and 1914, respectively.

In 1908 Agnes Fay married Arthur Ivason Morgan, a veteran of the Spanish-American War. He was a high school Latin and Greek teacher and a football coach. Eventually, when the family moved to California, he became a sales manager at Sperry Flour, a General Mills subsidiary located in San Francisco. They had one son, Arthur Ivason, born in 1923. He studied chemical engineering and did postgraduate work in Switzerland. Following in his mother's footsteps, he became an authority on foods.[1]

Agnes Morgan had a warm personality and was interested in and concerned about others. She was productive in her work, which centered around teaching, research, and administrative responsibilities. Her successes resulted in an extraordinary number of publications (approximately 300 articles, reports, and one book), other professional accomplishments, and many "firsts" in chemistry. An active life such as hers had no place for hobbies except for a colony of cocker spaniels, which were also research subjects in her nutritional studies. Morgan's main source of relaxation was reading mystery stories mostly by British authors such as Dorothy Sayers, Carter Dickinson, and Agatha Christie.[2]

Morgan wanted to work as a chemist. She knew, however, that as a woman she would not be taken seriously by her male peers. She chose to teach at the collegiate level, where prejudice against women in science was minimal and she would be able to conduct research. While acquiring her education she worked as an instructor of chemistry at Hardin College, at the University of Montana, and at the University of Washington. She became known as an innovative and successful teacher who encouraged women in studying science and who became a leader in nutritional education. Her professional career began in 1915 as an assistant professor of nutrition in the College of Agriculture at the University of California, Berkeley. By 1923 she was a full professor, and from 1938 she was a biochemist at the Agricultural Experiment Station.

At the University of California, Berkeley, Morgan established a curriculum in nutrition and dietetics. Through determined efforts she elevated the status of nutrition to a scientific level and developed home economics as a scientific discipline. Between 1916 and 1938, under her direction the Department of Home Economics underwent a number of changes. One of her achievements was the high standards of scientific training provided for students of home economics. She was instrumental in organizing an interdepartmental graduate study in nutrition. Students were taught biochemistry, physiology, anatomy, medicine, and household science. Later, similar interdisciplinary programs were established based on her prototype curriculum. She also directed the research of a large number of doctoral and master's degree students.

Morgan's first research projects focused on the nutrients of foods grown in California and on the changes in food values during processing. She studied the stability of proteins in wheat, almonds, walnuts, and meat during heat treatment and the mechanism of the damage that resulted. She demonstrated that heat-damaged proteins diminished the amount of amino acids essential for growth and maintenance of the body.

Much of her research was concerned with human and animal nutrition and food technology. She studied the effect of sulfur dioxide as a preservative on vitamins and found that it conserved vitamin C but damaged thiamine. She also was interested in children's nutrition and observed the effect of supplementary feedings on the growth of children. Morgan also conducted studies of serum cholesterol and diet and studied bone density in aging. During World War II, at the request of the Federal

Bureau of Nutrition, she researched methods of dehydration to maintain the nutritional value and the quality of food and to improve dehydrated food. The process she invented for dehydrating scrapple, a liver pudding, was patented.

Because animal research can be better controlled than human research, many of Morgan's projects involved rats, guinea pigs, hamsters, and cocker spaniels. She studied the effect of vitamins on hormone activities and noted that vitamin D and calcium influence the physiological activity of the parathyroid secretion and, consequently, bone growth. This study called attention to and demonstrated the danger of giving too much vitamin D to babies. Other research projects were concerned with the effect of vitamin A and carotene on thyroid activity and the effect of riboflavin and pantothenic acid on the adrenal gland. Researching adrenal function, Morgan found that vitamin B is essential for the pigmentation of hair and skin. The lack of this vitamin turns hair gray. For this work she received the Garvan and Borden Awards. For studying the vitamin B content of grapes and wine, she was given a research grant in 1962 from the Society of Medical Friends of Wine.

Morgan was a nutritional advisor to several California governors and worked on establishing standards for human nutrition. In 1939 she conducted a food study of the San Quentin Prison, and in 1960 an investigation of the toxic effects of agricultural chemicals.[3] Earlier, she worked for the U.S. Office of Scientific Research and Development. From 1946 to 1950 she served as the only woman on the Committee of Nine, established to guide the national program of research.

By all accounts Morgan, who had good written and spoken communications skills, was an excellent administrator and was respected for her powerful personality. Against many odds she established a home economics curriculum based on solid scientific foundations. In addition, she secured financial support for her research from outside the university. At the Federal Food Administration's request during World War I, she introduced new courses in Household Science: dietetics for nurses, a graduate course for hospital dieticians, and some extension courses.[4] She initiated and chaired the California State Nutrition Committee from 1940 to 1942. Morgan also established Alpha Nu, an undergraduate home economics honors society, at the University of California, Berkeley, in 1915; and she was responsible for the creation of Iota Sigma Pi, an honors society for women chemists, at the University of Washington in 1911.

As a pioneer in vitamin research, her accomplishments and discoveries were numerous. She was the first scientist who recognized the effects of preservatives on vitamins. She was the first to recognize that vitamin deficiency causes the graying of hair and that heat damages proteins in food. Furthermore, she was the first to observe the interaction of vitamin

D and parathyroid extract and to observe that pantothenic acid deficiency damages the adrenal glands.

Morgan's honors were just as numerous as her discoveries. She received the Garvan Medal from the American Chemical Society in 1949 and the Borden Award from the American Institute of Nutrition in 1954. In 1950 her colleagues elected her to deliver the Faculty Research Lectures (she was the first woman to be elected), and the University of California awarded her the honorary LL.D. degree in 1959. The new home economics building at the University of California, Davis, campus was named A. F. Morgan Hall in 1961; and in the next year she was elected a fellow of the American Institute of Nutrition. The *San Francisco Examiner* presented her with the Phoebe Apperson Herst Gold Medal in 1963 for being one of the ten outstanding women in the Bay Area. A commemorative symposium was held in her honor in 1965, and she was named an honorary member of the California Dairy Council in 1966. At the Iota Sigma Pi Society's 1969 convention, its research award was named for her.

Agnes Fay Morgan "was an active member of twenty-one professional societies."[5] These included the American Chemical Society, the American Association for the Advancement of Science, the American Institute of Nutrition, and the Society of Biological Chemists.

Morgan retired from the University of California in 1954 and died of a heart attack on July 20, 1968, in Berkeley, California. She was a versatile person: a chemist, a nutritionist, a teacher, and an administrator. She was also a consultant to private and government agencies. Her teaching, substantial research, and courage in a men's world were and continue to be an inspiration to women chemists.

Notes

1. M.A. Cavanaugh, "Agnes Fay Morgan (1884–1968)," in *Women in Chemistry and Physics: A Biobibliographic Sourcebook*, eds. L.S. Grinstein, Rose K. Rose, and M.H. Rafailovich (Westport, Conn.: Greenwood Press, 1993), pp. 434–435.

2. "Garvan Medal to Agnes Morgan," *Chemical and Engineering News* 27 (1949): 905.

3. E.N. Todhunter, "Biographical Notes from History of Nutrition, Agnes Fay Morgan, May 4, 1884–July 20, 1968," *Journal of the American Dietetic Association* 53 (December 1968): 599.

4. M. Nerad, "Gender Stratification in Higher Education: The Department of Home Economics at the University of California, Berkeley, 1916–1962," *Women's Studies International Forum* 10 (1987): 160.

5. M. Nerad, "Gender in Higher Education: The History of the Home Economics Department at the University of California at Berkeley," Ph.D. dissertation, University of California, Berkeley, 1988, p. 35.

Bibliography

Cavanaugh, M.A. "Agnes Fay Morgan (1884–1968)," in *Women in Chemistry and Physics: A Biobibliographic Sourcebook*. Edited by L.S. Grinstein, R.K. Rose, and M.H. Rafailovich, pp. 434–448. Westport, Conn.: Greenwood Press, 1993.

Emerson, G.A. "Agnes Fay Morgan and Early Nutrition Discoveries in California." *Federation Proceedings of the American Societies for Experimental Biology* 36 (1977): 1911–1914.

Fraenkel-Conrat, J. "Agnes Fay Morgan: Founder, Scientist, Person." *Iotan Newsletter* 34 (November 1983).

"Garvan Medal to Agnes Morgan." *Chemical and Engineering News* 27 (1949): 905.

Gorman, Mel. "Agnes Fay Morgan, 1884–1968," in *American Chemists and Chemical Engineers*, 2 vols. Edited by Wyndham D. Miles, pp. 348–349. Washington, D.C.: American Chemical Society, 1976–1994.

"Landmarks of a Half Century of Nutrition Research." *Journal of Nutrition* 91, suppl. 1, pt. 2 (1967): 1–67.

Morgan, Agnes Fay. Regional Oral History Office, Bancroft Library. *Centennial History Project*. Interview with Agnes Fay Morgan, for the Centennial History Office. Berkeley: University of California, 1959.

Morgan, Agnes Fay, and I.S. Hall. *Experimental Food Study*. New York: Farrar and Rinehart, 1938.

Nerad, M. "Gender in Higher Education: The History of the Home Economics Department at the University of California at Berkeley." Ph.D. dissertation, University of California, Berkeley, 1988.

———. "Gender Stratification in Higher Education: The Department of Home Economics at the University of California, Berkeley 1916–1962." *Women's Studies International Forum* 10 (1987): 157–164.

Nutritional Status U.S.A. Edited by Agnes F. Morgan. Berkeley, Calif.: Agricultural Experiment Station, 1959.

Okey, R. "Agnes Fay Morgan (1884–1968): A Biographical Sketch." *Journal of Nutrition* 104 (1974): 1103–1107.

Okey, R., et al. *Agnes Fay Morgan, 1884–1968, In Memoriam*. Berkeley: University of California, 1969.

The Progress of Science: A Review of 1941. Edited by S.E. Farguhar. New York: Grolier Society, 1941.

Raacke, I.D. "Morgan, Agnes Fay." In *Notable American Women: The Modern Period: A Biographical Dictionary*. Edited by Barbara Sicherman, Carol Hurd Green with Ilene Klanbrov, Harriette Walker. Cambridge, Mass.: Belknap Press, 1980.

Todhunter, E.N. "Biographical Notes From History of Nutrition, Agnes Fay Morgan, May 4, 1884–July 20, 1968." *Journal of the American Dietetic Association* 53 (December 1968): 599.

KATALIN HARKÁNYI

DOROTHY VIRGINIA NIGHTINGALE

(1902–)

Chemist

Birth	February 21, 1902
1922	A.B., University of Missouri
1923	A.M., University of Missouri
1923–39	Instructor, University of Missouri
1928	Ph.D., organic chemistry, University of Chicago
1938	Honors Fellow, University of Minnesota
1939–48	Assistant Professor of Chemistry, University of Missouri
1942–45	Civilian with Office of Scientific Research and Development
1943–45	Committee on Medical Research—antimalarial
1946–47	Research Associate, University of California, Los Angeles
1948–58	Associate Professor of Chemistry, University of Missouri
1958–72	Professor of Chemistry, University of Missouri
1959	Garvan Medal, American Chemical Society
1972–	Professor Emerita, University of Missouri, Columbia

Dorothy Nightingale's work contributed much to the knowledge of organic synthetic reactions. Her career spanned a time of rapid discovery in organic chemistry, but it was also a time when the field was not very accepting of women scientists. Nightingale not only had to be quite brilliant at chemistry, but also had to have a tough shell to succeed. One special thing about her research was that she did not just make new molecules, she actually studied how the process happened. This kind of quantification, termed physical chemistry, was rare at that time. She did high-quality, difficult chemistry and a lot of it, compiling a very extensive publication record. In addition, her teaching was inspirational. On receiving the Garvan Medal, she said she wanted to continue to be "not common."[1]

Dorothy Virginia Nightingale was born in 1902 to Jennie (Beem) and William David Nightingale of Fort Collins, Colorado. Jennie had taught in a country school and worked as a secretary in an Indianapolis orthopedic hospital before marrying rancher William in 1900. They home-

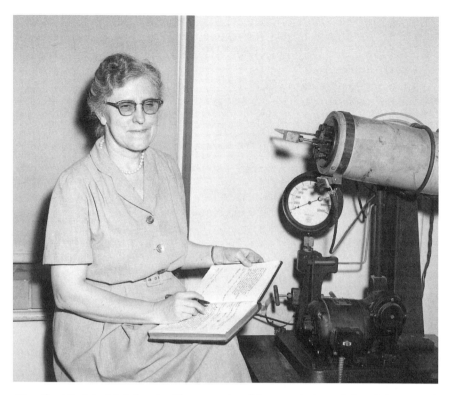

Dorothy Virginia Nightingale. Photo reprinted by permission of the University of Missouri Archives.

steaded outside of Fort Collins until 1919, when they moved to Columbia, Missouri.[2]

In about 1911, some of the college students staying at the boarding house that Jennie managed took Dorothy to the chemistry lab and entertained her with experiments. This was her first exposure to science. Early in college at the University of Missouri, foreign language and history interested her and she was a grader for the German department. She thought she would take just one class in chemistry. However, her chemistry professor, Herman Schlundt, sparked her interest to continue with chemistry. He encouraged her to go beyond her initial goal of teaching high school chemistry—to go to graduate school and then teach at the college level.[3]

After graduating in 1922, Nightingale went on to earn a master's degree at the University of Missouri. She determined the spectra of organomagnesium halides and other luminescent compounds for her master's thesis. She completed her degree in 1923 and became an instructor in the chemistry department at the University of Missouri, only the second woman to do so, and soon thereafter began to pursue a Ph.D. Her doctoral work, "Studies in the Merexide and Alloxantine Series," was supervised by Professor Julius Stieglitz and was completed in 1928, when she received her Ph.D.

An honors fellow at the University of Minnesota in 1938, Nightingale became an assistant professor at the University of Missouri the following year. During World War II she worked with the Office of Scientific Research and Development on the Committee on Medical Research performing antimalarial research. In 1946 she was a research associate at the University of California, Los Angeles. Nightingale was promoted to associate professor in 1948 and to full professor in 1958 at the University of Missouri. The American Chemical Society awarded her the Garvan Medal for distinguished service in chemistry in 1959.

Nightingale produced an amazing body of research. Her colleagues said she alone accounted for 40 percent of the organic chemistry publications from Missouri's chemistry department.[4] If one considers the quantity and significance of her research, the university was quite slow in promoting her.[5] Attaining full professorship ordinarily takes from ten to fifteen years, but it took Nightingale fully thirty years. Science, organic chemistry in particular, was quite an "old-boys" network that served as an obstacle to Nightingale's success and advancement.

Before she retired in 1972, Nightingale was director of graduate studies at Missouri. She believed teaching and research were intertwined. In fact, she said, "We teach as we direct a student's research."[6] She graduated 23 Ph.D. students and 27 master's students—a truly impressive record, since supervising several students at once requires both time and commitment. It was significant that so many graduate students wanted Nightingale to direct them, because having a woman scientist direct one's research was a stigma at that time. Nightingale encouraged her students to publish; in fact, many of her own publications had Ph.D. students as second authors. Several of her students, such as H. B. Hucker, J. A. Gallagher, R. H. Wise, and B. Sukornick, went on to contribute more to the literature. Nightingale must have been a fabulous teacher and mentor. Former students became friends. She corresponded with them and sent out newsletters among them reporting on their fellow alumni.[7]

Nightingale's early work involved the chemiluminescence of organomagnesium halides, which have to do with light emissions given off from a chemical reaction. She did a great deal of work with Friedel-Crafts reactions, which are Lewis acid-catalyzed organic reactions. The two main types of Friedel-Crafts reactions are alkylation and acylation of aromatic hydrocarbons catalyzed by anhydrous aluminum chloride. Alkylation is the introduction of alkyl groups (paraffinic hydrocarbon groups) into aromatic hydrocarbons. These processes are important in industry for producing such varied products as high-octane gasoline; cumene (a solvent used to produce phenol and acetone); ethylbenzene (a solvent used to produce styrene); better cleaning detergents; synthetic rubber; and plastics and elastomers. Many of these substances are toxic and flammable and, therefore, a little dangerous. Friedel-Crafts reactions began to be studied in 1877, but there are a large number of reactions. It took decades to study them all, and Nightingale contributed a good deal to the field through her extensive publications. Many of her articles are cited in the definitive work on the Friedel-Crafts reactions, *Friedel-Crafts and Related Reactions*, by George Olah.

Nightingale was concerned about the dearth of women in chemistry. She at one time urged women to teach chemistry at the junior college level to get a start in the field. She also collected statistics on women in college chemistry faculties. Nightingale also believed in being involved in professional organizations. She was a member of the American Chemical Society and past vice-president and treasurer of the Missouri section. She was a member of the American Association of University Women. Phi Beta Kappa could claim her as a member, and the local chapter elected her its vice-president. She was once also the secretary and vice-president for Sigma Xi and past vice-president and local chapter president of Sigma Delta Epsilon.

After retiring in 1972, Nightingale returned to Boulder, Colorado. She identifies herself as Christian and her hobbies as music and photography, especially the photography of wildflowers. She likes "motoring," seeing the countryside via car, horseback riding, and mountain climbing, and she is a member of the Colorado Mountain Club. She has traveled extensively in Mexico, Guatemala, and the mountainous areas of southern Europe. She has even cruised along the Antarctic peninsula.[8]

Notes

1. "Dorothy V. Nightingale," *Chemical and Engineering News* 37 (April 20, 1959): 117.

2. Adriane P. Borgias, "Dorothy Virginia Nightingale," in *Women in Chemistry and Physics: A Biobibliographic Sourcebook*, ed. L. S. Grinstein, R. K. Rose, and M. H. Rafailovich (Westport, Conn.: Greenwood Press, 1993), p. 449.

3. Ibid.

4. "Dorothy V. Nightingale," p. 117.

5. Martha J. Bailey, *American Women in Science: A Biographical Dictionary* (Santa Barbara, Calif.: ABC-CLIO, 1994), p. 272.

6. Borgias, "Dorothy Virginia Nightingale," p. 452.

7. "Dorothy V. Nightingale," p. 117.

8. Borgias, "Dorothy Virginia Nightingale," p. 451.

Bibliography

American Men and Women of Science. New York: R.R. Bowker, 1992.

American Women, 1935–1940: A Composite Biographical Dictionary. Edited by Durward Howes. Detroit: Gale, 1981.

Bailey, Martha J. *American Women in Science: A Biographical Dictionary.* Santa Barbara, Calif.: ABC-CLIO, 1994.

Borgias, Adriane P. "Dorothy Virginia Nightingale," in *Women in Chemistry and Physics: A Biobibliographic Sourcebook.* Edited by L. S. Grinstein, R. K. Rose, and M. H. Rafailovich. Westport, Conn.: Greenwood Press, 1993.

"Dorothy V. Nightingale." *Chemical and Engineering News* 37 (April 20, 1959): 117.

A History of the Department of Chemistry, University of Missouri-Columbia, 1843–1975. n.p., 1975.

Nightingale, Dorothy. "Alkylation and the Action of Aluminum Halides on Alkylbenzenes." *Chemical Reviews* 25 (1939): 329–376.

———. "Anomalous Nitration Reactions." *Chemical Reviews* 40 (1947): 117–140.

Nightingale, Dorothy, O.L. Wright, and H.B. Hucker. "The Anomalous Acylation of Some Monoalkylbenzenes." *Journal of Organic Chemistry* 18 (1953): 244–248.

JILL HOLMAN

CECILIA PAYNE-GAPOSCHKIN
(1900–1979)
Astronomer

Birth	May 10, 1900
1923	B.A., Newnham College; Pickering Fellow, Harvard Observatory
1924	Elected to membership in the American Astronomical Society
1925	Ph.D., Radcliffe College
1931	Became U.S. citizen
1934	First recipient of the Annie J. Cannon Prize, American Astronomical Society; married Sergei I. Gaposchkin
1938	Named Phillips Astronomer at Harvard Observatory
1939	Started the Harvard Observatory Philharmonic Orchestra

Cecilia Payne-Gaposchkin. Photo reprinted by permission of Radcliffe College Archives.

1956	Became the first woman to advance to the rank of professor at Harvard; appointed Chairman of the Dept. of Astronomy at Harvard
1967	Professor Emerita, Harvard University
1976	Henry Norris Russell Prize, American Astronomical Society
1977	Minor planet 1974 CA named Payne-Gaposchkin in her honor
Death	December 7, 1979

Cecilia Payne-Gaposchkin was an authority on stellar development, variable stars, and galactic structure. She demonstrated that the stars are similar to the earth and sun in their chemical makeup. She also proved that the apparent differences in their spectra are due to varying surface temperatures. Her later work, which dealt with the classification of stars of variable brightness, helped astronomers to understand stellar evolution.

Cecilia Payne was born in 1900 in England. Her father, Edward Payne, practiced law and wrote books on American history. Her mother, Emma Pertz, was a painter. Payne's first encounter with astronomy occurred when she saw her first meteorite at the age of 5. At age 6 she began to attend school and by age 12 could speak French and German and had a basic knowledge of Latin and arithmetic. She also studied geometry and algebra. Cecilia then moved to a school in London that provided no science lessons; but as Cecilia already knew that she wanted to be a scientist, she decided to study science on her own. The second year at the school proved better, as Payne learned algebra and Euclid. When Dorothy Dalglish came to the school to teach science, she recognized Cecilia's love of the subject and lent her books on physics.

Over the following years, Cecilia was trained to be a classical scholar. She found it necessary to fight for an education in science. Having taught herself botany, calculus, and coordinate geometry, she was finally assigned a mathematics teacher. Cecilia then moved to St. Paul's Girls' School, where she concentrated on science and music.

Cecilia's one desire was to go to Cambridge to study science. To her, chemistry was "an exercise in memory and manual dexterity, but physics opened up a whole new world."[1] Cecilia's instructor allowed her to discover the areas that excited her. By the end of her studies, physics was replacing botany as her primary interest. Cecilia received a scholarship for Cambridge and entered Newnham College in 1919. She was delighted with the study of physics. One particular guest lecturer had a very profound influence on her. Professor Arthur Eddington was to announce the results of the eclipse expedition to Brazil in 1918. Cecilia was fascinated by the lecture, and the direction of her interest changed. At Cambridge astronomy was considered a branch of mathematics, so she could not study it directly. She began to study physics and to attend as many lectures on astronomy as she could. She also read any book on astronomy that she could find. During a public night at the observatory, Cecilia had a chance encounter with Eddington, whom she told that she would like to become an astronomer. He, either then or later, stated that he "could see no insuperable objection."[2] This statement got her through many later rebuffs. She asked Eddington what she should read; when he found that she had read everything he suggested, he directed her to the *Monthly Notices* and the *Astrophysical Journal*. By the time she finished at Cambridge, Payne had already published one paper and been elected to the Royal Astronomical Society.

As college life drew to a close, Payne attended a lecture given by Harlow Shapley, the newly appointed director of the Harvard Observatory. After the lecture she met Shapley and told him that she would like to work for him. Putting all her efforts into getting enough money in grants and fellowships to support herself for a year in the United States, she arrived at the Harvard Observatory in 1923 as a Pickering fellow. Payne respected Shapley. In fact, she called him her "idol" in her autobiography.[3]

At Harvard, Payne had a chance to meet well-known women astronomers, including **Annie Jump Cannon** and **Antonia Maury**. When asked what she would like to study at the observatory, Payne indicated an interest in stellar atmospheres. Shapley approved and gave her access to the extensive Harvard plate collection. Payne said that she set out to "make quantitative the qualitative information that was inherent in the Henry Draper system."[4] After months of research, the first breakthrough came when she determined the temperatures of hotter stars. At Shapley's urging, she wrote her first paper on stellar spectra. As her work progressed, the relation between line intensity and temperature began to take shape. She found that one could infer the ionization potential of an atom from the behavior of its lines in the spectral sequence when the scale had been established by means of other atoms of known properties. After two years, she had determined a stellar temperature scale and had measured the abundance of chemical elements in the stars. In addition to her Ph.D. thesis, Shapley encouraged her to make her work into a book. This she did, calling it *Stellar Atmospheres: A Contribution to the Observational Study of Matter at High Temperatures*. Radcliffe College granted her the first Ph.D. in astronomy given to a student at the Harvard Observatory.

Payne was offered a position at the Lick Observatory, but when she told Shapley about it, he offered her a position at Harvard. She served as a National Research Council fellow and then as a staff member. Payne would remain at the Harvard Observatory for over fifty years.

Harvard did not have an advanced spectrograph, which would have made it possible for Payne to elaborate on her earlier work on the temperature and composition of stellar atmospheres. By 1928 she was spending more and more time on an ongoing observatory project of measuring stellar magnitudes and distances. Payne found this work tedious and unproductive, yet she continued to work at Harvard. In 1931 she became a U.S. citizen.

At the astronomical meetings in Göttingen, Germany, she met the Russian astronomer Sergei Gaposchkin. He had tried to get a position at the Harvard Observatory but had been turned down. He appealed to Payne at the Göttingen meetings, and she then campaigned for his appointment at Harvard. She called him a "good but not brilliant astronomer."[5] Cecilia offered to raise the salary for Gaposchkin herself, suggesting that he have a fellowship in photometry work. In October, Gaposchkin was offered the position of research assistant for one year.

Payne continued her research in photometry and spectroscopy. By 1934 her relationship with Gaposchkin had deepened. As Sergei would require financial assistance, Cecilia could continue working. They were married in New York. Payne was 35 years old. To aid those who wished to look up her scientific papers, she took Payne-Gaposchkin as her last name.

Payne-Gaposchkin continued to teach courses for undergraduate and graduate students. She supervised projects, edited observatory publications, and served on several commissions of the International Astronomical Union. She was a member of the American Philosophical Society and the American Academy of Arts and Sciences. She also worked with her husband on studies of eclipsing variable stars; later, the Gaposchkins studied the variable stars photographed on Harvard plates, work that continued into the 1950s. Although the Gaposchkins continued to work together on projects, Cecilia was more prominent. Otto Struve compared Payne-Gaposchkin with her husband, saying, "she is his equal in industry . . . and his superior in knowledge and judgment."[6]

The Gaposchkin family included three children: Edward Michael, Katherine Lenora, and Peter. Cecilia continued her work through her pregnancies. She presented a paper to the American Astronomical Society five months before Edward's birth in 1935 and directed research on photometry and variable stars the following summer. During the summer of 1937, after Katherine was born, Payne-Gaposchkin spent six weeks lecturing at the Yerkes Observatory in Wisconsin.[7]

In 1938, Payne-Gaposchkin was named Phillips Astronomer at the Harvard Observatory. In the mid-1940s, Payne-Gaposchkin became more involved in the running of the observatory as a member of the Harvard Observatory Council. She was also affiliated with the Smithsonian Astrophysical Observatory. In 1956 she was named professor of astronomy and was appointed chairman of the Department of Astronomy at Harvard University. Payne-Gaposchkin received the highest award of the American Astronomical Society—the Henry Norris Russell lectureship—in 1976. This honor was bestowed for "a lifetime of eminence in astronomical research."[8] Cecilia was the first woman to receive this honor since its establishment in 1946.

Payne-Gaposchkin was involved in and chaired the Forum for International Relations at the observatory with her husband. This became a vehicle for those in the university community to discuss events during World War II. Her personal interests showed her to be somewhat of a renaissance woman: she sang in the Lexington Choral Society and taught Sunday school in the Unitarian church. She sometimes took on sewing or knitting projects, and she enjoyed playing bridge. Cecilia also wrote poetry and was an unpublished writer of detective stories. She also started the Harvard Observatory Philharmonic Orchestra in 1949.

By 1979, Payne-Gaposchkin had completed her autobiography. During a trip with her husband, she returned early and was clearly ill. Lung cancer was diagnosed. She died in December 1979.

Notes

1. Cecilia Payne-Gaposchkin, *Cecilia Payne-Gaposchkin: An Autobiography and Other Recollections*, ed. Katherine Haramundanis (New York: Cambridge University Press, 1984), p. 108.

2. Ibid., p. 120.

3. Ibid., p. 156.

4. Ibid., p. 161.

5. Peggy A. Kidwell, "Cecilia Payne-Gaposchkin: Astronomy in the Family," in *Uneasy Careers and Intimate Lives: Women in Science 1789–1979*, eds. Pnina G. Abir-Am and Dorinda Outram (New Brunswick, N.J.: Rutgers University Press, 987), p. 227.

6. Ibid., p. 235.

7. Ibid., p. 232.

8. E. Margaret Burbidge, "Giving Women Astronomers Their Due," *Physics Today* 44 (1990): 92.

Bibliography

Burbidge, E. Margaret. "Giving Women Astronomers Their Due." *Physics Today* 44 (1990): 91–92.

Kidwell, Peggy A. "Cecilia Payne-Gaposchkin: Astronomy in the Family," in *Uneasy Careers and Intimate Lives: Women in Science 1789–1979*. Edited by Pnina G. Abir-Am and Dorinda Outram. New Brunswick, N.J.: Rutgers University Press, 1987.

———. "Three Women of American Astronomy." *American Scientist* 78 (1990): 244–251.

Payne-Gaposchkin, Cecilia. *Cecilia Payne-Gaposchkin: An Autobiography and Other Recollections*. Edited by Katherine Haramundanis. New York: Cambridge University Press, 1984.

———. *The Galactic Novae*. New York: Dover, 1964.

———. *Stars and Clusters*. Cambridge, Mass.: Harvard University Press, 1979.

———. *Stars in the Making*. Cambridge, Mass.: Harvard University Press, 1952.

Payne-Gaposchkin, Cecilia, and K. Haramundanis. *Introduction to Astronomy*. 2nd ed. Englewood Cliffs, N.J.: Prentice-Hall, 1970.

Rossiter, Margaret W. *Women Scientists in America: Struggles and Strategies to 1940*. Baltimore: Johns Hopkins University Press, 1982.

DOLORES FIDISHUN

MARY ENGLE PENNINGTON

(1872–1952)

Bacteriologist, Chemist

Birth	October 8, 1872
1892	Completed requirements for a B.S. degree, University of Pennsylvania; received Certificate of Proficiency in Biology
1895	Ph.D., chemistry, University of Pennsylvania
1895–97	Fellow, chemical botany, University of Pennsylvania
1897	Fellow, physiological chemistry, Yale University
1898	Director, Clinical Laboratory, Woman's Medical College of Pennsylvania
1901	Established the Philadelphia Clinical Laboratory with Elizabeth Atkinson
1904	Bacteriologist, Philadelphia Bureau of Health
1905	Bacteriological Chemist, Bureau of Chemistry, U.S. Department of Agriculture
1908–19	Chief, Food Research Laboratory
1908	U.S. Delegate, International Congress on Refrigeration, Paris
1910	U.S. Delegate, International Congress on Refrigeration, Vienna
1913	U.S. Delegate, International Congress on Refrigeration, Chicago and Washington
1919	Notable Service Medal
1922–52	Established and ran consulting business, New York
1923–31	Director, Household Refrigeration Bureau, National Association of Ice Industries
1940	Garvan Medal, American Chemical Society
1940–45	Consultant, R&D Branch, Military Planning Division, War Shipping Administration
1947	Fellow, American Society of Refrigerating Engineers; Fellow, American Association for the Advancement of Science; elected to the Hall of Fame, Poultry Historical Society

1948   U.S. Delegate, International Congress on Refrigeration,
              Copenhagen
Death   December 27, 1952

Mary Engle Pennington, whose pioneering work in food preservation continues to affect our daily lives, was born in Nashville, Tennessee, in 1872. She was the elder of two daughters of Sarah B. Molony (who came from a well-established Pennsylvania Quaker family) and Henry Pennington. When Mary was a few years old, the family moved to Philadelphia. Although her father was a southerner, her parents thought it was best to live near Mrs. Pennington's family.

Henry Pennington established a successful label manufacturing business in Philadelphia and bought a home near the University of Pennsylvania. He loved to garden and shared this hobby with his older daughter. Both parents were enlightened for their time and supported Mary when she decided to pursue a university education.

As a young girl, Mary liked to read and spent much time in the public library. At the age of 12 she discovered a medical chemistry book that awakened her interest in chemistry. Because she could not understand much of the terminology, she went to the University for help. She was encouraged to come back when she could better understand the subject and was invited to study chemistry at the University.[1]

In 1890, when she was 18 years old, Mary Pennington began her higher education studies. At the Towne Scientific School, University of Pennsylvania, she took chemistry, biology, and hygiene. By 1892 she had finished all the course requirements for a bachelor's degree but was never granted that diploma; instead, she was given a Certificate of Proficiency in biology. At that time the University did not grant bachelor's degrees to women. Her determination and her professors' encouragement prompted Mary to continue her studies. She attended the Electrochemical School of Edgar Fahs Smith for her graduate studies and, in 1895, was awarded a doctoral degree. Her dissertation discussed the derivatives of the elements columbium and tantalum. She spent two more years at the University of Pennsylvania studying chemical botany and one year at Yale University as a fellow in physiological chemistry. At Yale, together with Russell Chittenden, she investigated the effect of colored light on plants and their growth.[2]

In 1898, after the year at Yale and to her family's delight, she returned to Philadelphia, where she remained until 1922. During that time she was very active and held a number of positions. Pennington established the Philadelphia Clinical Laboratory with Elizabeth Atkinson in 1901. The laboratory performed chemical and bacteriological analyses for physicians. At the same time, she accepted directorship of the Clinical Laboratory of the Woman's Medical College of Pennsylvania. Later, in 1904, she was asked to reorganize the Philadelphia Bacteriological Laboratory, eventually becoming a bacteriologist with the Philadelphia Bureau of Health. By 1905 she was the bacteriological chemist at the Bureau of Chemistry, U.S. Department of Agriculture. Her work greatly impressed Harvey W. Wiley, chief of the Bureau of Chemistry, who inspired Pennington to become head of the newly created Food Research Laboratory. To qualify for this position, candidates had to pass a civil service examination. Wiley feared that Pennington's gender would hinder her appointment and urged her to sign the exam with her initials. She received the highest scores and was appointed before her true identity was discovered.[3]

After World War I, Pennington left civil service employment and became the director of research and development at the American Balsa Company, a manufacturer of refrigerated boxcars. In 1922 she established a consulting firm in New York City and eventually moved her office to the prestigious Woolworth Building. The firm concentrated on problems associated with the perishable food industry. As a result of her research, she developed food preservation techniques and thus improved storage of eggs, poultry, and frozen fish. This research enabled her to formulate workable refrigeration techniques.

Mary Pennington had a remarkable career. She was educated as an analytical chemist but soon became a bacteriologist. While working at the Food Research Laboratory in Philadelphia, she and other chemists under her leadership conducted research on food preservation. She was concerned about spoiled milk and other dairy products, such ice cream, that often made children ill. This particular work resulted in new procedures for milk inspection. By scientifically demonstrating the existence of harmful bacteria in ice cream to dealers, Pennington convinced them to adopt sanitary methods of handling food.

Pennington was the first to recognize the important relationship between humidity and temperature in storing eggs. By solving the technical problems of humidity control, Pennington was able to prevent the drying or molding of food products at freezing temperatures. Before World War I she invented a new and superior method of slaughtering and storing chickens. She also did considerable work with the frozen egg industry in improving the transportation of eggs. Based on her long study of this problem, 29°F–31°F became the accepted temperature range for shipping eggs. To conserve eggs, she also designed a packing case to reduce breakage.

During World War I, Pennington worked for the Perishable Products Division, U.S. Food Administration. One of the Division's tasks was to increase the amount of unspoiled food exported to Europe. In this venture she was given responsibility for uniting the efforts of private industries and government agencies concerned with the preservation of food during transportation.

Pennington's interest in frozen foods and their transportation led her to improve the design and construction of household and commercial refrigerators, refrigerated warehouses, and cooling rooms. Pennington also devised standards for railroad refrigerator cars. Because the transportation of perishable foods involves insulation and technical applications, she attempted to resolve this problem as well. With A. B. Davis in 1927 and again in 1928, she applied for two patents for strawboard insulating materials.

Pennington was a prolific author, with numerous articles and reports to her credit. She co-authored *Eggs* in 1933, published and co-authored *Food and Food Products*, and wrote the handbook entitled *The Care of Perishable Food Aboard Ship*.

Pennington received numerous honors and held memberships in many professional organizations. In 1919, President Hoover awarded her the Notable Service Medal for her innovations in refrigeration. She received the Garvan Medal in 1940 from the American Chemical Society. In 1947 she was the first woman to be elected to the Hall of Fame of the Poultry Historical Society. In the same year she became a fellow in the American Association for the Advancement of Science and was elected a fellow of the American Society of Refrigerating Engineers. She was a member of the American Chemical Society, the Society of Biological Chemists, the Society for Bacteriology, the Institute of Food Technologists, the Poultry Science Association, and the American Institute of Refrigeration. She also held memberships in Sigma Xi, Kappa Kappa Gamma, and Iota Sigma Pi. Pennington was also a U.S. delegate to several international congresses on refrigeration held in Paris, Vienna, Chicago, Washington, and Copenhagen.

Mary Pennington became a successful woman scientist in an era when career opportunities were very limited for her and her female colleagues. She was a chemist, a bacteriologist specializing in food science, and in essence an engineer in the field of refrigeration. Her theoretical knowledge, ability to apply this knowledge, and technical skills helped her achieve numerous successes. Her excellent public relations skills made her effective in working with the food industry and various government agencies.

Pennington lived a quiet life in New York City. Her home, overlooking the Hudson River, was furnished with Early American furniture and had an all-electric kitchen, which was rare for that time period. Since her lifetime work was concerned with food preservation and frozen food, she bought much of her own food frozen or canned. Her Persian cat and the flowers she grew on the terrace kept her company.

Although she was nicknamed the "ice woman," Pennington was

known for her warm, friendly nature, her hospitality, and her loyalty to family. Despite her scientific work, Pennington remained interested in sewing and knitting, and she had the reputation of being a gracious hostess and a good entertainer.[4] On December 27, 1952, she died of a heart attack in St. Luke's Hospital. She was buried in Philadelphia.

Notes

1. A. Pierce, "Mary Engle Pennington: An Appreciation," *Chemical and Engineering News* 18 (November 10, 1940): 941.
2. A. C. Goff, *Women Can Be Engineers* (Youngstown, Ohio: n.p.), p. 184.
3. Mary R. Creese and Thomas M. Creese, "Mary Engle Pennington (1872–1952)," in *Women in Chemistry and Physics: A Biobibliographic Sourcebook*, eds. Louise S. Grinstein, Rose K. Rose, and Miriam H. Rafailovich (Westport, Conn: Greenwood Press, 1993), p. 462.
4. Pierce, "Mary Engle Pennington," p. 942.

Bibliography

Creese, Mary R., and Thomas M. Creese. "Mary Engle Pennington (1872–1952)," in *Women in Chemistry and Physics: A Biobibliographic Sourcebook*. Edited by Louise S. Grinstein, Rose K. Rose, and Miriam H. Rafailovich. Westport, Conn.: Greenwood Press, 1993.

"Dr. Mary Engle Pennington." *IceRef* (February 1953): 58.

Heggie, B. "Ice Woman." *New Yorker* 17 (September 6, 1940): 23–30.

"Mary Pennington, Engineer, 80, Dead." *New York Times* 102 (December 28, 1952): 48.

Miles, Whyndham D. *American Chemists and Chemical Engineers*. Washington, D.C.: American Chemical Society, 1976–1994.

Ogilvie, Marilyn Bailey. *Women in Science; Antiquity through the Nineteenth Century: A Biographical Dictionary with Annotated Bibliography*. Cambridge, Mass: MIT Press, 1986.

Pennington, Mary E. *The Care of Perishable Food aboard Ship*. New York: Ehlenberger, 1942.

Pennington, Mary E., and D.D. Tressler. "Food Preservation by Temperature Control," in *The Chemistry and Technology of Food and Food Products*, Vol. 3. Edited by M.B. Jacobs, pp. 1822–1857. New York: Interscience Publishers, 1951.

Pennington, Mary E., et al. *Eggs*, 2 vols. Edited by Paul Mandeville. Chicago: Progress Publications, 1933.

Pierce, A. "Mary Engle Pennington: An Appreciation." *Chemical and Engineering News* 18 (November 10, 1940): 941–942.

Siegel, Patricia Joan, and Kay Thomas Finley. *Women in the Scientific Search*. Metuchen, N.J., and London: Scarecrow Press, 1985.

Vare, Ethlie Ann, and Greg Ptacek. *Mothers of Invention; From the Bra to the Bomb: Forgotten Women and Their Unforgettable Ideas*. New York: Quill, 1987.

Wiser, Vivian. "Pennington, Mary Engle," in *Notable American Women: The Mod-*

ern Period. Edited by Barbara Sicherman and Carol Hurd Green, with Ilene Kantrov and Harriette Wolker. Cambridge, Mass.: Belknap Press, 1980.

Yost, Edna. *American Women of Science.* Philadelphia: Frederick A. Stokes, 1943.

KATALIN HARKÁNYI

MARGUERITE PEREY
(1909–1975)
Chemist

Birth	November 19, 1909
1929	Diplôme d'État de Chimiste, École d'Enseignement Technique Féminine; joined Marie Curie's staff at the Institut du Radium
1939	Discovered the element francium (87)
1946	Completed doctoral dissertation defense on March 2; appointed Maître de Recherches at the Institut du Radium
1949	Appointed to new chair of nuclear chemistry, Université de Strasbourg
1958	Director, Nuclear Research Center, Université de Strasbourg, which she founded (Laboratoire de Chimie Nucléaire of the Centre de Recherches Nucléaires at Strasbourg-Cronenbourg)
1960	Grand Prix de la Ville de Paris; Officier of the Légion d'Honneur
1962	Became first woman member of the Académie des Sciences on March 12
1964	Lavoisier Prize, Académie des Sciences; Silver Medal, Société Chimique de France
1974	Commandeur of the Ordre National du Mérite
Death	May 13, 1975

Marguerite Catherine Perey, the first woman to enter the French Academy of Science in three hundred years, was born in Villenoble, France, in 1909. She had three brothers and a sister. Their father, Emile, owned a flour mill; their mother, Anne Jeanne, gave piano lessons. They lived a solidly middle-class, Protestant life until a stock market crash and Em-

Marguerite Perey. Photo courtesy of Professor Jean-Pierre Adloff, Université Louis Pasteur, Strasbourg.

ile's death just before World War I. From that time, the family lived practically hand-to-mouth and higher education was out of the question for the five children.

Marguerite trained at the École d'Enseignement Technique Féminine, a woman's school for technicians, and earned a certificate in chemistry in 1929. Like many women who entered scientific fields during the first half of the twentieth century, Perey probably benefitted more from attending a somewhat inferior school that was open only to women than she would have from attending a "better" school where her male colleagues and instructors would not have taken her seriously. In any case, she emerged from technical school a bright, competent, and confident chemist.

Her first job after completing school was with the Institut du Radium in Paris, directed by the Nobel Prize winner **Marie Sklodowska Curie**. Perey began her employment rather badly. Entering the Institute for the first time, she met Curie in the waiting room and mistook her for the laboratory secretary. Both women were embarrassed but went on to form a close bond in later years. Curie rewarded Perey for her intelligence and willingness to learn by personally supervising her work and education and by making Perey her assistant. Only Curie's own children, who also worked in the laboratory, received more attention and praise from her.

Under the mentorship of Curie, Perey continued to learn and grow as a chemist. She contributed a generous portion of her small salary to help her family, but her closest relationship was to Curie. When Curie died in 1934, Perey felt the loss greatly, both personally and professionally. She never married but dedicated herself to her work, her colleagues, and her students.

Since the late nineteenth century, scientists had tried to isolate an element they called "eka-cesium," an alkali metal that would be the most electropositive of the elements. They believed that its atomic number (the number of its protons) would be 87. Several scientists had claimed to discover the new element, but their claims were later shown to be fraudulent. In 1939, at the age of 29 and without even the equivalent of the American bachelor's degree, Perey succeeded where many more experienced chemists had failed. She isolated the element while she was attempting to purify another element, actinium. As actinium decays it gives off beta-rays and produces other radioelements. Perey was thorough and quick enough to notice that one of these short-lived elements was a new one, with a half-life of approximately 21 minutes. Others had observed actinium's decay over the previous forty years but had not detected the presence of the long-sought element. It turned out that the new element, which Perey called actinium K, did have the expected atomic number 87. This was the last naturally occurring element ever found; all new elements found later were created by bombarding known elements with atoms. Scientists believe that no other elements will ever be found in nature.

Perey's discovery was announced to the world at a session of the Académie des Sciences. Even her colleague, Jean Perrin, who presented the information to the Academy and who had himself presented one of the fraudulent claims, did not fully believe it. It must have been a shock to the all-male Academy to learn that a young, relatively inexperienced chemist—and a woman, at that—had indeed made this important discovery.

Following her discovery of the new element, Perey turned her attentions to examining its properties. As she continued her research at the Institut du Radium, she was encouraged by her colleagues to begin formal studies in chemistry at the Sorbonne. She earned a diploma there, and then went on to earn a doctorate. Her doctoral thesis, defended in 1946, was entitled "L'élément 87: Actinium K." In it she proposed the name "francium" for the new element, in honor of her country. Perey's mother attended the defense, expecting her daughter's work to be rejected. She never believed in Perey's brilliance or expected her to be successful.

But Perey continued to demonstrate to her peers her extreme competence. In 1949 she was appointed to a new position teaching nuclear chemistry at the University of Strasbourg in France, where she developed a program in radiochemistry and nuclear chemistry. She also opened a small laboratory of her own, which became an important center for nuclear chemistry research, the Laboratoire de Chimie Nucléaire of the Centre de Recherches Nucléaires at Strasbourg-Cronenbourg. The results of her work as well as the work of her students and collaborators were widely published in national and international journals.

With her colleague Jean-Pierre Adloff, Perey used chromatography to isolate a nearly pure sample of francium, but it took another scientist, Earl K. Hyde, to finally succeed in purifying it. (Adloff later became an unofficial biographer of Perey, sharing his knowledge and personal reminiscences in several journal articles after her death.) Perey also studied the relationships between francium and cancerous organs to see whether francium could have use in treating certain types of cancer, and she patented methods for obtaining and studying francium.

Perey received many honors and awards during her career, including the Grand Prix de la Ville de Paris in 1960 and the Silver Medal of the Société Chimique de France in 1964. She was made an *officier* of the Legion of Honor in 1960 and a *commandeur* of the National Order of Merit in 1974. Perhaps most significant, she was elected a corresponding member of the Académie des Sciences in 1962. The Academy had been founded in 1662 under King Louis XIV and had expressly forbidden women members ever since. Not even Perey's mentor, Marie Curie, had been admitted to this official academy, which had systematically excluded even Nobel Prize–winning women for three hundred years.

In 1960, Perey noticed a burn mark on her left hand. She feared it was the first sign of radiation cancer, which had killed Marie Curie, her daughter **Irène Joliot-Curie**, and several others as a result of their years of working with radioactive substances. Her family did not take the threat seriously and tried to persuade her that it was merely an acid burn.[1] But it was cancer, and there was no good treatment for it. Perey was forced to spend more and more of her time in the hospital and had to give up many of her responsibilities. She resigned from the University of Strasbourg and moved to Nice. She maintained contact with her own laboratory for as long as she could, but eventually her periods of illness were too long and too severe to continue the contact.

In 1967, Perey traveled to Warsaw to participate in a celebration of the one hundredth anniversary of Marie Curie's birth. She was well enough to join her friends in a thirtieth anniversary observance of the discovery of francium in 1969, but the years of illness were taking their toll. She spent the last two years of her life in hospitals and clinics; according to her collaborator J. P. Adloff, "she faced with admirable courage and stoicism the exorable evolution of the illness."[2] She died in the hospital in 1975 at the age of 65.

Notes

1. George B. Kauffman and Jean-Pierre Adloff, "Marguerite Catherine Perey," in *Women in Chemistry and Physics*, eds. Louise S. Grinstein et al. (Westport, Conn.: Greenwood Press, 1993), p. 471.

2. J.P. Adloff, "Marguerite Perey, 1909–1975," *Radiochemical and Radioanalytical Letters* 23 (1975): 193.

Bibliography

Adloff, J.P. "Marguerite Perey, 1909–1975." *Radiochemical and Radioanalytical Letters* 23, no. 4 (1975): 189–193.

Davis, H.M. *The Chemical Elements*. Revised by G.T. Seaborg, pp. 26–28. Washington, D.C.: Science Service, 1959. [English translation of sections of "Sur un élément 87, dérivé de l'Actinium." *CRHSAS* 208 (1939): 97–99.]

Kauffman, George B., and Jean-Pierre Adloff. "Marguerite Catherine Perey (1909–1975)," in *Women in Chemistry and Physics*. Edited by Louise S. Grinstein et al., pp. 470–475. Westport, Conn.: Greenwood Press, 1993.

———. "Marguerite Perey and the Discovery of Francium." *Education in Chemistry* 26 (September 1989): 135–137.

Perey, Marguerite. "L'Élément 87: Actinium K." Thèse Doctoratès sciences physiques, Université de Paris, 1946.

Weeks, Mary Elvira. "Francium," in *Discovery of the Elements*. Revised by H.M. Leicester, pp. 838–839. Easton, Pa.: Journal of Chemical Education, 1968.

CYNTHIA A. BILY

GERTRUDE PERLMANN
(1912–1974)
Biochemist

Birth	April 20, 1912
1936	D.Sc., chemistry and physics, German University of Prague

Gertrude Perlmann. Photo courtesy of the Rockefeller Archive Center.

1937–39	Carlsberg Foundation and Carlsberg Laboratory, Copenhagen
1939–41	Harvard Medical School
1941–45	Massachusetts General Hospital
1945	U.S. citizenship
1947–51	Assistant Visiting Investigator, Rockefeller Institute
1951–57	Associate Visiting Investigator, Rockefeller Institute
1957–58	Assistant Professor of Biochemistry, Rockefeller Institute
1958–73	Associate Professor of Biochemistry, Rockefeller Institute
1965	Garvan Medal, American Chemical Society
1973–74	Professor of Biochemistry, Rockefeller Institute
Death	September 9, 1974

Gertrude Perlmann's research contributed to the identification of proteins in body fluids, which may indicate pathological conditions. Understanding proteins' structure and function helps researchers to know how biological systems work and to detect sources of problems in abnormal situations.

Perlmann was born in Liberec (Reichenberg), Czechoslovakia, in 1912. At age 25 she fled her homeland for Denmark after the German invasion. Perlmann received her doctoral degree in chemistry and physics from the German University of Prague in 1936. In 1939 she left Denmark before World War II began and went to Harvard Medical School and Massachusetts General Hospital to join the physical chemistry staff. She moved in 1945 to the Rockefeller Institute in New York City, where she did the most notable work of her career in biochemistry.

Perlmann's area of scientific study was protein chemistry. She measured the structure of proteins and changes that take place as a result of particular chemical interactions. Proteins are extremely important because they are the chemical building blocks of biological systems, which serve structural and functional roles. For example, proteins are the basis of physical components of soft tissues (like muscle) and of important functional molecules (like antibodies) that aid in disease resistance. Proteins are formed by amino acids. Two hundred amino acids have been discovered; they can appear in numbers from 50 to 1,000, arranged in specific order to form a particular protein.

Special proteins have structural and functional roles in cells; among them are hormones, enzymes, and antibodies. The proteins called enzymes catalyze chemical reactions that take place in processes such as digestion and formation of molecules such as RNA and DNA. Proteins that serve functional roles are said to be biologically active. This aspect enables scientists to use various analytical techniques to study how the proteins change, either in ways that make the proteins themselves readily measurable or in a manner that facilitates processes in cells, which produce chemical products that can be measured.

Perlmann was among the first to recognize the crucial role that phosphate ions play in the structure of some proteins that form a class referred to as phosphoproteins. Many proteins have phosphate groups attached to their chemical structure, and the removal of such groups can make a dramatic difference in molecular activity. Perlmann began working with egg albumin because it contains phosphate in amounts corresponding to one or two groups per molecule. She was able to detect the presence of phosphate and its removal (known as dephosphorylation) by using an analytical technique called electrophoresis.

Electrophoresis allows researchers to detect specific characteristics of protein molecules. The system employs an electric current that attracts components in a solution causing proteins in a mixture to migrate at different rates according to their structure and electrical charge. This results in separation that enables the biochemist to isolate a relatively pure sample of an individual protein in the mixture. Scientists use specialized marker substances that attach to molecules and make it possible to follow their movement in a solution and to determine which ones move faster than others. The result is called an electrophoretic pattern. These patterns have been determined for many biological substances and can serve as a way to identify proteins.

Working with egg albumin, which was known to be a phosphorus-containing protein, or phosphoprotein (a class that researchers found interesting but where little work had been done), Perlmann monitored patterns that were evident through electrophoretic analysis. She discovered that some enzymes called phosphatases could remove the phosphate from the ovalbumin protein. She reported that three native forms of ovalbumin exist with one, two, or no phosphate groups.

Making use of similar studies to monitor phosphate attachment and removal, Perlmann did her most significant work on pepsin and its inactive precursor protein, pepsinogen. Pepsin plays an important role in the human digestive process. It is secreted in the form of pepsinogen by cells in the stomach known as chief cells. Digestion breaks down food proteins, liberating the amino acids that form them. The process begins in the stomach where there exists an acidic environment, a condition required to initiate pepsinogen's chemical transformation into pepsin. Digestion of protein into amino acids occurs in the duodenum, a portion of the small intestine. Once the proteins are broken down, their component amino acids are absorbed and distributed among tissues in the body to form amino acid source pools available for protein synthesis as needed by various organs.

Perlmann performed amino acid analysis of pepsin and determined the sequence of amino acids that forms the segment of pepsinogen that comes off the molecule when it changes into pepsin. Optical studies enabled her to determine the three-dimensional structures of pepsin and pepsinogen. She measured the effects of changing chemical conditions and temperature to demonstrate the distinct means by which the proteins maintain their structure.

Perlmann won the American Chemical Society's Garvan Medal in 1965, which honors distinguished service to chemistry by women. She died in New York City in 1974.

Bibliography

"Garvan Medal." *Chemical and Engineering News* 43, no. 15 (April 12, 1965): 94.
McGraw-Hill Modern Scientists and Engineers 2 (1980): 409–410.
Perlmann, Gertrude E. "Correlation between Optical Rotation, Fluorescence, and

Biological Activity of Pepsinogen." *Biopolymers*, Symp. No. 1 (1964): 383–387.

——. "Enzymic Dephosphorylation of Pepsin and Pepsinogen." *Journal of General Physiology* 41 (1958): 441–450.

——. "Relation of Protein Conformation to Biological Activity." *Biochemical Journal* 89, no. 1 (1963): 45.

——. "The Specific Refractive Increment of Some Purified Proteins." *Journal of the American Chemical Society* 70 (1948): 2719–2724.

FLORA SHRODE

MARY LOCKE PETERMANN
(1908–1975)
Biochemist

Birth	February 26, 1908
1929	A.B., chemistry, Smith College
1939	Ph.D., physiological chemistry, University of Wisconsin
1939–42	Postdoctoral fellow, physical chemistry, University of Wisconsin
1944–45	Postdoctoral fellow, University of Wisconsin
1945–46	Research Chemist, Memorial Hospital, NY
1946–48	Finney-Howell Foundation Fellow, Sloan-Kettering Institute
1948–60	Associate, Sloan-Kettering Institute
1951–66	Associate Professor of Biochemistry, Sloan-Kettering Division, Graduate School of Medical Sciences, Cornell University
1960–63	Associate Member, Sloan-Kettering Institute
1963	Sloan Award in Cancer Research
1963–73	Member, Sloan-Kettering Institute
1966	Garvan Medal, American Chemical Society; D.Sc., Smith College; Distinguished Service Award, American Academy of Achievement
1966–73	Professor of Biochemistry, Sloan-Kettering Division, Graduate School of Medical Sciences, Cornell University

1973–75   Member Emerita, Sloan-Kettering Institute

Death   December 13, 1975

A pioneer in cellular chemistry, Mary Locke Petermann studied the serum proteins of patients with cancer but is best known for isolating and
characterizing animal ribosomes. Ribosomes are particles made of proteins and joined with nucleic acid that occur in the nuclei of cells. They
are a basic ingredient of genes and are vital to protein synthesis.

The daughter of first-generation Americans, Petermann was born in
1908 in Laurium, Michigan. Her father, Albert Edward Petermann, had
been born in Laurium in 1877. He won a scholarship from Cornell University and graduated in 1900 with a bachelor's degree in philosophy.
After graduation he moved to Calumet, Michigan, where he studied law.
In 1902 he married Anna Mae Grierson. Albert Petermann practiced law
in Calumet for many years. From 1940 until his death in 1944, he was
president and general manager of Calumet and Hecla Consolidated Copper Company.

Mary had two brothers, Albert Edward Jr. and Paul. Her older brother,
Albert, like his father, became a successful corporate lawyer. Paul Petermann suffered brain damage from a fall down a flight of stairs when he
was a toddler. He died of pneumonia in 1928.

Mary's interest in mathematics began in high school. When she told
her high school advisor she wanted to become a mathematician, he said:
"Why? The only thing you could ever do with mathematics would be
to teach at a woman's college."[1] She attended a preparatory school in
Massachusetts before entering Smith College as a chemistry major. Most
of her friends were majoring in humanities and refused to discuss science
with her. "This was lucky for me," Petermann said. "I had to read the
books they were interested in to follow their debates. I gained a more
liberal education than I would if I had closeted myself exclusively with
science students."[2] Petermann was elected to Phi Beta Kappa and graduated with high honors in chemistry in 1929.

From 1929 to 1930, Petermann worked as a chemical technician for the
Department of Physiology at the Yale University Medical School studying liver function in dogs. During the next four years she worked at the
Boston Psychopathic Hospital where patients with mental diseases were
undergoing carbon dioxide inhalation, a forerunner to shock treatment
that aimed at changing brain chemistry. Petermann began graduate work
at the University of Wisconsin in 1936. She received a Ph.D. in physiological chemistry in 1939 for research focusing on the function of the
adrenal gland. After completing her doctorate, Petermann remained at
the University of Wisconsin working in the Department of Physical

Chemistry as a research associate. She was the first woman chemist on the staff. At this time she worked on pioneering studies on the splitting of protein antibodies by enzymes. Her research provided the basis for Dr. Rodney Porter's work on the structure of immunoglobulins, for which he won the Nobel Prize in 1972. Porter cited her work in his Nobel address.

Petermann did seminal work in cellular chemistry. Her interest in protein chemistry developed over the six years she spent at the University of Wisconsin after completing her doctoral program. She investigated the physical chemistry of a number of proteins using the techniques of ultracentrifugation, the use of a very high-speed centrifuge, and electrophoresis, moving colloidal particles through fluid under the action of an electrical field. She began her study of submicroscopic ribonucleoprotein particles in the late 1940s. At the same time, George E. Palade of the Rockefeller Institute was looking at electron micrographs of liver sections. He observed small, dense, spherical particles in the cytoplasm. The particles initially were known by a number of names: Petermann particles, Palade's particles, microsomal particles, and ribonucleoprotein particles. At a Biophysical Society Meeting in 1958, the particles were named ribosomes. Palade was introduced at the meeting as the "father of the particles." He told the audience that if he was the father, then Mary Petermann was the "mother of the particles."[3] In 1974, George Palade, Albert Claude, and Christian De Duve shared the Nobel Prize in physiology and medicine for their discoveries concerning the structural and functional organization of the cell. Palade cited Petermann in his Nobel lecture.

Of her work on ribosomes, Petermann said:

> I am often reminded of a book I read as a child. It was about walled gardens and every time a door in a wall opened one saw another wall and another door. In seeking to learn the connections between a gene and its expression as a protein, molecular biology has opened many doors, but new and higher walls always appear and often no one is quite sure where to look for the next door. Nature defends her mysteries well but subcellular biochemistry has made remarkable progress in the last two decades and should continue to find and to open the significant doors.[4]

During World War II, Petermann worked as a research associate on contract with the Committee on Medical Research for the analysis of blood substitutes. In 1945, Dr. Cornelius D. Rhoades, director of Memorial Hospital and newly appointed head of the Sloan-Kettering Institute for Cancer Research in New York City, decided to explore the role of nucleoproteins in cancer. He began looking for a physical chemist specializing in proteins. He hired Petermann as a research chemist. Rhoades, who published with Petermann, referred to her as "that girl from Wisconsin."[5]

In 1948, Petermann was appointed associate at the Sloan-Kettering Institute. When she began working at Sloan-Kettering, she was assigned to work under George B. Brown, who needed someone to work on the metabolism of nucleic acids. She was told to look into RNA. Brown told her to "work in cytoplasm because he had the nucleus under control."[6] This is where her work on ribosomes began.

In 1960, Petermann was promoted to associate member at Sloan-Kettering. Three years later she became a full member, the first woman to attain that position. She concurrently held the title of associate professor of biochemistry, Sloan-Kettering Division, Graduate School of Medical Sciences, Cornell University, from 1951 to 1966. In 1966 she was appointed professor, the first woman to hold the title of full professor at that institution. Petermann was the only woman member of Sloan-Kettering for a very long time. Like other women working in science during this era, Petermann did not find ready acceptance by male peers. The men she worked with made disparaging remarks about her, claiming she was "difficult to work with."[7]

At Sloan-Kettering, Petermann ran a small lab, working with two or three students. She made it a practice to hire women. She said she did not hire men because she believed they would "find it difficult working for a woman."[8] Under her tutelage a number of women received their Ph.D. degrees. Her first student was Dr. Mary G. Hamilton. Hamilton is currently associate professor of chemistry at Fordham University. Amalia Pavlovec, a Czech refugee, worked with Petermann for years. Pavlovec has remained at Sloan-Kettering as a member of the research staff.

During her years at Sloan-Kettering, Petermann was paid substantially less than her male peers. Petermann came from a wealthy family that owned copper mines in Michigan. The administration cited the fact that she was independently wealthy as an excuse to keep her salary low. The policy was that an associate member could not be paid more than a member, so her low salary held other people back.

Upon retirement in 1973 Petermann, as member emerita at Sloan-Kettering Institute, organized a chapter of the American Women in Science organization (AWIS) at the Institute. She then went to work on correcting the inequities in women's salaries. She was elected president pro tem and later president of Memorial Sloan-Kettering Cancer Center Association for Professional Woman (MSKCCAPW) in 1974.

During her 46–year career in science research, Mary Locke Petermann garnered several prestigious awards. In 1963 she received the Alfred P.

Sloan Award for her "basic and distinguished contribution to knowledge of the relevance of proteins and the nucleoproteins to abnormal growth."[9] Recipients of the Sloan Award received $10,000 in addition to their salary, travel, and other expenses during a year at another university or research institute. Petermann used her award money to work in the laboratory of Arne Tiselius, a Nobel laureate of 1948, in Uppsala, Sweden. While in Europe she gave lectures in Sweden, England, the Netherlands, Israel, and France.

The year 1966 was a stellar one for Petermann. She received the American Chemical Society's Garvan Medal "for her outstanding contributions to the isolations from mammalian tissues and for precise physical chemical characterization of those nucleoprotein particles, the ribosomes, which play a crucial role in the sites of protein synthesis."[10] (The Francis P. Garvan award was established in 1936 to recognize distinguished service to chemistry by women chemists in the United States. The award included $5,000 and an inscribed gold medal.) That same year Petermann received an honorary D.Sc. degree from Smith College for her pioneering research in cellular chemistry. She also received the Distinguished Service Award from the American Academy of Achievement.

Petermann was the author of over eighty scientific publications and a classic book in scientific research entitled *The Physical and Chemical Properties of Ribosomes*. She was a member of several scientific societies including Sigma Xi, Sigma Delta Epsilon, the American Chemical Society, the American Society of Biological Chemists, the American Association for the Advancement of Science, the American Association for Cancer Research, the Biophysical Society, the Harvey Society, and the Enzyme Club. She was a fellow of the New York Academy of Science. Although Petermann never married, she stayed close to her brother Edward and his family. She joined the Fifth Avenue Presbyterian Church in New York in 1951 and is remembered as a faithful and active member.

Petermann died in 1975 of intestinal cancer at the home of her niece, Dorothy Jamison, in Philadelphia. Although she had been successfully treated for cervical cancer in 1955, she was misdiagnosed at Sloan-Kettering as suffering from a "nervous stomach." By the time she was correctly diagnosed and underwent surgery to remove the tumor, it was too late.

From her work with antibodies to her work identifying and characterizing ribosomes, Petermann treaded new water. As a research scientist, she continued to open doors.

Notes

1. B. H. Jennings, "Petermann's Particles" (unpublished manuscript, 1982), p. 1.

2. Ibid.

3. Mary L. Moeller, "Mary Locke Petermann (1908–75)," in *Women in Chemistry and Physics*, eds. Louise Grinstein, Rose Kay Rose, and Miriam Rafailovich (Westport, Conn.: Greenwood Press, 1993), p. 482.

4. Mary Locke Petermann, "How Does a Ribosome Translate Linear Genetic Information?" *Subcellular Biochemistry* 1 (1971): 73.

5. Mary G. Hamilton, telephone interview with K. C. Benedict, November 8, 1995.

6. Ibid.

7. Mary L. Moeller, telephone interview with K.C. Benedict, November 2, 1995.

8. Hamilton, interview.

9. "Mary Locke Petermann, Ph.D., Recipient of Alfred P. Sloan Award in Cancer Research," letter of citation, March 27, 1963, in Memorial Sloan-Kettering Cancer Center Archives, R. G. 225.3, Petermann Collection, boxes 2 and 3, Rockefeller Archives Center, Tarrytown, New York.

10. "1966 Recipients of ACS Awards," program, 1966, in American Chemical Society Archives, Petermann Collection, Washington, D.C.

Bibliography

Moeller, Mary L. "Mary Locke Petermann (1908–75)," in *Women in Chemistry and Physics*. Edited by Louise Grinstein, Rose Kay Rose, and Miriam Rafailovich. Westport, Conn.: Greenwood Press, 1993.

Petermann, Mary Locke. "How Does a Ribosome Translate Linear Genetic Information?" *Subcellular Biochemistry* 1 (1971): 73.

———. *The Physical and Chemical Properties of Ribosomes.* New York: Elsevier, 1964.

K. C. BENEDICT

LUCY WESTON PICKETT

(1904–)

Chemist

Birth	January 19, 1904
1925	B.A., Mt. Holyoke College
1927	M.A., Mt. Holyoke College
1927–28	Instructor, Goucher College
1929	First published article
1930	Ph.D., University of Illinois
1930–35	Instructor, Mt. Holyoke College
1932–33	Fellowship, Royal Institution, London
1935–40	Assistant Professor of Chemistry, Mt. Holyoke College
1938–39	Fellowship, Liège, Belgium
1940–45	Associate Professor of Chemistry, Mt. Holyoke College
1945–68	Professor of Chemistry, Mt. Holyoke College
1947–48	Visiting Professor, University of California
1954–62	Chair, Dept. of Chemistry, Mt. Holyoke College
1957	Garvan Medal, American Chemical Society
1962	Last published article
1968	Retired from Mt. Holyoke College

Lucy Weston Pickett, born in 1904 in Beverly, Massachusetts, was a pioneer in the use of spectroscopy for investigating the structure of molecules. Her mother, Lucy Weston, was a schoolteacher and the principal of an elementary school until her marriage in 1902. Her father, George Ernest Pickett, the descendant of several sea captains, was a seaman himself in his youth. Lucy's brother, Thomas Austin Pickett, was born in 1907. Lucy and Thomas both became chemists and led similar academic and professional lives while maintaining a close personal relationship.

Lucy Weston Pickett. Photo reprinted by permission of The
Mount Holyoke College Archives and Special Collections.

Lucy attended high school in Beverly and entered Mount Holyoke
College in South Hadley, Massachusetts, in 1921. She had planned to
major in Latin but changed her mind during her first semester of college.
Attracted to the sciences by her freshman course in chemistry, she explored other sciences in her sophomore year and finally decided on a
double major in chemistry and mathematics in her junior year.[1] She graduated from Mount Holyoke College *summa cum laude* in 1925.

At Mount Holyoke, Lucy was not only a dedicated student but also a
member of the debating team. Her friend and senior-year roommate, Margaret Chapin, wrote of her in 1924: "She's a budding absent-minded professor. Yesterday, being much thrilled over something new in organic, she
walked halfway from Shattuck [chemistry/physics building] to Mead
[dormitory], in full sunlight with an open umbrella."[2] She also found some
time for social events, as Margaret records: "Lucy went to the Llamie

[yearbook] Dance last night with a chem assistant from Amherst Aggie [University of Massachusetts] whom she never saw before who was imported by her brother. He seems to have quite bewitched her."[3]

Despite her youthful social interests, Pickett eventually determined that a career in chemistry was more important: "In the '20s and '30s, when I was in school, we were all earnest young people and we felt at that time we had to decide between marriage and a career. That was characteristic of the early times."[4]

After receiving her B.A., Pickett stayed on for two more years at Mount Holyoke and obtained an M.A. in chemistry in 1927. Her first position was as an instructor of chemistry at Goucher College in Maryland in 1927–1928. In 1928 she enrolled in the graduate program in chemistry at the University of Illinois at Urbana. With a major in analytical chemistry and minors in physical chemistry and physics, Pickett received her Ph.D. in May 1930. Her dissertation was on the effect of X-rays on chemical reactions and the X-ray structure of organic compounds.

While completing her doctoral program and considering various job offers, Pickett received a letter from the Mount Holyoke chemistry department inviting her back to join the faculty. "And so," Pickett decided, "that's what I wanted to do."[5] Except for several leaves of absence, Pickett remained at Mount Holyoke from 1930 until her retirement in 1968.

In 1932–1933, on leave from Mount Holyoke with a fellowship from the American Association of University Women, Pickett went to London to work with the famous X-ray crystallographer Sir William Bragg at the Royal Institution. Upon her return, although she would have liked to continue her work in X-ray crystallography, Pickett realized that the necessary equipment could not be obtained at Mount Holyoke. Instead, she joined Mount Holyoke's active team of researchers (including **Emma Perry Carr** and **Mary Sherrill**) who were investigating molecular structures through spectroscopy.

In 1939, on an Educational Foundation fellowship, Pickett went to the University of Liège in Belgium to work with spectroscopist Victor Henri and, in the same year, to Harvard to work with molecular spectroscopist George Kistiakowsky. In 1942, Pickett and Carr traveled to Chicago for a conference on spectroscopy organized by Robert S. Mulliken, where they presented a paper on the spectra of dienes (organic compounds with two double bonds). Mulliken later organized the Laboratory of Molecular Structure and Spectra at the University of Chicago. In 1947, Pickett was a visiting professor at the University of California at Berkeley in the fall and spent the subsequent January at the California Institute of Technology in the laboratories of Linus Pauling. Pickett spent the summer of 1952 in Chicago, working with Mulliken on theoretical interpretations of the spectrum of the benzenium ion.

Pickett was promoted from instructor to assistant professor at Mount Holyoke in 1934, to associate professor in 1940, and to full professor in 1945. She received the Camille and Henry Dreyfus endowed chair in chemistry in 1955 and became the first Mary Lyon professor of chemistry in 1958. She served as chair of the chemistry department from 1954 to 1962.

In 1957, Pickett was named the recipient of the prestigious Garvan Medal, awarded by the American Chemical Society, for her research in molecular spectroscopy. She also received honorary D.Sc. degrees from Ripon College in Wisconsin in 1958 and from Mount Holyoke in 1975.

Upon her retirement from Mount Holyoke, the Lucy Pickett fund was established by her colleagues and students to bring noted speakers to the department. The first of these was Robert S. Mulliken, 1966 recipient of the Nobel Prize in chemistry, with whom Pickett had collaborated on a paper in 1954. During the 1970s she requested that the Lucy W. Pickett speakers be women; and in a letter written in 1993 she expressed her delight in the list of women scientists who had spoken under the auspices of the fund named for her.[6]

Pickett retired to Bradenton, Florida, where she said, "I lead a rather quiet but pleasant life."[7] During her retirement years she traveled to Africa, South America, Russia, and Greece and also did volunteer work tutoring disadvantaged students.

Of her life as a chemist at Mount Holyoke, Pickett recalled, "I just felt that I was with a group of remarkable people. . . . I really felt a part of a dedicated and hard-working group that was having fun together."[8]

Notes

1. Author's interview with Lucy Weston Pickett, May 1990; tape in Mount Holyoke College Library/Archives.
2. Letter from Margaret Chapin to her mother, December 1, 1924, in Chapin files, Mount Holyoke College Library/Archives.
3. Letter from Margaret Chapin to her mother, April 26, 1925.
4. Author's interview with Pickett, May 1990.
5. Author's interview, May 1990.
6. Letter dated October 26, 1993, in Pickett files, Mount Holyoke College Library/Archives.
7. Letter dated October 26, 1993.
8. Author's interview, May 1990.

Bibliography

Fleck, George. "Lucy Weston Pickett," in *Women in Chemistry and Physics: A Bio-bibliographic Sourcebook.* Edited by L.S. Grinstein, R.K. Rose, and M. Rafailovich. Westport, Conn., Greenwood Press, 1993.

Jennings, Bojan. "The Professional Life of Emma Perry Carr." *Journal of Chemical*

Education 63 (1986): 923–927.

Pickett, Lucy. "Developments in the Teaching of Analytical Chemistry." *Journal of Chemical Education* 20 (1943): 102.

———. "Some New Experiments on the Chemical Effects of X-Rays and the Energy Relations Involved." *JACS* 52 (1930): 465–479.

———. "An X-Ray Study of Substituted Biphenyls." *JACS* 58 (1936): 2299–2303.

Pickett, Lucy, et al. "Molecular Complexes of Tetracyanoethylene with Tetrahydrofuran, Tetrahydropyran and P-Dioxane." *Journal of Physical Chemistry* 66 (1962): 1754–1755.

CAROLE B. SHMURAK

HELEN W. DODSON PRINCE

(1905–   )

Astronomer

Birth	December 31, 1905
1927	B.A., mathematics, Goucher College
1927–31	Statistician, Maryland Department of Education
1932	M.A., University of Michigan
1932–33	Assistant, Dept. of Astronomy, University of Michigan
1933–45	Assistant Professor of Astronomy, Wellesley College
1934	Ph.D., astronomy, University of Michigan
1934–35	Summer Observer, Maria Mitchell Observatory, Nantucket Island
1938–39	Summer Research Assistant, Observatoire de Paris, Section d'Astrophysique at Meudon
1943–45	Staff, Radiation Laboratory, Massachusetts Institute of Technology
1945–50	Assistant Professor of Mathematics and Astronomy, Goucher College
1947–62	Faculty, McMath-Hulbert Observatory, University of Michigan
1952	Sc.D., Goucher College

Helen W. Dodson Prince. Photograph by Acme Photo Service. Courtesy of Wellesley College Archives.

1955	Annie J. Cannon Prize
1956	Married Edmond Lafayette Prince
1957–76	Professor of Astronomy, University of Michigan
1962–76	Associate Director, McMath-Hulbert Observatory
1974	Faculty Distinguished Achievement Award, University of Michigan

Helen W. Dodson Prince made the observation and analysis of solar activity, with an emphasis on solar flares, her life-long vocation. She knew as an undergraduate at Goucher College in her native Baltimore, Maryland, that she had a talent for mathematics and physics. She made the conscious choice to go into astronomy, a field that had always been dominated by men, and she never regretted it.

Prince entered the graduate school of the University of Michigan in 1931. She earned her master's degree in 1932 and her doctorate in astronomy in 1934. Her doctoral thesis, "A Study of the Spectrum of 25

Orionis," gives some indication of exactly how detailed her interests would become.

From 1933 through 1945, Prince served as an assistant professor of astronomy at Wellesley College. A significant portion of the women in graduate astronomy in the United States during this period took classes with her. Because her summers were free, Prince spent them at various observatories, continuing her own research work. The summers of 1934 and 1935 found her at the Maria Mitchell Observatory on Nantucket Island. She continued her studies of 25 Orionis, eventually publishing her spectral observations in the *Astrophysical Journal*. The summers of 1938 and 1939 found her at the Observatoire de Paris at Meudon, France, a world center for the observation and analysis of solar activity. Here her interest in solar activity became dominant, leading to the eventual total lifetime publication of over 100 journal articles.

During the later years of World War II, from 1943 to 1945, Prince was a staff member of the Radiation Laboratory at the Massachusetts Institute of Technology. There she contributed to the further mathematical development of radar. After the war she returned to her native Baltimore and alma mater Goucher College, where she was a full-time professor of astronomy from 1945 to 1947, and a part-time professor from 1947 to 1950. In order to continue the studies of the sun that she had started at Meudon, Prince had accepted a shared-time appointment at McMath-Hulbert Observatory at the University of Michigan, where motion pictures of the sun could be taken in the light of a single spectrum line. She continued on a shared-time basis at McMath-Hulbert until 1950, when she was promoted to associate professor of astronomy at the University of Michigan and came to McMath-Hulbert full-time.

McMath-Hulbert Observatory was founded on the shores of Lake Angelus outside of Pontiac, Michigan, in 1935 by Robert R. McMath, a former University of Michigan engineering student who became an auto industry executive, and Judge Hulbert, a local judge with an interest in astronomy and movies. It began as an amateur effort with little ready money, competing with the established observatories like Mount Wilson where there were many big-name researchers, plenty of cash, and the cachet of the Carnegie Institution. The Observatory was eventually drawn into the University of Michigan orbit and became a valuable asset to the astronomy department. Robert McMath was an amazing organizer and an exacting boss. Prince wound up working as one of the few female solar astronomers in the country at a cutting-edge institution from which nobody had originally expected anything of value. She was to thrive there for 29 years.

Although Prince never abandoned teaching students, it was her exhaustive research that made McMath-Hulbert Observatory an internationally recognized source of solar-terrestrial interaction observations during the 1950s and 1960s. In 1951, Prince began a collaboration with Arthur Covington, Canada's pioneering solar radio astronomer, that lasted into the 1970s. Their work was an effort to coordinate radio and optical studies of solar flares for better classification and explanation. In 1955, Prince was awarded the Annie J. Cannon Medal by the American Astronomical Society for her work in solar flare investigations.

Prince was appointed associate director of McMath-Hulbert in 1962, a post in which she served until her retirement in 1976. She continued her observations at the Observatory after her appointment as professor emerita until 1979.[1]

Note

1. Much of the information contained in this essay was gathered from Dr. Prince's resumé and her retirement memoir. The author is indebted for additional information to Rudi Lindner and Dr. Richard Teske of the University of Michigan; Dr. Richard Jarrell of York University in North York, Ontario; and Dorrit Hoffleit of Yale University.

Bibliography

Dodson, Helen W. "The Spectrum of 25 Orionis, 1933–1939." *Astrophysical Journal* 91 (1940): 126.

——. "The Sun—The Earth's Near and Disturbing Neighbor." *Michigan Alumnus Quarterly Review* 59 (1952): 62.

Dodson, Helen W., and E. Ruth Hedeman. "The Frequency and Position of Flares within a Spot Group." *Astrophysical Journal* 110 (1949): 242.

Dodson, Helen W., E.R. Hedeman, and R.R. McMath. "Photometry of Solar Flares." *Astrophysical Journal*, supp. 20 (1956): 241–270.

Prince, Helen W. Dodson, and E.R. Hedeman. "Solar Minimum and the International Years of the Quiet Sun." *Science* 143 (1964): 237.

Prince, Helen W. Dodson, E.R. Hedeman, et al. *Catalogue of Solar Particle Events, 1955–1969.* Dordrecht, Holland: Reidel Publishing, 1975.

Prince, Helen W. Dodson, and O.C. Mohler. "McMath-Hulbert Observatory of the University of Michigan." *Solar Physics* 5 (1968): 417–422.

Who's Who of American Women, 1966–67. Chicago: Marquis Who's Who, 1967.

REBECCA L. ROBERTS

KAMAL J. RANADIVE

(1917–   )

Cancer Researcher

Birth	November 8, 1917
1939	B.Sc., biology, Ferguson College, Pune, University of Bombay
1941	M.Sc., cytology, University of Bombay
1949	Ph.D., microbiology, University of Bombay
1949–50	Rockefeller Postdoctoral Fellow, Johns Hopkins University
1950–51	Rockefeller Postdoctoral Fellow, Columbia University
1952	Senior Research Officer, Indian Cancer Research Institute, Bombay
1952–70	Head, Biology Division, Indian Cancer Research Institute, Bombay
1958	Dr. Basantidevi Amirchand Award, Indian Council of Medical Research, for cancer research
1966	Award for Cancer Research, Medical Council of India; Watumull Award for work in leprosy-isolation of "ICRC" bacillus
1966–70	Acting Director, Indian Cancer Research Institute
1970–78	Head, Biology Division, Tata Memorial Hospital and Cancer Research Institute
1976	Sandoz Award and Gold Medal for work on environmental carcinogens specific to India
1978	Retired with emerita status, Indian Council of Medical Research
1982	"Padmabhushan" Award, Government of India
1991	Distinguished Woman Award, Benaras Hindu University; Tata Memorial Hospital Jubilee Award for outstanding work in cancer control

Kamal Ranadive, one of India's most distinguished women scientists, is the pioneer who introduced tissue culture technology into India after her postdoctoral training under Rockefeller Foundation fellowships in the United States during 1950–1951. She founded the first Indian cancer research center with Professor V. R. Khanolkar, an eminent Indian scientist

Kamal J. Ranadive. Photo courtesy of M. N. Bartakke.

in cancer research, and she developed a new discipline of growing animal cells in vitro. During 1952–1953, Ranadive's leadership led to the establishment of the first experimental biology laboratory and tissue culture laboratory in India. As a result of her leadership, the technology spread throughout India. Having been in the cancer research field for more than 30 years, she received "Padmabhushan," the highest award from the government of India, in recognition of her contributions in cancer and leprosy research.[1]

Ranadive was born in 1917. Having received her bachelor of science degree with honors, she went on to earn her master's degree in cytology from the University of Bombay. She then continued her professional training in cytology and applied research by pursuing her studies in cancer research. In 1943 she took her first job at Tata Memorial Hospital in Bombay, India. In 1949 she received her Ph.D. in microbiology from the University of Bombay. Her dissertation, "Experimental Studies in Breast Cancer," was supervised and approved by Dr. Georgina Bonser, University of Leeds, England.

From 1949 to 1951, Ranadive was in the United States under Rockefeller Foundation fellowships to do postdoctoral research on tissue culture at Johns Hopkins under George O. Guy and also at Columbia University Medical Center under Margaret Murray. These experiences helped Ranadive gain confidence in her cancer research. In 1952 she began her professional career as a cancer research scientist at the Indian Cancer Research Center, Bombay, as a senior research officer. In the same year she initiated the Experimental Biology Laboratory and Tissue Culture Laboratory in Bombay. From 1966 to 1970 she was acting director of the Indian Cancer Research Center. Ranadive is to be commended both for being selected for this position and for demonstrating her leadership in administering the Center; but it is perhaps equally notable that at the time, there were very few foreign-trained women scientists in cancer research in India. She broke the barriers of the time and inspired Indian women scientists to enter the research field, especially cancer research related to cancer of women and children.

In 1970 the Tata family, renowned for its philanthropy, donated funding for the Indian Cancer Research Institute to promote continuing research in cancer. The Tata Memorial Hospital and the Center merged with a new name—Tata Memorial Hospital and Cancer Research Institute—and Ranadive was appointed head of the Biology Division. She worked in this capacity until her retirement in 1978, when she was awarded the status of emerita scientist.

Ranadive has won many awards and honors for her contributions to cancer and leprosy research, including the most prestigious award of "Padmabhushan" from the government of India. Her other awards for cancer research include the Dr. Basantidevi Amirchand Award and the Medical Council of India Award. She won the Watumull Award for her work in leprosy research, and the Sandoz Award and Gold Medal for her work on environmental carcinogens specific to India. In 1991, Ranadive received both the Distinguished Woman Award from Benaras Hindu University and the Tata Memorial Hospital Jubilee Award for outstanding work in cancer control. She has also received several international fellowships, including Rockefeller Foundation fellowships and World Health Organization fellowships to attend meetings abroad.

Ranadive has produced over 200 publications in cancer and leprosy research. She has advised and guided students during their master's and doctoral degree programs as well as postdoctoral training programs at the University of Bombay; and she introduced the discipline of applied biology (the application of biology to medical research) into the science curriculum there. As a member of the Steering Group of Science and Technology Committee of the Seventh and Eighth Five Year Plan (1987–1992), Government of India, she made a strong effort to focus the Indian government's attention on funding for cancer research and women's and children's health and medical issues.

Ranadive remains active in cancer research. She has undertaken some major projects and published the research results because of her special interest in the medical health of tribal populations in India. Her project on "Immunohematology of Tribal Blood" involved the study of infant deaths and the nutritional and biosocial aspects of tribal cultures in western India. The project was sponsored in part by the Indian Women Scientist Association, the Sir Tata Trust, the Government of India, and the Maharashtra State Government. Another project, "Analysis of Cytokines and Oncogene Expression in Tribal Blood Samples," involved collaborative research with Dr. Gabriel Ferandes, University of Texas Health Science Center, San Antonio.

As a career professional scientist, Ranadive has delivered papers at the International Cancer Research meetings held in Germany, the Netherlands, eastern Europe, the United States, China, Japan, and Russia. As a member of the International Scientific Committee, she has received invitations to chair sessions and to act in various capacities at meetings organized by international bodies including UNESCO-ICRO (International Cell Research Organization); European Group of Breast Cancer; International Tissue Culture Association; and DEPCA (Detection, Prevention and Control of Cancer)—American Group. Ranadive has also served on the editorial boards of several international journals, including *International Journal of Cancer, Year Book of Cancer, Indian Journal of Cancer, Indian Journal of Medical Research*, and *Indian Journal of Experimental Biology*.

As a pioneer woman of science, Ranadive took a leading role in establishing the Indian Women Scientist Association and served as its president in 1973. She takes an active part in the Association and continues to serve as a consultant. Dr. V. R. Khanolkar's early identification of and belief in her abilities, as well as the inspiration and support he continued to lend her, helped Ranadive to overcome social barriers and make contributions to the betterment of the lives of the people of India.

Note

1. Most of the information for this essay is based on Dr. Ranadive's resumé and conversation with her friend, Dr. M.N. Bartakke.

Bibliography

Directory of Indian Women Today. Edited by Ajit Cour and Arpana Cour. New Delhi: International Publications, 1976.

SAROJINI LOTLIKAR

SARAH RATNER

(1903–   )

Biochemist

Birth	June 9, 1903
1924	B.A., Cornell University
1927	M.A., Columbia University
1930–34	Assistant biochemist (then Associate biochemist), Columbia College of Physicians and Surgeons
1934–36	Geis Fellow, Columbia College of Physicians and Surgeons
1937	Ph.D., Columbia University
1937–39	Macy Resident Fellow, Columbia College of Physicians and Surgeons
1939–46	Instructor (then Assistant Professor), Columbia College of Physicians and Surgeons
1946–53	Assistant Professor of Pharmacology, New York University College of Medicine
1953–54	Associate Professor, New York University College of Medicine
1954–	Adjunct Associate Professor of Biochemistry, New York University College of Medicine
1954–57	Associate Member, Division of Nutrition and Physiology, Public Health Research Institute of New York
1956	Schoenheimer Lecturer
1957–	Member, Dept. of Biochemistry, Public Health Research Institute of New York
1959	Carl Neuberg Medal, American Society of European Chemists and Pharmacists
1961	Garvan Medal, American Chemical Society
1975	L. and B. Freedman Foundation Award, New York Academy of Sciences
1978–79	Fogarty Scholar-in-Residence, National Institutes of Health

Sarah Ratner was born, raised, and worked throughout her professional career in the New York City area. Educated in an era when few women were found in the science labs, her work from the 1930s through the

1950s was instrumental in opening the vistas of scientific knowledge to our understanding of human physiology and biochemistry. To her and her colleagues' credit, Ratner's scientific expertise and scholarly methods were well respected in the scientific communities of chemistry, physiology, enzymology, and the newly established biochemistry. Her lifework has mainly centered around the intricacies of nitrogen metabolism. Her work has impacted the practice and research of medicine and the biological and physical sciences.

Sarah Ratner was born in 1903 to Russian Jewish immigrant parents, Aaron and Hannah Ratner. Sarah and her twin brother were the youngest of five children. Upon coming to the United States, her father had built a manufacturing business. Aaron Ratner was a self-educated man who loved books, inventions, inventing, and learning in general. Sarah grew up surrounded by classic literature and her father's active pursuit of knowledge. Unusual for his time and culture, but consistent with his progressive nature, Aaron Ratner educated his only daughter as well as his four sons. Indeed, his thirst for knowledge seems to have particularly borne fruit in Sarah, as she was the only child in the family to pursue academics.

Sarah says that in high school, "my choice of courses gave preference to sciences and mathematics."[1] She was guided by older friends to continue her education. In the 1920s, not many colleges were available for women to study the sciences. Nonetheless, she was accepted and awarded a scholarship to study chemistry at Cornell.

Ratner states that she suffered from shyness during her college years. The emphasis in the chemistry department at Cornell was on industrial chemistry, but she was not personally interested in pursuing it as a career. After finishing her bachelor's degree, she returned to New York City to gain practical experience in both analytical chemistry and pediatric research. She says that the latter is what attracted her, and she began to explore the literature and research of physiological chemistry. Hers was a timely choice, as this field of study continued to mature through the many discoveries of the day. Ratner worked under Dr. Hans T. Clarke at the Columbia College of Physicians and Surgeons. She was involved there in one of the centers of scientific research that opened up the new field of biochemistry.[2]

Ratner emphasizes the role of a mentor when she describes how Clarke impressed her as a graduate assistant. She speaks of his extensive scientific knowledge, his respect for and trust of his students, and his interpersonal insights. She is regarded in much the same way by her own students. Several of them speak of how Ratner has positively affected their lives with her advice, moral standards, and scholarly wisdom.[3] She illustrates, in her life as a student and later as a professor, how sincere guidance and instruction from an interested educator/researcher may open personal, vocational, and educational opportunities for students.

As Ratner worked and studied in Clarke's laboratory, she became involved in pioneering work identifying the role of prostaglandins (a uterine hormone) in uterine contractions. This proved to be a significant discovery for the medical profession.[4]

In 1937, when she finished her doctorate, Ratner continued to experience some of the hindrances of being a woman in the sciences. Even though her peers were being placed in research positions, opportunities for her were limited to nonresearch, teaching positions at women's colleges. She persisted, however, and found an out-of-town research position, which she had to relinquish a few months later because of illness and death in her family.

Fortunately, she was able to return to Columbia's College of Physicians and Surgeons to work with Dr. Rudolf Schoenheimer. There she was thrust into the exciting study of amino acid metabolism with isotope tracers. It was 1937, and this lab was breaking new ground in biochemical research. Combined with the organic chemistry roots of her studies under Dr. Clarke, the methods she now learned to trace the metabolic pathways of nitrogen isotopes were to be cornerstones of her next 50 years of research: nitrogen metabolism in the urea cycle.

From 1942 to 1947, Ratner worked with Dr. David Greene in the research of amino acid oxidases, one particular area of interest in the larger field of enzymology. Her research with Greene brought her new understanding of the role of enzymes. This would later influence her own research in nitrogen metabolism.

Regarding the impact of her womanhood on her career, Ratner states:

Though I was much in favor of spending some postdoctoral years in association with gifted investigators, and I had been fortunate in this way, my hopes for an opportunity to explore new areas of nitrogen metabolism ran together with the awareness that appointments in medical schools were extremely rare for women. . . . In the 1940s, a faculty appointment represented one of the few ways of providing such opportunities.[5]

Her appointment came in 1946, to the Department of Pharmacology at the New York University Medical School. In addition to teaching pharmacology, she was able to pursue her research of the mechanism of urea synthesis. In what was to be her characteristically meticulous and complete manner, she pieced together the biochemical evidence of this urea cycle. She conducted experiments to test her hypotheses and was ultimately successful in proposing a new intermediate for urea synthesis—

the compound argininosuccinate. This was a tremendous discovery for scientific research, theory, and practice.

Ratner continued her research on new problems in nitrogen metabolism in 1954 at the Department of Biochemistry in the Public Health Research Institute of the City of New York. Her work helped elucidate the now famous Krebs citric cycle, which is a cornerstone of biochemical study. She greatly contributed to the understanding of human metabolic disorders of urea synthesis, which have far-reaching medical effects from brain to kidney disorders.

An important thread that runs throughout Sarah Ratner's professional endeavors as a scientific researcher is her intense involvement with her pursuit. She is depicted by others and describes herself as thriving on the challenge of scientific inquiry:

> Experimentation is for me a very human, personal, and intense activity. In experimental design, and in responding to the urge to strengthen proof by reducing the number of experimental variables, there are creative and aesthetic as well as intellectual aims and pleasures. All of this is what keeps investigators in their laboratories.[6]

Ratner's eloquent statement surely voices the inner desire and urgency that spurs on laboratory researchers in numerous disciplines.

Another important aspect of Ratner's professional research was her many contributions to the research literature. A meticulous writer and scientist, she would apparently thrash out the inconsistencies of opposing theories and use new findings to elucidate the details of the metabolic pathways to which she contributed. On at least two occasions, she wrote reviews of research for the journal *Advances in Enzymology*. Her love of intellectual stimulation is attested to in her own writings and in others' accounts of her endeavors.

Ratner's expertise in the research literature is a result of the high standards she held herself accountable to in both her writing and research methods. Her research was always presented in a lucid and precise manner. She also guided her students by admonishing them to be concise, accurate, and careful in their research publications.[7] In research, her attention to detail brought many previously unnoticed factors and compounds to the attention of researchers through both the literature and the laboratory.

Ratner was at once a brilliant scientist, lucid writer, lover of the arts, and a genuinely personable individual. She was trained in and contributed to a field of research dominated by men. Her numerous honors and awards attest to the quality and significance of her contribution during the dawning of modern biochemistry. She was awarded both the Neuberg (1959) and Garvan (1961) medals. She received honorary doctorates of science (D.Sc.) from three universities: the University of North Carolina, 1981; Northwestern University, 1982; and the State University of New York at Stony Brook, 1984. She is a member of the National Academy of Sciences; a fellow of the American Academy of Arts and Sciences, and of the Harvey Society, and of the New York Academy of Sciences. As of this writing, she remains a member of the Department of Biochemistry at the Public Health Research Institute of New York.

Notes

1. Sarah Ratner, "A Long View of Nitrogen Metabolism," *Annual Review of Biochemistry* 46 (1977): 4.
2. Ibid., p. 5.
3. C.T. Nuzum, "Consulting Dr. Ratner," in *An Era in New York Biochemistry: A Festschrift for Sarah Ratner*, ed. Maynard E. Pullman (New York: New York Academy of Sciences, 1983), pp. 129–131; Efraim Racker, "The Laughter of Sarah," in *An Era in New York Biochemistry*, pp. 183–185.
4. Ratner, "A Long View of Nitrogen Metabolism," p. 6.
5. Ibid., pp. 11–12.
6. Ibid., p. 17.
7. Nuzum, "Consulting Dr. Ratner," pp. 130–131.

Bibliography

An Era in New York Biochemistry: A Festschrift for Sarah Ratner. Edited by Maynard E. Pullman. New York: New York Academy of Sciences, 1983.

Ratner, Sarah. "A Long View of Nitrogen Metabolism." *Annual Review of Biochemistry* 14 (1977): 1–24.

———. "Urea Synthesis and Metabolism of Arginine and Citrulline." *Advances in Enzymology* 15 (1954): 319–387.

Ratner, Sarah, and Hans T. Clarke. "The Action of Formaldehyde upon Cysteine." *Journal of the American Chemical Society* 59 (1937): 200–206.

KATHLEEN PALOMBO KING

ELLEN SWALLOW RICHARDS

(1842–1911)

Chemist, Ecologist, Home Economist

| Birth | December 3, 1842 |
| 1870 | A.B., Vassar College (one of its first graduates) |

Ellen Swallow Richards. Photo reprinted by permission of the Sophia Smith Collection, Smith College.

1873	B.S., Massachusetts Institute of Technology (first woman graduate); M.A., Vassar College
1873–78	Teacher, Dept. of Chemistry, MIT
1875	Married Robert Hallowell Richards on June 4
1878–84	Teacher, MIT Women's Science Laboratory
1879	First woman member, American Institute of Mining and Metallurgical Engineers
1884–1911	First woman teacher, MIT Dept. of Sanitary Chemistry
1910	Honorary doctorate, Smith College
Death	March 30, 1911

Although Ellen Swallow Richards's contributions to science were numerous, perhaps her most significant contribution was the improvement of the standard of living by applying her knowledge of chemistry to sanitary conditions. She was also a pioneer in opening up the fields of scientific inquiry for other women, developing the home economics movement, and analyzing water and mineral samples. Because of her work in environmental and sanitary engineering, one of her biographers, Robert Clarke, deemed her the "woman who founded ecology."

An only child, Ellen Henrietta Swallow was born in Dunstable, Massachusetts, in 1842. She attended the village school, Westford Academy, from 1859 to 1863 after her parents moved the family from Dunstable. She also received instruction from her parents, Fanny Taylor and Peter Swallow, who were both schoolteachers. Upon leaving Westford Academy she taught school, worked in her father's store, and performed housekeeping chores. During this time she also attended a school in Worcester, Massachusetts.[1] Not content to cease her education, at age 25 Richards went to fledgling Vassar College.

Her admission to Vassar with a rating near the junior year followed a six-year struggle with health problems and hours of self-study. She graduated in 1870 with an A.B. degree and in 1873 with an M.A. Among her Vassar professors were **Maria Mitchell**, who was her astronomy professor, and Charles Farrand, her chemistry professor. After leaving Vassar, Ellen became a special student at the Massachusetts Institute of Technology during a period when MIT did not accept women students. MIT admitted Ellen for free to avoid the scandal of enrolling a woman. Her special enrollment was deemed the "Swallow experiment."[2] Eventually, she became the first woman to be enrolled in a U.S. scientific institution. Upon earning a B.S. in chemistry in 1873, she also became MIT's first woman graduate. Women were allowed to enroll officially when the MIT governing board agreed on May 11, 1876, to allow special students into chemistry without regard to sex. Though she stayed two additional years, MIT never granted her a doctorate. However, Smith College awarded her an honorary doctorate in science in 1910.

From 1873 to 1878, she taught in the MIT chemistry department without a title or salary as the first woman teacher. After a two-year engagement, she married Robert Hallowell Richards—then head of the MIT Department of Mining Engineering and her professor of mineralogy—on June 4, 1875. Robert H. Richards became one of his wife's most ardent supporters.

In 1878, MIT asked Ellen Swallow Richards to provide instruction in a women's laboratory she had developed and to serve as an instructor in chemistry and mineralogy. She had established the laboratory in 1876 with funds given to MIT from the Boston Women's Education Association. She went on to contribute to its maintenance, a practice she followed for several projects she considered important. The laboratory did not provide classes for women as a part of the regular curriculum, but it proved so successful that MIT abolished it in 1883 and allowed women to enroll in regular classes. From 1884 to 1911, Richards taught in the MIT Department of Sanitary Chemistry. After introducing a course on sanitary engineering, she remained for approximately 27 years teaching analysis of food, water, air, and sewage. As a teacher, she trained many of the men who became authorities in the field of sanitary engineering.[3]

Throughout her career, Richards made a conscious choice to put her schooling to practical use. She wrote more than 30 books and pamphlets in addition to contributing to journals, and she was a highly desired public lecturer. She set about developing proper food preparation and cooking techniques, analyzing water samples, and improving basic living conditions. She deemed her ideas "euthenics," or the science of the controlled environment for right living. As a result, many also named her the developer of sanitary engineering. Richards attended the Chicago World's Fair in 1893 to serve as the superintendent of the Rumford Kitchen with Mary Hinman Abel. Abel had earlier studied the German *Volksküchen*, or soup kitchens for the poor.[4] Together, she and Richards provided demonstrations of proper sanitary procedures and the cooking of nutritional yet low-cost meals at the Rumford Kitchen. Richards also helped establish the Boston School Lunch Program with the help of staff at the New England Kitchen. Located in Boston, the New England Kitchen served a purpose similar to that of the Rumford Kitchen.

In her own home, Richards followed and demonstrated proper sanitation and ventilation procedures. She completely remodeled her home to improve the heating and ventilation systems. She redesigned a water heater to make it more efficient, and she rerouted the plumbing so that water went behind the furnace and could be heated in the winter to save fuel.[5] Among the first to own a vacuum and gas stove, Richards is also given credit for discovering the process of cleaning wool with naphtha and thereby revolutionizing the dry cleaning industry.[6]

As a student of MIT professor William R. Nichols, Richards executed a complete survey of Massachusetts drinking water and sewage for the Massachusetts Board of Health in 1872, taking 40,000 samples in all. A state laboratory assumed responsibility for the testing in 1897. A singular contribution for posterity from this sampling was Richards's Normal Chlorine Map, which warned of early inland water pollution. She also contributed the first Water Purity Tables and the first state water quality standards in the United States. Richards served as a consultant primarily in food analysis and published several textbooks on this topic. Her *Food Materials and Their Adulterations* (1885) influenced the passage of the first Pure Food and Drug Act in Massachusetts. She even studied spontaneous combustion to prevent fires. With Edward Atkinson she helped design the Aladdin Oven; their work led to designing fire-resistant factories, designs that became models for industry. One of Richards's co-authors, Atkinson appointed her the first woman science consultant to the Manufacturers Mutual Insurance Company.[7]

Richards also became a specialist in mineral analysis. As an undergraduate she had studied samarskite, and at the time she had believed that the substance might contain undiscovered elements. Although she did not explore the mineral herself, as she predicted, samarium and gadolinium were later found in this material. She distinguished herself further in metals analysis for discovering vanadium in small amounts in ore metals. Her skills in metals analysis propelled her into becoming the first woman elected to the American Institute of Mining and Metallurgical Engineers in 1879.

After 1890, Richards devoted much of her time to forming the home economics movement. At a public lecture in Boston on November 30, 1892, Richards coined the term "oekology" when she called for the christening of this science.[8] Her branch of ecology, later called "domestic sciences" and then "home economics," arose from her work in sanitary chemistry to improve the environmental problems caused by industrialization. She said of this new discipline in 1897, "The science of household economics is in what chemists call a state of supersaturated solution; it needs only the insertion of a needle point to start crystallization."[9]

Richards helped to found the American Home Economics Association in 1908. Actually, the association had begun with a series of lectures at Lake Placid, New York, in the 1900s. Richards served as the first president of the American Home Economics Association and remained in that post until 1910. As president, she helped establish the *Journal of Home Economics* by underwriting its production. Richards was also appointed to the National Education Association's Council in 1910. In this position, she supervised home economics teaching in schools across the country.

On the forefront of promoting women's education, especially in science, in 1882 Richards organized the science section of the Society to Encourage Studies at Home.[10] In Boston around 1896 she attempted to start a science and arts high school for girls, but her efforts fell short. Richards was more successful in organizing a school of housekeeping in Boston in 1899 at the Women's Educational and Industrial Union. This school developed into the home economics department of Simmons College. Richards was also involved in establishing the American Association of University Women (AAUW) with Marion Talbot. The AAUW had begun as the Association of Collegiate Alumnae in Boston in 1881.[11] Richards has also been credited with helping to establish the Seaside Laboratory, later the Woods Hole Marine Biological Laboratory.[12] She died at age 69 on March 30, 1911, in Jamaica Plain, Massachusetts.

Notes

1. Margaret Ogilvie, *Women in Science: Antiquity through the Nineteenth Century: A Biographical Dictionary with Annotated Bibliography* (Cambridge, Mass.: MIT Press, 1986), p. 150.

2. Joan Marlow, "Ellen Swallow Richards (1842–1911), Founder of Home Economics," in *The Great Woman* (New York: A and W Publications, 1979), p. 130.

3. G. Kass-Simon, "Ellen Swallow Richards and the Ecology Movement: Righting the Record," in *Women of Science: Righting the Record*, eds. G. Kass-Simon and Patricia Farnes (Bloomington: Indiana University Press, 1990), p. 153.

4. Hamilton Cravens, "Establishing the Science of Nutrition at the USDA: Ellen Swallow Richards and Her Allies," *Agricultural History* 64, no. 2 (1990): 127.

5. G. Kass-Simon, "Ellen Swallow Richards," p. 152.

6. Joan Macksey and Kenneth Macksey, *Book of Women's Achievements* (New York: Stein and Day, 1975), p. 16.

7. Ann Wintriss, "Profile: Ellen Swallow," *Human Ecology Forum* 7, no. 3 (1977): 16.

8. G. Kass-Simon, "Ellen Swallow Richards," p. 153.

9. Quoted in Wintriss, "Profile," p. 16.

10. Ogilvie, *Women in Science*, p. 151.

11. Margaret Rossiter, "Women Scientists in America before 1920," *American Scientist* 62 (May–June 1974): 321.

12. G. Kass-Simon, "Ellen Swallow Richards," p. 155.

Bibliography

Clarke, Robert. *Ellen Swallow: The Woman Who Founded Ecology*. Chicago: Follett, 1973.

Cowan, Ruth Schwartz. "Ellen Swallow Richards: Technology and Women," in *Technology in America: A History of Individuals and Ideas*. Edited by Carroll W. Pursell. Cambridge, Mass.: MIT Press, 1990.

Douty, Esther (Morris). *America's First Woman Chemist, Ellen Richards*. New York: Messner, 1961.

Frankfort, Roberta. "Epilogue: Ellen H. Richards and the New Professional Education for Women," in *Collegiate Women: Domesticity and Career in Turn-of-the-Century America*. New York: New York University Press, 1977.

Hunt, Caroline Louisa. *The Life of Ellen H. Richards*. Boston: Whitcomb & Barrows, 1912.

Kass-Simon, G. "Ellen Swallow Richards and the Ecology Movement: Righting the Record," in *Women of Science: Righting the Record*. Edited by G. Kass-Simon and Patricia Farnes. Bloomington: Indiana University Press, 1990.

Ogilvie, Margaret. *Women in Science; Antiquity through the Nineteenth Century: A Biographical Dictionary with Annotated Bibliography*. Cambridge, Mass.: MIT Press, 1986.

Richards, Ellen Henrietta (Swallow). *Air, Water, and Food from a Sanitary Standpoint*. 2nd ed. New York: J. Wiley & Sons, 1904.

———. *Conservation by Sanitation: Air and Water Supply Disposal of Waste*. New York: J. Wiley & Sons, 1911.

—————. *Euthenics: The Science of Controllable Environment.* New York: Arno Press, 1910.

—————. *Food Materials and Their Adulterations.* Boston: Estes and Lauriat, 1886.

—————. *Sanitation in Daily Life.* Boston: Whitcomb & Barrows, 1907.

Richards, Ellen Henrietta (Swallow), and Edward Atkinson. *The Science of Nutrition: Treatise upon the Science of Nutrition.* Boston: Damrell & Upham, 1896.

Ridley, Agnes Fenster. "Richards, Ellen Henrietta Swallow," in *Biographical Dictionary of American Educators.* Westport, Conn.: Greenwood Press, 1978.

Rossiter, Margaret. "Women Scientists in America before 1920." *American Scientist* 62 (May–June 1974): 312–323.

Yost, Edna. "Ellen H. Richards," in *American Women of Science.* Philadelphia: Frederick A. Stokes, 1943.

FAYE A. CHADWELL

JANE RICHARDSON
(1941–　)
Crystallographer

Birth	January 25, 1941
1958	Third place in national Westinghouse Science Talent Search
1962	B.A., Swarthmore College
1963	Married David C. Richardson
1966	M.A., philosophy of science, Harvard University; M.A.T., natural sciences, Harvard University
1970–84	Associate, Dept. of Anatomy, Duke University
1984–88	Medical Research Associate Professor, Depts. of Biochemistry and Anatomy, Duke University
1985	MacArthur Fellow
1986	D.Sc. (Honorary), Swarthmore College
1988–91	Medical Research Associate Professor, Dept. of Biochemistry, Duke University
1991–	James B. Duke Professor, Dept. of Biochemistry, Duke University; elected to National Academy of Sciences and the American Academy of Arts and Sciences
1994	D.Sc., University of North Carolina, Chapel Hill

Jane and David Richardson. Photograph by Les Todd, Duke University Photography.

Jane S. Richardson was born in 1941 in Teaneck, New Jersey. As a teenager in 1958, her early interest in science won her third place in the nation at the Westinghouse Science Talent Search. She went on to study philosophy, mathematics, and physics at Swarthmore College, graduating cum laude with a B.A. degree in 1962. She also became a Woodrow Wilson Fellow in the same year. In 1963, Jane married David C. Richardson; together they have pursued the use of X-ray crystallography to determine the molecular structures of individual proteins.[1] Jane and David have two children: Robert was born in 1973 and Claudia in 1980.

Jane began her career as a technical assistant in the Department of Chemistry at the Massachusetts Institute of Technology in Cambridge, Massachusetts. She worked at this job for about five years, from 1964 through 1969. While working at MIT, Jane decided to pursue a graduate degree in science and enrolled at Harvard. She received her M.A. in Philosophy of Science and M.A.T. in Natural Sciences from Harvard University in 1966.

In 1969, Jane Richardson moved to Bethesda, Maryland, to work as a general physical scientist at the Laboratory of Molecular Biology, National Institutes of Health. In that year she also co-authored her first paper, which was published in the *Proceedings of the National Academy of*

Sciences. Within one year she moved to Duke University in North Carolina to work as an associate in the Department of Anatomy. She has since been associated with Duke University and currently holds the James B. Duke professorship in the Department of Biochemistry, where she teaches biochemistry at the graduate and undergraduate levels. She is also serving as co-director of the Comprehensive Cancer Center at Duke University.

Richardson is a highly regarded researcher and an expert on protein folding. In an interview that appeared in the *New York Times*, she half-whimsically compares the protein-folding problem with the paper-folding problems of the Japanese art of origami.[2] Richardson is a crystallographer who studies the three-dimensional structures of proteins, emphasizing the underlying principles of their architecture, aesthetics, interrelationships, and folding mechanism. Her other research interests are protein crystallography, design of new proteins for synthesis, and comparison and classification of protein structures. Many of her colleagues and fellow researchers regard her work on protein folding as a unique and completely new style of representing a protein. According to a 1994 *UNESCO Courier* article, "no protein actually looks like Richardson's Ribbon. . . . Richardson's work is the design of completely new protein molecules based on information she has gleaned from existing natural proteins. Part of her contribution to her field has been her ability to capture certain qualities of proteins and depicting them in certain ways."[3]

Jane Richardson's professional activities and honors are numerous. She is a member of several learned societies, having served on advisory boards and executive boards for several of them. Her honors include membership in the nation's most prominent scientific organizations, the National Academy of Sciences and the American Academy of Arts and Sciences. She was a recipient of the MacArthur "Genius" Award in 1985, and her work was also chosen for *Science Digest* One Hundred Best Innovations. In 1986 her alma mater, Swarthmore College, awarded her an honorary doctor of science degree. In 1990 Duke University also presented her with an honorary degree, and she received another from the University of North Carolina in 1994.

In her long and continuing tenure at the Duke University, Jane Richardson has authored and co-authored more than 60 publications in well-known magazines such as *Nature* and *Science*. She has lectured extensively and participated in conferences nationally and internationally. She has visited Russia, Israel, Germany, Canada, and Switzerland among other nations as an invited speaker.

Richardson understands the challenges and barriers that women face as they try to succeed in a male-dominated profession. Commenting on the small number of women elected to the high-profile positions in the

National Academy of Sciences, Jane Richardson said, "It's an old club, elected by its own members."[4]

Notes

1. "Brief Biography on Jane Richardson," Public Information Office, The John D. and Catherine T. MacArthur Foundation, July 1985.
2. James Gleick, "Secret of Proteins Is Hidden in Their Folded Shapes," *New York Times* (June 14, 1988): C1.
3. John Hodgson, "Art and the Science of Life; Biotechnology: A Resourceful Gene," *UNESCO Courier* (June 1994): 36.
4. Natalie Angier, "Academy's Choices Don't Reflect the Number of Women in Science," *New York Times*, Late Edition, Section 1 (May 10, 1992): 24.

Bibliography

Angier, Natalie. "Academy's Choices Don't Reflect the Number of Women in Science." *New York Times*, late edition, sect. 1 (May 10, 1992): 24.
"Brief Biography on Jane Richardson." Public Information Office, The John D. and Catherine T. MacArthur Foundation, July 1985.
Gleick, James. "Secret of Proteins Is Hidden in Their Folded Shapes." *New York Times* (June 14, 1988): C1.
Hodgson, John. "Art and the Science of Life; Biotechnology: A Resourceful Gene." *UNESCO Courier* (June 1994): 36.

USHA MEHTA

DOROTHEA KLUMPKE ROBERTS
(1861–1942)
Astronomer

Birth	August 9, 1861
1886	B.S., University of Paris
ca. 1887	Became Paris Observatory's first woman student and staff member
1889	Earned first Prix des Dames, Astronomical Society of France
1891–1901	Directed Paris Bureau of Measurements, supervised Paris Observatory's portion of *Carte du Ciel* (Chart of the Heavens) project.

Dorothea Klumpke Roberts. Photo courtesy of the Mary Lea
Shane Archives of the Lick Observatory.

1893	Elected first woman officier of Paris Academy of Sciences; became first woman to receive D.Sc. degree in mathematics from University of Paris
1899	First woman to undertake astronomical observations from above earth's surface
1901	Married pioneer astronomical photographer Isaac Roberts
1932	Helene-Paul Helbronner Prize, French Academy of Sciences
1934	Elected Chevalier of the Legion of Honor and received Cross of the Legion
Death	October 5, 1942

The inventions of photography and spectroscopy in the late nineteenth
century revolutionized astronomy, stimulating the expansion of observatories and the birth of astrophysics. Larger telescopes and the new

technologies enabled astronomers to study the chemical composition, temperature, and evolution of stars through the quality of their light, the spectrum that resulted when light from a star was passed through a prism and photographed. Women astronomers at the American women's colleges undertook research projects involving the observation of astronomical objects; others found employment as human "computers" in the observatories of America and Europe, examining photographic plates for novel astronomical objects, calculating their positions and orbits, and classifying myriad stars.[1] Dorothea Klumpke Roberts, one of the first American women to earn a doctorate in mathematics, was internationally renowned for both her observational and computational work. She was elected to membership in twelve learned societies, including the Astronomical Society of France, Royal Astronomical Society, American Astronomical Society, Astronomical Society of the Pacific, International Astronomical Union, British Astronomical Association, Comité National Français d'Astronomie, and American Association for the Advancement of Science.[2]

Klumpke Roberts inherited her pioneer spirit from her family. Her father, John Gerard Klumpke, was a German immigrant who headed to California during the Gold Rush, became a successful boot and shoe maker (gold miners paid up to $100 for a pair of his boots), and later prospered as a realtor. He married Dorothea Matilda Tolle, an artistically talented woman. The couple had five daughters and two sons. Dorothea was born in San Francisco on August 9, 1861.[3]

Determined that their gifted daughters receive the same educational opportunities as men, Mrs. Klumpke enrolled them in schools in Germany, Switzerland, and France, where the sisters distinguished themselves. Anna became a noted artist, the protégé and biographer of Rosa Bonheur. Julia became a concert violinist and composer; Matilda, a pianist; and Augusta, a neurologist. Dorothea initially studied languages, but her fascination with astronomy led her to pursue mathematics and mathematical astronomy at the University of Paris, from which she received her bachelor of science degree in 1886.[4]

In 1887, Klumpke joined the staff of the Paris Observatory, where she worked primarily on the measurement of star positions on photographic plates. When the International Astronomical Congress convened in Paris to discuss the proposed Carte du Ciel (Chart of the Heavens) project—an ambitious venture to photograph the entire sky down to the fourteenth magnitude and prepare a corresponding catalogue of stars down to the eleventh magnitude—Klumpke translated the papers of the scientists attending the conference into French for the official record.[5]

By 1891, the plates awaiting measurement had become so numerous that the program administrator created a special Bureau of Measurements to handle the Paris Observatory's portion of the project. Klumpke

applied for the position of director. She was selected over 50 male applicants and held that post from 1891 to 1901. The quality of her work
earned her the first award of the Prix des Dames (Women's Prize) from
the Astronomical Society of France in 1889 and election as Officier of the
Paris Academy of Sciences in 1893, the first such honor accorded a
woman. The paper she presented on the mapping of the heavenly bodies
at the Congress of Astronomy and Astro-Physics, held in Chicago during
the 1893 World's Columbian Exposition, netted her a $300 prize from
the French Academy of Sciences.[6] During this period she frequently lectured on astronomy and published the results of her research on comets,
meteorites, the spectra of stars, and other celestial phenomena.[7]

Klumpke became the Paris Observatory's first woman student and, on
December 14, 1893, the first woman to be awarded the doctor of science
degree in mathematics from the University of Paris. Her doctoral dissertation, a mathematical study of the rings of Saturn, completed Sonya
Kovalevskaya's work on that topic. The eminent Gaston Darboux, who
chaired her examination committee, complimented her on her brilliant
dissertation defense: "Your thesis is the first which a woman has presented and defended with success before our Faculty to obtain the degree
of Doctor of Mathematical Sciences. You deservedly open the way, and
the Faculty is eager to declare you worthy to obtain the degree of Doctor."[8]

Klumpke had an intrepid spirit. She was thrilled to be chosen by the
Société Française de Navigation Aérienne to observe from a balloon the
anticipated return of the Leonid meteors on November 16, 1899. She
spent the day of the expedition sewing warm clothes and addressing
small souvenirs "in case of no return." As the balloon "La Centaure"
climbed to an altitude of 1,640 feet and drifted westward, Klumpke and
her traveling companions marveled over the dazzling expanse of Paris
and meteors trailing blue, green, and red light. Only 11 of the 24 meteors
they observed emanated from the Lion, but the trip left an indelible
impression: "It seemed to me that, in the absence of earth's jar and grind,
the eye was clearer, the heart more awake, and the soul filled to its brim
with divine, with reverent adoration. . . . Never before had nature
seemed to me so grand, so beautiful."[9]

The balloon landed 176 miles from its starting point. The astronomical
community hailed Klumpke as the leading woman aeronaut of France
and the first woman to undertake astronomical observations from above
the earth's surface. She published an account of her exhilarating experience in the *Century Magazine*.[10]

In 1901, Klumpke married Isaac Roberts, then age 72, a retired builder
and pioneer in astronomical photography who had built a private observatory in Crowborough, Sussex. The observatory was equipped with
a 20–inch reflector and camera; with these, it developed the first good

representations in England of the Andromeda galaxy and other nebulae. Klumpke and Roberts collaborated on his extensive celestial photography program. After his death in 1904, Klumpke Roberts continued the work, publishing the results of their collaboration. She then moved to France to live with her mother and sister Anna, remaining there after their mother died in 1922.[11]

Klumpke Roberts published *Isaac Roberts' Atlas of 52 Regions, a Guide to William Herschel's Fields of Nebulosity* in 1929 as a memorial to her husband and issued a supplement in 1932. The *Celestial Atlas* contained fine enlargements of plates taken with the 20–inch reflector, which greatly facilitated the identification of faintly luminous areas of the sky and the motions of the stars involved. For this work, she received the Helene-Paul Helbronner prize from the French Academy of Sciences in 1932. Her greatest honor came in 1934 when she was elected Chevalier of the Legion of Honor and the President of France bestowed upon her the Cross of the Legion, in recognition of 48 years of service to French astronomy.[12]

Shortly thereafter, Klumpke Roberts returned with Anna to San Francisco, where their house became a magnet for artists, scientists, and musicians. In retirement she maintained her interest in astronomy and the careers of young astronomers. For years she gave annual donations to the Paris Observatory, to be awarded to a young staff member. She generously supported the Astronomical Society of France and revived its Prix des Dames, which had been discontinued. Her gift to the Astronomische Gesellschaft recognized the author of the best paper on Herschel's 52 Areas. She bequeathed money to the American Astronomical Society and the Astronomical Society of the Pacific, the latter for support of the Klumpke-Roberts Lecture Fund, named in memory of her parents and husband. In 1937 she established and endowed the Dorothea Klumpke Roberts Prize Funds, also named in honor of her parents and husband. The income from this fund supported respective awards in astronomy and mathematics presented to outstanding students at the University of California at Berkeley.[13]

Klumpke Roberts had long been committed to furthering women's advancement in astronomy. In lively addresses delivered at the International Congress of Women, held in London in 1899, and at a 1919 meeting of the Astronomical Society of the Pacific, she paid homage to the women who from antiquity to her own time had made significant contributions to astronomy. She predicted that women's future work in the field would eclipse even those accomplishments:

> When woman becomes aware of what she owes unto astronomy, when she understands that all sciences contribute to the furthering of astronomy, when she but catches a glimpse of the wealth of

information and the treasures of reward held in store for her by astronomy, then she will not rest until every child in every school thruout the world be taught with the a b c's the rudiments of astronomy.[14]

Dorothea Klumpke Roberts died in San Francisco on October 5, 1942. In 1974, as a fitting tribute to this premier scientist, the Astronomical Society of the Pacific redirected the Klumpke-Roberts Lecture Fund to the Klumpke Roberts Award, named in her honor, "for outstanding contributions to education or popularization in astronomy." Carl Sagan was the first recipient; others have included Isaac Asimov and astronomer and newspaper columnist **Helen Sawyer Hogg**.[15]

Notes

1. Pamela E. Mack, "Straying from Their Orbits: Women in Astronomy," in *Women of Science: Righting the Record*, eds. G. Kass-Simon and Patricia Farnes (Bloomington: Indiana University Press, 1990), pp. 72–75, 88, 93, 111–112; Peggy Aldrich Kidwell, "Women Astronomers in Britain, 1780–1920," *Isis* 75 (September 1984): 539. The assistance of Mr. Pat Galloway of the Astronomical Society of the Pacific in identifying photo sources and materials on Klumpke Roberts and women in astronomy was invaluable.

2. Deborah Jean Warner, "Women Astronomers," *Natural History* 88 (May 1979): 20; James T. White, ed., "Roberts, Dorothea Klumpke," *National Cyclopaedia of American Biography* 31 (1944): 406.

3. James T. White, ed., "Klumpke, John Gerard," *National Cyclopaedia of American Biography* 31 (1944): 403; Katherine Bracher, "Dorothea Klumpke Roberts: A Forgotten Astronomer," *Mercury* 10 (September–October 1981): 139.

4. Bracher, "Dorothea Klumpke Roberts," p. 139; Lori Kenschaft, trans., "*Les Femmes dans la Science* by A. Rebiere," *Association for Women in Mathematics Newsletter* 13 (September–October 1983): 10.

5. Bracher, "Dorothea Klumpke Roberts," pp. 139–140.

6. Kenneth Weitzenhoffer, "The Triumph of Dorothea Klumpke," *Sky and Telescope* 72 (August 1986): 109–110; Bracher, "Dorothea Klumpke Roberts," p. 140; White, "Roberts, Dorothea Klumpke," p. 405.

7. White, "Roberts, Dorothea Klumpke," p. 405; Kenschaft, "Les Femmes," p. 10.

8. White, "Roberts, Dorothea Klumpke," p. 405; Weitzenhoffer, "Triumph," p. 109; Kenschaft, "Les Femmes," p. 10.

9. Dorothea Klumpke, "A Night in a Balloon: An Astronomer's Trip from Paris to the Sea in Observation of the Leonids," *Century Magazine* 60 (June 1900): 276, 278, 281–282.

10. Weitzenhoffer, "Triumph," pp. 109–110.

11. Robert G. Aitken, "Dorothea Klumpke Roberts—An Appreciation," *Publications of the Astronomical Society of the Pacific* 54 (December 1942): 219–220; Bracher, "Dorothea Klumpke Roberts," p. 140.

12. Aitken, "Dorothea Klumpke Roberts," p. 220; Bracher, "Dorothea Klumpke Roberts," p. 140.

13. Aitken, "Dorothea Klumpke Roberts," p. 221; C.H. Adams, "General Notes. The Dorothea Klumpke-Roberts Prizes at the University of California," *Publications of the Astronomical Society of the Pacific* 49 (June 1937): 172–173.

14. Dorothea Klumpke Roberts, "Woman's Work in Astronomy," *Publications of the Astronomical Society of the Pacific* 31 (July 1919): 216–218.

15. Bracher, "Dorothea Klumpke Roberts," p. 140; Katherine Bracher, *The Stars for All: A Centennial History of the Astronomical Society of the Pacific* (San Francisco: Astronomical Society of the Pacific, 1989), p. 34; Andrew Fraknoi, "The Klumpke-Roberts Award," *Sky and Telescope* 72 (August 1986): 110.

Bibliography

Aitken, Robert G. "Dorothea Klumpke Roberts—An Appreciation." *Publications of the Astronomical Society of the Pacific* 54 (December 1942): 217–222.

Bracher, Katherine. "Dorothea Klumpke Roberts: A Forgotten Astronomer." *Mercury* 10 (September–October 1981): 139–140.

Klumpke, Dorothea. "The Bureau of Measurements of the Paris Observatory." *Astronomy and Astro-Physics* 12 (November 1893): 783–788.

———. "A Night in a Balloon: An Astronomer's Trip from Paris to the Sea in Observation of Leonids." *Century Magazine* 60 (June 1900): 276–284.

———. "The Work of Women in Astronomy." *Observatory* 22 (August 1899): 295–300.

Lankford, John, and Rickey L. Slavings. "Gender and Science: Women in American Astronomy, 1859–1940." *Physics Today* 43 (March 1990): 58–65.

Mack, Pamela E. 1990. "Straying from Their Orbits: Women in Astronomy in America," in *Women of Science: Righting the Record*. Edited by G. Kass-Simon and Patricia Farnes, pp. 72–116. Bloomington: Indiana University Press, 1990.

Ogilvie, Marilyn Bailey. "Roberts, Dorothea Klumpke," in *Women in Science; Antiquity through the Nineteenth Century; A Biographical Dictionary with Annotated Bibliography*. Cambridge, Mass.: MIT Press, 1986.

Siegel, Patricia Joan, and Kay Thomas Finley. "Dorothea Klumpke Roberts," in *Women in the Scientific Search: An American Biobibliography*, pp. 50–51. Metuchen, N.J.: Scarecrow Press, 1995.

Weitzenhoffer, Kenneth. "The Triumph of Dorothea Klumpke." *Sky and Telescope* 72 (August 1986): 109–110.

White, James T., ed. "Roberts, Dorothea Klumpke." *National Cyclopaedia of American Biography* 31 (1944): 405–406.

MARIA CHIARA

Elizabeth Roemer

(1929–)

Astronomer

Birth	September 4, 1929
1950	B.A., University of California, Berkeley; Bertha Dolbeer Scholar; Dorothea Klumpke Roberts Prize, University of California, Berkeley
1955	Ph.D., astronomy, University of California, Berkeley; Lick Observatory Fellow
1955–56	Research Astronomer, University of California, Berkeley
1956	Research Associate, Yerkes Observatory, University of Chicago
1957–66	Astronomer, Flagstaff Station, U.S. Naval Observatory
1959	*Mademoiselle* Merit Award
1964–79	Chair, Working Group on Orbits and Ephemerides of Comets Commission, Positions & Motions Minor Planets, Comets & Satellites, 20 International Astronomical Union (also 1985–88)
1966–69	Associate Professor of Astronomy, Lunar and Planetary Laboratory, University of Arizona
1969–	Professor of Astronomy, Lunar and Planetary Laboratory, University of Arizona
1971	Benjamin Apthorp Gould Prize, National Academy of Sciences
1973–75	Member, Space Science Review Panel Associateship Program, Office of Scientific Personnel, National Research Council (Chair, 1975)
1979–85	Commission 20: Vice President, 1979–82; President, 1982–85
1980–	Astronomer, Stewart Observatory
1986	NASA Special Award
1988–	President, Commission 6 Astronomical Telegrams (also 1976–79)

Elizabeth Roemer is both a researcher and author of numerous publications on astronomy and the astrophysics of comets and minor planets.

She is a premiere recoverer of "lost" comets (i.e., comets whose "planned rediscovery" is based on predictions from previous returns). Her lifetime study of comets includes at least 79 recoveries of returning periodic comets, visual and spectroscopic binary stars, and computation of orbits of comets and minor planets. In addition, her publications have covered a wide variety of topics, such as comets and minor planets, astronomy and practical astronomy, computation of orbits, astrometric and astrophysical investigations of comets, minor planets and satellites, and dynamical astronomy.

Born in 1929 in Oakland, California, Roemer began her career as a teacher of adult classes in the Oakland public school system from 1950 to 1952 while she pursued her degree in astronomy at the University of California, Berkeley. During that time she also served as assistant astronomer and later, from 1954 to 1955, as lab technician at the Lick Observatory, a well-respected institution for astronomical observations. Roemer joined the faculty of the University of California at Berkeley as research astronomer in the 1955–1956 academic year and also served as research associate at the Yerkes Observatory, University of Chicago, in 1956. In 1957 she joined the staff of the U.S. Naval Observatory, Flagstaff Station, as astronomer and remained there until 1966. It was at the U.S. Naval Observatory that Roemer made her first major recoveries. In addition, her photographic records of comets began to earn her national recognition. At the U.S. Naval Observatory, she used a 40–inch astronomic reflecting telescope of very high definition to show that comet nuclei could not be larger than 3 to 4 kilometers for short-period comets and 7 to 8 kilometers for long-period comets. In 1965, Roemer was named acting director of the Observatory. Roemer joined the faculty of the University of Arizona as associate professor from 1966 to 1969 and became a full professor at that institution's Lunar and Planetary Laboratory in 1969. Since 1980, she has been a senior astronomer at the Stewart Observatory in Arizona.

Elizabeth Roemer is recognized by her peers as one of the best recoverers of comets and as a contributor to many scientific and astronomical discoveries. "Her innumerable precise photographic observations of comets have been the basis for a great many cometary orbits of importance, and her notes on the physical characteristics of comets have also been invaluable."[1] In 1965, asteroid 1657 was named Roemera for her by a colleague, P. Wild. Roemer has served on many committees, national and international, for the study of comets and asteroids and has been awarded a number of prizes for her work, including the Dorothea Klumpke Roberts Prize from the University of California at Berkeley in 1950; the Benjamin Apthorp Gould Prize from the National Academy of Sciences in 1971; and the NASA Special Award in 1986. She is a member of Phi Beta Kappa, Sigma Xi, the International Astronomical Union, the

British Astronomical Association, and the Royal Astronomical Society (London).

Note

1. Fred Whipple, *The Mystery of Comets* (Washington, D.C.: Smithsonian Institution Press, 1985), p. 84.

Bibliography

American Men and Women of Science, 1992–93. 18th ed. New Providence, N.J.: R.R. Bowker, 1992.

Brown, Peter Lancaster. *Comets, Meteorites and Men*. New York: Taplinger Publishing, 1974.

Chapman, Robert, and John C. Brandt. *The Comet Book: A Guide for the Return of Halley's Comet*. Boston: Jones & Bartlett, 1984.

Roemer, Elizabeth. "Activity in Comets at Large Heliocentric Distance." *Publications of the Astronomical Society of the Pacific* 74 (October 1962): 351–365.

———. "Astronomic Observations and Orbits of Comets." *Astronomical Journal* 66 (October 1961): 368–371.

———. "Comet Notes." *Publications of the Astronomical Society of the Pacific* 83 (June 1971): 370–371.

———. "An Outburst of Comet Schwassmann-Wachmann I." *Publications of the Astronomical Society of the Pacific* 70 (1958): 272.

Whipple, Fred. *The Mystery of Comets*. Washington, D.C.: Smithsonian Institution Press, 1985.

Who's Who in the West, 1989–90. 22nd ed. Wilmette, Ill.: Marquis Who's Who, 1989.

Who's Who of American Women, 1979–80. 11th ed. Chicago: Marquis Who's Who, 1979.

The Women's Book of World Records and Achievements. Edited by Lois Decker O'Neill. Garden City, N.Y.: Anchor Press, 1979.

STEFANIE BUCK

NANCY GRACE ROMAN
(1925–   )
Astronomer

Birth	May 16, 1925
1946	B.A., Swarthmore College

Nancy Grace Roman. Photo courtesy of NASA.

1949	Ph.D., University of Chicago
1949–52	Research Associate, stellar astronomy, Yerkes Observatory, University of Chicago
1952–55	Instructor, Yerkes Observatory
1955	Assistant Professor, Yerkes Observatory
1955–56	Physicist, Radio Astronomy Branch, U.S. Naval Research Lab (USNRL)
1956–58	Astronomer, Head of Microwave Spectroscopy Section, USNRL
1960–64	Chief, Astronomy and Solar Physics, Geophysics and Astronomy Programs Office, Space Flight Development, NASA
1964–79	Chief, Astronomy and Relativity Programs, NASA
1966	D.Sc., Russell Sage College
1969	D.Sc., Hood College
1971	D.Sc., Bates College
1976	D.Sc., Swarthmore College

1979–80	Program Scientist for space telescope
1981–	Senior Scientist, Astronomical Data Center, NASA
1994–	Head, Astronomical Data Center, NASA

For as long as I can remember, I wanted to be an astronomer. In fifth grade I started an astronomy club among my friends. I then read every astronomy book I could find in our local library. In high school I happily took the two and a half years of math that were offered. In college I majored in astronomy but also took five years each of math and physics. I believe that the only way I managed to get through history and German in my freshman year was that I did not need to spend much time studying the math and astronomy I was taking. I received a B.A. in astronomy from Swarthmore College in 1946 and a Ph.D., also in astronomy, from the University of Chicago in 1949.

In my Ph.D. thesis, on the Ursa Major Cluster (which includes all but the end stars in the Big Dipper), I studied the motions of many other stars that had been suggested as members to determine which ones were likely really to be. I then used the membership in the cluster to determine distances and, from these, the brightness each star would have if it were at a standard distance. The latter is called the star's absolute magnitude. Astronomers send the light from a star through a prism to create a rainbow, called a spectrum. Two stars for which the spectrum looks the same will have the same absolute magnitude. Astronomers use the appearance of its spectrum to estimate the distance to a star that is too far away to measure its apparent change in direction as the earth orbits the sun. I then studied other clusters to determine the distance of each cluster and to calibrate the absolute magnitudes of stars with spectra not found in the Ursa Major Cluster.

From 1949 to 1955 I worked at the University of Chicago, first as a research associate and eventually as an assistant professor. I became interested in stars with abnormally large velocities. It was apparent that stars that were far from the plane of the Milky Way tended to move more rapidly and also exhibited evidence of fewer elements heavier than helium in their spectra. In my work I showed that even among the bright stars near the sun, stars with fewer heavy elements moved more rapidly. Later work indicated that the faster stars were older. I also taught graduate courses and did a great deal of observing, of both the spectra and the brightness of stars.

I enjoyed teaching very much but did not believe that I had a chance for tenure. Hence I left the University of Chicago when I had the opportunity to get an interesting position in the new field of radio astronomy. (No women had tenure in a major astronomy department at that time, and very few had faculty positions.) At the Naval Research Laboratory from 1955 to 1959, I worked on various projects including measuring the distance to the moon by radar and mapping the sky at 67 cm. I also continued my research on stellar motions. In 1956 an astronomer who had been born in Russia was invited to the dedication of a new observatory in Soviet Armenia. He declined, fearing that he might not be able to leave the country if he went. At the last minute I was invited in his place, because the observatory's director was interested in a short paper I had written.

When NASA was formed in 1958, most of its science staff came from Project Vanguard and other groups at the Naval Research Laboratory that were active in rocket astronomy. Although I had had few scientific contacts with these groups, I was well known as a result of the trip to the Soviet Union and I was asked if I knew anyone who would like to set up a program in space astronomy. The challenge of starting with a clean slate to define a program that would influence astronomy for decades was too great to turn down, even though I questioned getting into administration. I was not happy about giving up research, but I found that I actually enjoyed management. People are more difficult and complex than stars, but also often as interesting. At NASA I had the scientific responsibility for investigations in astronomy, solar physics, geodesy, and relativity that took advantage of the capability of making observations from above the atmosphere and of doing experiments in space. In this role I organized a coherent program of scientific investigations, taking account of both the technical capabilities available and the projects that members of the scientific community wished to pursue. I then convinced senior administrators that the program should be supported, and I worked with engineers to help them understand the requirements of the scientific instruments. I also prepared testimony for Congress, but I never testified myself. Among the many programs for which I was responsible were the Orbiting Solar and Astronomical Observatories, several geodetic satellites, the International Ultraviolet Explorer, and the Hubble Space Telescope as well as numerous rocket and balloon programs.

Science is an international activity, particularly in astronomy with its small number of professionals and little commercial value. I met frequently with representatives of the European Space Research Organization to coordinate our programs to avoid duplication. Some satellites were designed and built by Europeans but launched by the United States with U.S. experiments included. Europeans also flew experiments on our satellites. For many years I was the highest ranking woman in NASA. The only other female scientist was another woman astronomer whom I hired, who worked part-time when her children were young and eventually increased her hours to full-time.

I retired from NASA in 1980 to maintain a home for my elderly mother. Since then I have enjoyed keeping active with part-time positions. I provided support to various NASA programs, particularly the Hubble Space Telescope and the Earth Observation System. I have also worked since 1981 at the NASA Space Science Data Center in the Astronomical Data Center. There, I edit and document astronomical catalogues for electronic archiving. These can be retrieved easily by astronomers throughout the world. In this, I work closely with my counterparts in France with whom we exchange both catalogues and techniques. In 1993 I spent a week observing for the first time in almost 20 years. I enjoyed it as much as ever.

Throughout history there have been a comparatively large number of women in astronomy. Several made major contributions to the field, but, until my generation, they were usually restricted to low-level positions in which they did a great deal of "drudge" work such as is now assigned to electronic computers. Being a pioneer sometimes presented challenges, but it was exciting.

Bibliography

Roman, Nancy G. "A Catalogue of High Velocity Stars." *Astrophysical Journal* 118, suppl. 2 (1955): 195.

———. "A Correlation between the Spectroscopic and Dynamical Characteristics of the Late F- and Early G-Type Stars." *Astrophysical Journal* 112 (1950): 554.

———. "High Velocity Stars," in *Stars and Stellar Systems*, Vol. 5: "Galactic Structure," p. 345. Chicago: University of Chicago Press, 1965.

———. "High Velocity Stars as Population I Objects." *Astrophysical Journal* 62 (1957): 146.

———. "The Spectra of the Bright Stars of Types F5–K5." *Astrophysical Journal* 116 (1952): 122.

———. "A Study of the Concentration of Early-Type Stars in Cygnus." *Astrophysical Journal* 114 (1951): 492.

———. "The Ursa Major Group." *Astrophysical Journal* 110 (1949): 205.

Roman, Nancy G., and W.W. Morgan. "The Moving Cluster in Perseus." *Astrophysical Journal* 111 (1950): 426.

NANCY G. ROMAN

VERA COOPER RUBIN

(1928–　)

Astronomer

Birth	1928
1948	B.A., Vassar College; married Robert Rubin
1951	M.A., Cornell University
1954	Ph.D., Georgetown University
1955	Research Astronomer, Georgetown University
1960	Assistant Professor, Georgetown University
1965	Joined Dept. of Terrestrial Magnetism, Carnegie Institution, as Research Associate; became Staff Member in about 1970
1981	Elected to National Academy of Sciences
1993	National Medal of Science

Astronomers long believed that visible matter—the stars and gas observed in galaxies—was essentially all the matter in the universe. Vera Rubin, however, has demonstrated that what can be seen is only about 10 percent of what really exists. An amazing 90 percent of our universe consists of dark, invisible matter. We know it is there because we can measure its effect on the orbits of visible stars and gas. Rubin's research has focused on the study of galaxies: their movement, their internal rotation, and their distribution. Her career has followed an unusual pattern; she has repeatedly done path-breaking work in fields decades before they were popular, moving on as the fields became crowded. As her work has made her famous, she has used her position to work for the advancement of women and minorities in science.

As a child, Vera Cooper fell in love with astronomy watching the night sky through her bedroom window. "I would prefer to stay up and watch the stars than go to sleep," she remembers. "There was just nothing as interesting in my life as watching the stars every night."[1] With the help of her father, an electrical engineer, she built a small telescope through which she tried to photograph the moon. Because her telescope had no driving motor, the photographs were a "total flop," but the project was fun. At age 17 she went to Vassar College, where she knew that **Maria Mitchell**, the first American to discover a comet, had taught astronomy.

When she graduated she married Robert Rubin, a graduate student in physical chemistry at Cornell.

To join her husband, Rubin turned down an acceptance from Harvard's large and prominent graduate department of astronomy in favor of Cornell's minute astronomy program, which had only two instructors. One was a former naval navigator, who actually told students not to study astronomy; but the other was Martha Stahr Carpenter, from Berkeley, who taught galaxy dynamics. Cornell's strength was in its physics department, and Rubin studied with prominent physicists Philip Morrison, Richard Feynman, and Hans Bethe. For her master's thesis she analyzed the motion of 108 galaxies. She discovered that they shared a large-scale, systematic motion in addition to motion due to the expansion of the universe: galaxies in one direction seemed to approach us, and galaxies in the other direction receded. Because the galaxy dynamics she had studied had taught her the formalism to analyze rotational motions, she used it, entitling her first paper the "Rotation of the Universe."

In 1950 the astronomical community was not prepared to believe in large-scale motions, especially not coming from an unknown 22–year-old. The data to which she had access were barely sufficient for the project. Rubin's first presentation of her work, at a meeting of the American Astronomical Society, left her "somewhat less than mashed to the ground."[2] Astronomically isolated at Cornell, she knew no one else at the meeting; she was not an AAS member; she was nursing her first child; and with a single exception, the audience's comments were hostile. She gave her talk and left. The editor of the meeting's proceedings insisted on changing the title of the published abstract to the much less ambitious "Differential Rotation of the Inner Metagalaxy," and the paper itself was rejected from both major journals to which Rubin submitted it.[3] "It got an enormous amount of publicity, almost all negative," recalls Rubin. "But at least from then on, astronomers knew who I was."[4] Rubin has since been vindicated: several years later the cosmologist Gérard de Vaucouleurs realized that the motion Rubin had discovered demarcated the plane of the local supercluster of galaxies. After an initially hostile reaction, astronomers accepted this view.

From Cornell, the Rubins went to Washington, D.C., where Robert had a job at the Applied Physics Laboratory. Vera was temporarily out of science and terribly frustrated. Robert, however, shared an office with Ralph Alpher, who introduced Vera's work to his colleague, George Gamow. Gamow, then a physics professor at George Washington University and already famous for applying nuclear physics to Big Bang cosmology, invited Vera Rubin to do a Ph.D. thesis. By then, Rubin had already entered the Georgetown University astronomy program. Motivated by Gamow's questions, Rubin did a dissertation showing that instead of being randomly distributed, galaxies tend to clump together.[5]

Once again, Rubin's work was ahead of its time. Since the 1970s, the clustering of galaxies has been an important area of research because it provides evidence for how galaxies form.

After earning her Ph.D. in 1954, Rubin spent ten years as a research astronomer and assistant professor of astronomy at Georgetown. The Rubins divided their time between science and their four children. Science and children left time for nothing else, and Rubin emphasizes that the combination worked because she had the support and cooperation of her husband and her parents, who lived in the Washington area. All four Rubin children have since followed their parents into science.

Having depended during the first part of her career on published observations made by other astronomers, Rubin took up observational astronomy herself in 1963. In 1965 she became a staff member at the Department of Terrestrial Magnetism (DTM) of the Carnegie Institution. At that point America's preeminent observatory, Mount Palomar, operated jointly by the Carnegie Institution and Caltech, was still reserved for male use on the grounds that there were no toilet facilities for women. In 1965, Rubin became the first woman officially permitted to observe there. (**Margaret Burbidge**, already a famous observational astronomer, had previously observed there by slipping in under the name of her husband, a theoretical astronomer.)

Rubin's DTM colleague, Kent Ford, specialized in making fast image tubes that made it possible to collect the spectra of extremely faint objects. With Ford, Rubin worked briefly on quasar redshifts, a measurement that allowed astronomers to trace the expansion of the farthest reaches of the universe. Rubin, who previously worked on topics she had to herself, found herself working for the first time on a topic everyone considered hot, making the atmosphere very competitive. "I found it personally very distasteful," Rubin recalls.

> I just didn't like the pressure of other astronomers calling and asking me if I had observed this and if I knew what the redshift was. I didn't get to a telescope very often and it meant that I either had to give out answers I was uncertain of, or say I hadn't done it and someone else would then go do it. I just decided that wasn't the way I wanted to do astronomy. I feel that I already have enormous numbers of internal pressures and I don't need external pressures on top of them. I really like to be left alone while I'm working.[6]

In the mid-1970s, Rubin and Ford returned to the study of the systematic motions of galaxies, the subject of Rubin's master's thesis. Now an observer, Rubin had more and better data than she had used as a student. When she and Ford found evidence that a large group of galaxies, including our own Milky Way, are moving rapidly with respect to the rest of the universe, astronomers took the work seriously.[7] Like Rubin's previous research, it was immediately controversial. Called the Rubin-Ford effect, it has come to be seen as strong support for a once radical point of view now gaining wide acceptance: matter is distributed through the universe in clumps, not smoothly, and the gravitational pull of the clumps accelerates groups of galaxies into motion on an immense scale. (For an example of related work, see the profile of **Sandra Faber** in this volume.)

Yet again seeking a less controversial field, Rubin and Ford began to study galaxy rotation. Planning a systematic, long-term program of observation, they hoped to measure differences in the rotation of various kinds of spiral galaxies, which would account for the differences in the galaxies' shapes. But in their first observations, made in 1978, they were surprised by a feature that proved common to all the galaxies they studied. Astronomers had previously assumed that stars at the outer margins of a galaxy travel slowly, just as the outermost planets in the solar system move most slowly in their orbits around the sun. Instead, Rubin and Ford found that stars at the outer margins of galaxies travel as fast as stars closer to the galaxy center.[8] This means that there must be lots of invisible matter even at the fringe of the galaxy where the number of visible stars dwindles, because that matter is necessary to accelerate the outer stars in their rapid orbits. Of all Rubin's unexpected and far-reaching results, this one has done most to change astronomers' view of the universe, because the amount of additional matter required to explain galaxy rotation is enormous: ten times the previously estimated mass of galaxies. What this mass is—exotic particles, cold planets, dead stars—has become one of the most pressing questions in astronomy.

Rubin has received many awards for her work: honorary doctorates from Creighton, Harvard and Yale Universities, and Williams College; and distinguished visiting professorships and fellowships at the Institute for Advanced Study, Cerro Tololo Inter American Observatory (Chile), the University of California at Berkeley and Williams College, and the Beatrice Tinsley Visiting Professorship at the University of Texas at Austin. In 1994 she was selected to deliver the Henry Norris Russell Lecture to the American Astronomical Society. She is a member of the National Academy of Sciences, and in 1993 she received a National Medal of Science, the highest scientific honor awarded by the president of the United States. In spite of her own success, she remains keenly aware of the barriers facing younger scientists, especially women. She has been a forceful spokesperson and a cherished teacher and mentor. As Rubin recently told the Association for Women in Science, "I spend an enormous amount of my time running, on the side, a business called 'trying to help young women.' "[9]

Notes

1. Vera Rubin, interviewed in Alan Lightman and Roberta Brawer, *Origins: The Lives and Worlds of Modern Cosmologists* (Cambridge, Mass.: Harvard University Press, 1990), p. 286.

2. Lightman and Brawer, *Origins*, p. 290.

3. Vera Rubin, "Differential Rotation of the Inner Metagalaxy," *Astrophysical Journal* 56 (1951): 47.

4. Vera Rubin in Marcia Bartusiak, "The Woman Who Spins the Stars," *Discover* 11 (October 1990): 90.

5. Vera Rubin, "Fluctuations in the Space Distribution of the Galaxies," *Proceedings of the National Academy of Sciences* 40 (1954): 541.

6. Vera Rubin in Sally Stephens, "Vera Rubin: An Unconventional Career," *Mercury* 21 (January/February 1992): 42.

7. V.C. Rubin, W.K. Ford Jr., and J. S. Rubin, "A Curious Distribution of Radial Velocities of ScI Galaxies with 14.0 m 15.0," *Astrophysical Journal* 183 (1973): L111; V.C. Rubin, W.K. Ford Jr., N. Thonard, and M.S. Roberts, "Motion of the Galaxy and the Local Group Determined from the Velocity Anisotropy of Distant ScI Galaxies I. The Data," *Astronomical Journal* 81 (1976): 687; and V.C. Rubin, W.K. Ford Jr., N. Thonard, and M.S. Roberts, "Motion of the Galaxy and the Local Group Determined from the Velocity Anisotropy of Distant ScI Galaxies II. The Analysis for Motion," *Astronomical Journal* 81 (1976): 719.

8. V.C. Rubin, W.K. Ford Jr., and N. Thonard, "Extended Rotation Curves of High-Luminosity Spiral Galaxies, IV. Systematic Dynamical Properties Sa-Sc," *Astrophysical Journal Letters* 225 (1978): L107; V.C. Rubin, W.K. Ford Jr., and N. Thonard, "Rotational Properties of 21 Scl Galaxies with a Large Range of Luminosities and Radii, from NGC 4605 (R = 4 kpc) to UGC 2885 (R = 122 kpc)," *Astrophysical Journal* 238 (1980): 471.

9. Barbara Mandula and Resha Putzrath, "AWIS Members Awarded Medal of Science," *AWIS [Association for Women in Science] Magazine* 23 (January/February 1994): 12.

Bibliography

Bartusiak, Marcia. "The Woman Who Spins the Stars." *Discover* 11 (October 1990): 88–94.

Lightman, Alan, and Roberta Brawer. "Vera Rubin" [interview], in *Origins: The Lives and Worlds of Modern Cosmologists*, pp. 285–305. Cambridge, Mass.: Harvard University Press, 1990.

Overbye, Dennis. *Lonely Hearts of the Cosmos: The Story of the Scientific Quest for the Secret of the Universe*. New York: HarperCollins, 1991.

Rubin, V.C. "Dark Matter in Spiral Galaxies." *Scientific American* 248 (June 1983): 96. [For a general audience.]

———. "The Dynamics of the Andromeda Nebula." *Scientific American* 228 (1973): 30. [For a general audience.]

———. "The Rotation of Spiral Galaxies." *Science* 220 (1983): 1339. [For a general audience.]

Rubin, Vera. *Bright Galaxies—Dark Matters*. Woodbury, N.Y.: American Institute of Physics, 1996. [A collection of Rubin's writings for the general reader.]
————. "Women's Work: For Women in Science, a Fair Shake Is Still Elusive." *Science* 86 (July/August 1986): 58–65.
Stephens, Sally. "Vera Rubin: An Unconventional Career." [Interview] *Mercury* (January/February 1992): 38–45.

JOANN EISBERG

FLORENCE B. SEIBERT
(1897–1991)
Biochemist

Birth	October 6, 1897
1918	A.B., Goucher College
1921–22	Van Meter Fellow, Yale University
1922–23	Porter Fellow, American Physiological Society, Yale University
1923	Ph.D., Yale University
1923–24	Porter Fellow, University of Chicago
1924–32	Sprague Memorial Institute, Chicago
1924	Ricketts Prize, Chicago
1932–37	Assistant Professor of Biochemistry, Henry Phipps Institute, University of Pennsylvania
1937–55	Associate Professor of Biochemistry, Henry Phipps Institute, University of Pennsylvania
1937–38	Guggenheim Fellow, University of Uppsala, Sweden
1938	Trudeau Medal, National Tuberculosis Association
1941	LL.D., Goucher College
1942	Garvan Gold Medal, American Chemical Society
1943	First Achievement Award, American Association of University Women
1944	National Achievement Award
1945	D.Sc., University of Pennsylvania
1946	Gimbel Philadelphia Award

1947	John Scott Award
1950	Distinguished Daughters of Pennsylvania Medal; D.Sc., Woman's Medical College
1955–59	Professor of Biochemistry, Henry Phipps Institute, University of Pennsylvania
1962	John Elliott Award
1990	Inducted into the National Women's Hall of Fame
Death	August 23, 1991

In the preface to her autobiography, *Pebbles on the Hill of a Scientist*, Florence B. Seibert summed up the story of her life's achievements in science. As a biochemist she made important contributions in the method of intravenous feeding, in the development of standards to diagnose tuberculosis, and in the fight against cancer. Seibert wrote her autobiography because she believed she had the responsibility to tell the public in terms they would understand what she had done as a scientist and how she had done it. She also hoped that her story would stimulate youth to pursue what she called a "rewarding endeavor."

Born in 1897 in Easton, Pennsylvania, Seibert contracted poliomyelitis at age 3. This disability and other bouts of ill health during her lifetime, including ongoing attacks of tachycardia, shaped her life but did not limit it. Her love of science, along with the encouragement she received from others, enabled her to persist and achieve in her chosen profession. Seibert had a strong support system, which included her family (especially her sister Mabel, who served as her assistant for many years), her professors, and a number of her colleagues.

Seibert wrote that "the necessary emphasis placed upon the physical aspects of life from childhood" had sowed in her "a seed which germinated interest in the biological sciences."[1] Her interest in science was nurtured at Goucher College, a college for women, where she learned that she "was not an invalid but was able to stand on [her] own two feet with a chance to make a contribution in the world."[2] As a graduate student in Chicago, she further developed her independence when she received "a new pair of legs" by learning to drive a specially fitted automobile. She could finally go all over the city and the countryside on her own, which she did with her little Pomeranian dog at her side. She continued to use the automobile as a way of getting around throughout her life.

Upon graduation from Goucher College in 1918, Seibert followed her chemistry professor, Dr. Jessie Minor, to Garfield, New Jersey, to work in the chemistry laboratory of a paper mill. By 1920 she had decided she wanted to study biochemistry. With the encouragement of Minor and another professor from Goucher, Seibert went to Yale University on a scholarship. It was at Yale, while working on her doctoral dissertation under Lafayette B. Mendel, that Seibert made the first of a number of discoveries for which she is remembered.

Seibert had chosen protein fevers as the subject of her doctoral research. At the time, protein injections such as milk were used to treat arthritis and other diseases. Sometimes these injections caused fevers, and Seibert's task was to determine which proteins caused them. As a result of her work on protein fevers both as a graduate student and as a postdoctoral fellow, she discovered that it was not the proteins but bacterial contamination in the distillation process that caused the fevers. She then developed a method to distill the solutions so that they were bacteria-free. This discovery had a tremendous impact on the safety of intravenous feeding.

Seibert was fortunate to have been involved, as a young scientist, "in a rare experience of immeasurable inspiration."[3] While a graduate student, she served as an assistant in research on diabetes. She wrote in her autobiography that "the thrill of seeing at first hand the emergence of such a discovery proved to me that there could be no greater challenge in life than to help track down these mysteries of our human bodies."[4]

For Seibert, the great challenge of her scientific life was in the field of tuberculosis research. In 1924 she began what was to become her life's work for the next 35 years when she became an assistant to Dr. Esmond R. Long, who had just received a grant from the National Tuberculosis Association to isolate the active substance that caused skin reactions in people who had tuberculosis. She worked with Long first in Chicago, and then she moved with him in 1932 to the Henry Phipps Institute at the University of Pennsylvania in Philadelphia.

As a result of her research, Seibert invented the first reliable test to diagnose tuberculosis by isolating the protein from the tuberculin bacillus in pure crystalline form. She also did important research on the question of immunity to the disease. Her work on tuberculosis was aided by the year she spent in Sweden (1937–1938) on a Guggenheim Fellowship, for it was there that she learned to use electrophoresis equipment that was not available in the United States. The instrument used electrical current to separate components; with this equipment, she could study tuberculoprotein molecules and purify them in a way that would not have been possible without the apparatus.

Seibert was honored in her lifetime for her work on tuberculosis. In 1938 she received the Trudeau Medal from the National Tuberculosis Association. She was the first woman to receive this award, and she wrote that it made her feel "most humble and most seriously committed to every possible further effort of which I was possible."[5] In 1942, Seibert received the Francis P. Garvan Gold Medal, an award that honored women chemists.[6] The award was for isolating the active substance in tuberculin and for preparing the International Standard of Tuberculin. Seibert also received numerous other honors and awards for her work, including five honorary degrees.

Seibert retired from the University of Pennsylvania near the end of 1958. She and her sister moved to Florida. After a five-year hiatus from active research, she found that she still had other challenges to meet. The next phase of her life was centered upon the isolation and elimination of bacteria in cancer.

Although Seibert advanced in her field as a researcher, she found early on that she and Dr. Florence Sabin were the only women present at most of the National Tuberculosis Association meetings they attended. That she was not the only woman, however, did make a difference in her life. It was to Sabin, who was on the Guggenheim board, that Seibert turned for support when applying for her Guggenheim. Seibert's mentor from her graduate student days had wanted to put her forward for this award but died before he could help her achieve the fellowship.

An article on Seibert in the September 6, 1942, edition of the *New York Times*, entitled "Her Battlefield Is a Laboratory," noted that she had become "somewhat of a legend among those who seek a preventive and an antidote for tuberculosis." The same article emphasized that Seibert did not fit the stereotype of a "scientist." At 4 feet 9 inches in height, she was a short woman who weighed less than 100 pounds. The article described her as being "deceptively different from the average idea of a distinguished scientist. She is not absentminded. . . . Her voice is . . . soft . . . , but when she talks her statements are as clearly outlined as the facets of a diamond."[7]

Throughout most of her life, Seibert worked as a scientist. Her views about who could do science were summed up in an interview: " 'Science has a lot of big men in it. . . . And big men are quick to give opportunities to women as well as to men if they see the kind of ability a scientific problem calls for and a willingness to put into it the kind of work it needs. But . . . science is not a lazy man's job—or a lazy woman's, either.' "[8]

Near the end of her life, Seibert was still being honored for her work. In 1990 she was inducted into the National Women's Hall of Fame in Seneca Falls, New York. She died in 1991 in St. Petersburg, Florida, survived by her sister Mabel.[9]

Notes

1. Florence B. Seibert, *Pebbles on the Hill of a Scientist* (St. Petersburg, Fla.: St. Petersburg Printing, 1968), p. 4.

2. Ibid., p. 9.

3. Ibid., p. 3.

4. Ibid., pp. 28–29.

5. Ibid., p. 72.

6. Kathleen McLaughlin, "Her Battlefield Is a Laboratory," *New York Times* (September 6, 1942): 15.

7. Ibid.

8. Edna Yost, *American Women of Science* (Philadelphia: Frederick A. Stokes, 1943), p. 185.

9. "Dr. Florence B. Seibert, Inventor of Standard TB Test, Dies at 93," *New York Times* (August 31, 1991): 12.

Bibliography

"Dr. Florence B. Seibert, Inventor of Standard TB Test, Dies at 93." *New York Times* (August 31, 1991): 12.

McLaughlin, Kathleen. "Her Battlefield Is a Laboratory." *New York Times* (September 6, 1942): 15.

O'Hern, Elizabeth Moot. *Profiles of Pioneer Women Scientists.* Washington, D.C.: Acropolis Books, 1985.

Seibert, Florence B. *Pebbles on the Hill of a Scientist.* St. Petersburg, Fla.: St. Petersburg Printing, 1968.

World Who's Who in Science. Edited by Allen G. Debus. Chicago: Marquis Who's Who, 1968.

Yost, Edna. *American Women of Science.* Philadelphia: Frederick A. Stokes, 1943.

ELLEN F. MAPPEN

MARY LURA SHERRILL
(1888–1968)
Chemist

Birth	July 14, 1888
1909	B.A., chemistry, Randolph-Macon College
1909–16	Instructor, Randolph-Macon College
1911	M.A., physics, Randolph-Macon College
1917–18	Adjunct Professor, Randolph-Macon College
1918–20	Associate Professor, North Carolina College for Women
1920–21	Associate Chemist, Chemical Warfare Service
1921–24	Assistant Professor of Chemistry, Mt. Holyoke College

Mary Lura Sherrill. Photo reprinted by permission of The
Mount Holyoke Archives and Special Collections.

1923	Ph.D., University of Chicago
1924–31	Associate Professor of Chemistry, Mt. Holyoke College
1931–54	Professor of Chemistry, Mt. Holyoke College
1946–54	Chair, Dept. of Chemistry, Mt. Holyoke College
1947	Garvan Medal, American Chemical Society
1954	Retired from Mt. Holyoke College
Death	October 27, 1968

Mary Sherrill continued the award-winning research tradition of Mount
Holyoke College, taking it in a new direction: the synthesis of antimalarial drugs. She was recognized not only as a consummate researcher
but also as an excellent teacher.

Mary Sherrill was born in Salisbury, North Carolina, in 1888. She was the youngest of seven children of a prominent southern family. Her mother, Sarah (Bost) Sherrill, was the daughter of a captain in the Confederate army who had been killed in the Civil War. Her father, Miles Sherrill, who had also been a soldier in the Confederate army, became a member of the state Senate, a trustee of the Davenport College for Women, and the state librarian. One of her brothers, Joseph Garland Sherrill, became a distinguished surgeon and a founding fellow of the American College of Surgeons. Another brother, Clarence, was an engineer who supervised the reconstruction of the Lincoln Memorial in Washington, D.C., and served as an aide to Presidents Theodore Roosevelt, Warren Harding, and Calvin Coolidge. He later became the city manager of Cincinnati, Ohio. Clarence humorously told the press how proud he was of his sister, Mary: "It sure is great to have somebody in the family get ahead."[1]

Mary attended local public schools in North Carolina and then entered Randolph-Macon Women's College in Lynchburg, Virginia, in 1906. Her interest in chemistry was awakened during her freshman year by Professor Fernando Wood Martin. After receiving her B.A. in chemistry in 1909, she stayed on at Randolph-Macon as an assistant in chemistry, continuing to take courses and receiving an M.A. degree in physics in 1911. She then taught at Randolph-Macon from 1911 to 1916.

Sherrill entered the Ph.D. program in chemistry at the University of Chicago in 1916. Owing to the presence of Julius Stieglitz in the chemistry department there, the program was especially welcoming to women students. Spending the academic year 1916–1917 at the University of Chicago, Sherrill returned there every summer from 1917 to 1920. During the rest of the year she taught at Randolph-Macon (1917–1918) and North Carolina College for Women (1918–1920). Her graduate research focused first on the synthesis of barbiturates and later on methods of synthesizing esters of methylenedisalicylic acid.

When the United States entered World War I in 1917, Stieglitz, then president of the American Chemical Society, helped organize American chemists for the war effort. Stieglitz was appointed consultant to the Chemical Warfare Service (CWS) in 1918, and Sherrill became a research associate for the CWS. She worked full-time for the CWS during 1920–1921. During this time she worked on the synthesis of a gas that would cause sneezing. According to one account, this gas "mixed in small quantities with illuminating gas would prevent accidental asphyxiation because of its immediate distasteful effects."[2] According to another account, the purpose of the research was to develop a less toxic sneeze gas than the Germans had used in World War I. The German sneeze gas penetrated ordinary gas masks and caused violent sneezing, compelling soldiers to remove their masks in an atmosphere that might contain lethal war gases.[3] Whatever the purpose of the gas, Sherrill held the patent on its commercial production.

Stieglitz continued to serve as her mentor. In 1921 he recommended her to **Emma Perry Carr** for a position at Mount Holyoke College in South Hadley, Massachusetts. Carr, then chair of the Mount Holyoke chemistry department, had also been mentored by Stieglitz in her years at the University of Chicago. Sherrill came to Mount Holyoke as an assistant professor and in her first years there finished her doctoral dissertation, receiving her Ph.D. from the University of Chicago in 1923. She was promoted to associate professor in 1924 and to professor in 1931. In 1946, Sherrill succeeded Carr as department chair. She and Carr were close friends. They carried out joint research, shared a home, and traveled together in the summers and after Sherrill's retirement in 1954. One of their colleagues recalled: "She was very devoted to Miss Carr and to carrying out anything that Miss Carr wanted to carry out and they worked very well together. She was a good teacher and a good scientist in her own right, but somehow we thought of them together."[4]

The Mount Holyoke chemistry department, under Carr's leadership, had organized as a group research team including undergraduates, master's candidates, and faculty. The research focused on the ultraviolet spectroscopy of organic molecules. In order for the spectroscopic data to be useful, the organic compounds had to be synthesized and purified. This was Sherrill's contribution. In 1928–1929, Sherrill received a fellowship to work in the laboratories of Jacques Errera in Brussels and Johannes van der Waals in Amsterdam, learning new purification techniques and studying how the dipole moments of organic compounds could be used to determine molecular structures. During World War II, Sherrill directed a new area of research at Mount Holyoke: the synthesis of new antimalarial drugs to replace quinine, which was no longer available.

Carr had been the first recipient of the American Chemical Society's Garvan Medal in 1937. Sherrill received the Garvan Medal in 1947 for her work on antimalarials as well as for her teaching. (**Lucy W. Pickett**, a third Mount Holyoke chemist, whose senior research thesis had been directed by Sherrill, was awarded the Garvan Medal in 1957.) After the awarding of the medal, one of the local newspapers described the Carr-Sherrill team in these words: "How they have been able to stand the pace has always been a topic of interesting debate. . . . Maybe the answer lies in a gorgeous sense of humor and an ability to get rest by jumping from one task to another like lads on the flying trapeze with 'the greatest of ease.' "[5]

Sherrill lived in South Hadley after her retirement but moved back to North Carolina when her health began to fail in 1961. She died on October 27, 1968, in High Point, North Carolina.

Notes

1. *Cincinnati Post* (September 17, 1929): n.p. (In Sherrill files in Mount Holyoke College Library/Archives.)

2. Ibid., n.p.

3. George Fleck, "Mary Lura Sherrill," in *Women in Chemistry and Physics: A Biobibliographic Sourcebook*, eds. L.S. Grinstein, R.K. Rose, and M. Rafailovich (Westport, Conn.: Greenwood Press, 1993), p. 534.

4. Interview with Lucy Weston Pickett by Carole B. Shmurak, May 1990. (Tape in Mount Holyoke College Library/Archives.)

5. *Holyoke Transcript-Telegram* [Holyoke, Massachusetts] (April 17, 1947), n.p. (In Sherrill files in Mount Holyoke College Library/Archives.)

Bibliography

Fleck, George. "Mary Lura Sherrill," in *Women in Chemistry and Physics: A Biobibliographic Sourcebook*. Edited by L.S. Grinstein, R.K. Rose, and M. Rafailovich. Westport, Conn.: Greenwood Press, 1993.

Jennings, Bojan. "The Professional Life of Emma Perry Carr." *Journal of Chemical Education* 63 (1986): 923–927.

Sherrill, Mary L. "Group Research in a Small Department." *Journal of Chemical Education* 34 (1957): 466–470.

———. "The Reaction of Bromonitromethane with Aromatic Compounds in the Presence of Aluminum Chloride." *Journal of the American Chemical Society* 46 (1924): 2753–2758.

———. "The Relation of Research to Teaching in a Liberal Arts College." *Journal of Chemical Education* 25 (1948): 512–514.

Sherrill, Mary L., et al. "Quinazoline Derivatives I." *Journal of the American Chemical Society* 68 (1946): 1299–1301. [This volume contains five articles by Sherrill and her team on their wartime efforts to synthesize antimalarial drugs.]

CAROLE B. SHMURAK

VANDANA SHIVA

(1952–)

Physicist, Environmentalist

Birth	1952
1972	B.Sc., physics, Punjab University, India

1974	M.Sc., physics, Punjab University, India
1979	Ph.D., Western Ontario University, Canada
1980	Joined Indian Institute of Management, Bangalore
1982	Consultant, United Nations University
1990–	Director, Research Foundation for Science, Technology, and Natural Resource Policy, Dehradun, India
1992	Global 500 Award, United Nations Environmental Program
1993	Earth Day International Award

Vandana Shiva is one of the most renowned women scientists in the world. "She is an Indian physicist and feminist militant who works with many community action groups fighting against environmental destruction."[1] She holds the position of director of the Research Foundation for Science, Technology, and Natural Research Policy in Dehradun, India. She is also the Science and Environment Advisor of the Third World Network. Shiva might be called the Rachel Carson of India. "Vandana Shiva is one of the world's most prominent radical scientists."[2]

Shiva, born Garewal in the Himalayan region in 1952, describes briefly her family background: "I was lucky to have been born the daughter of a forester in India and to have grown up in the Himalayan forest."[3] Two personal episodes in her life changed her career and outlook on science. As a young student she wanted to be a nuclear scientist, but her sister made her aware of nuclear hazards. As a young female scientist at the Atomic Research Institute, India, Shiva found only male domination rather than answers to her inquiries. She left her job and went to Canada, enrolling in the Foundation of Physics Program, where her questions were answered. The second episode took place at her first childbirth when the medical doctor advised a cesarean section because she was too old to have a normal delivery despite her excellent health. She left the doctor's office, consulted another doctor, and had a normal, natural delivery with a healthy child. She thereby experienced the "dominant knowledge." As a result of these personal experiences she became more involved with ecology, preservation of Himalayan forests, and the Chipko movement.

Shiva's name is closely related to ecofeminism and the Chipko movement. In the words of Judith Hall, "Shiva and Mies claim to have constructed a global ecofeminist discourse."[4] The Chipko movement was started by women in the Himalayan region of India to stop deforestation by the timber industry and soil erosion, to protest the lack of cattle feed, and to earn a livelihood making handicrafts. Women hug the trees that contractors arrive to fell. They are in the forefront of the movement. The Chipko movement accomplished a logging ban in the Himalayan area, where, above 1,000 meters, no commercial logging is now allowed. Shiva credits the Chipko movement with raising her awareness:

> It was the Chipko that taught me also about the politics of diversity. Maintenance of diversity in nature and society is the only survival politics . . . suicidal forces. It is a matter of gender conflict, the men will always want to cultivate cash crops, and the women will always want the subsistence crop. Women want plants that feed people and feed the soil, men want plants that will feed the market.[5]

Shiva received the 1994 Right Livelihood Award of $2 million, shared with four women from other countries. The citation reads: "Vandana Shiva . . . is an effective activist on a wide range of issues throughout India. She has argued against the mode of unsustainable development pushed by the World Bank and promoted a participatory process that articulates and endorses native people's knowledge. She has also been involved in initiating a national campaign to save indigenous seeds and resist the patenting of life forms."[6] This international award began in 1980 and is considered an alternative Nobel Prize. Its purpose is to celebrate and financially support people who have rejected the notion that nothing can be done about the global slide toward injustice and destruction. The award offers inspiration to social and environmental activists everywhere.

Shiva describes how the World Bank policies and

> "Miracle" seeds from the West destroyed the agricultural diversity and the social stability of the Punjab state in India. Farmers became overdependent on fertilizers and pesticides to protect their few strains of wheat and rice. Conflicts with Rajastan and Harayana over water resources increased because these "Miracle" crops needed intensive irrigation. The worsening of the lot of Punjab peasants contributed to the development of Punjab nationalism, discontent and violence.[7]

Conservation of biodiversity in tropical forests is Shiva's focus. She refutes the notion that biodiversity can be achieved if commercial interests are used to value genetic resources. She asserts that forest dwellers and indigenous workers in the forests have maintained biodiversity for thousands of years. The emerging biotechnologists will erode biodiversity by increasing uniformity in production. She expresses her concern about the conservation of diversity and the 25 percent disappearance of the world's flora and fauna species over the next generation in her book *Biodiversity: Social and Ecological Perspectives* (1991).

Shree Venkatram, while describing the Western multinational corporations' struggle for a patent to market Indian Neem leaves as natural pesticides, quotes Vandana Shiva's words: "The West does not see technologies that are developed by non-Western cultures and indigenous communities as technologies evolved by human societies, but as part of nature."[8] Shiva, as India's delegate to the 1992 International Earth Summit in Rio de Janeiro, Brazil, gave the message to the attendees about the interrelation of development, ecology, gender, race, and class at the global level.

Shiva's strong criticism of environmental degradation issues was clear in her presentation at a 1992 Chicago meeting:

> Environmentalists of the North need to learn that the poor of the world should not be made scapegoats for ecological degradation. It is not the South's overpopulation that is taxing the Earth's resources, so much as the World Bank development policies and the unwillingness of governments of the North to make corporations accountable for the environmental damage they do. Are we going to move into an era of environmental apartheid, where the North becomes clean and stays rich while the South stays poor and becomes the toxic dump of the world?[9]

Shiva has contributed to environmental protection, ecology, feminism, and the Chipko movement through numerous books, articles, and presentations at national and international conferences. She continues to be an active voice for the conservation of biodiversity, offering intelligent criticism of the hegemony of male-dominated, rich, northern hemisphere corporations and societies, a criticism based on a fundamental belief in nature itself.

Notes

1. Judith Bizot, "Vandana Shiva, Indian Scientist, Discusses Environmental Protection" (Interview) *UNESCO Courier* (March 1992): 8.

2. Vandana Shiva, *The Violence of the Green Revolution* (London and Atlantic Highlands, N.J.: Zed Books, 1991), cover.

3. Bizot, "Vandana Shiva," p. 8.

4. Judith K. Hall, "Ecofeminism and Geography," Master's thesis, Virginia Polytechnic and State University, 1994, pp. 13–16.

5. Vandana Shiva, "Greening of India: Environmental Activism," *India Currents* (July 3, 1992): 31.

6. Andrea Honebrink, "Right Livelihood Award Winners: Alternative Nobel Prize Honors Five Courageous Winners (Five Women)," *Utne Reader* 62 (1994): 47.

7. Vandana Shiva et al., *Biodiversity: Social and Ecological Perspectives* (London and Atlantic Highlands, N.J.: Zed Books, 1991), cover.

8. Shree Venkatram, "Neem Tree Is Closely Linked to the Life of Most Indians: Claim to Neem," *India Worldwide* (February 28, 1995): 38.

9. Vandana Shiva, "Greening of India: Environmental Activism," p. 31.

Bibliography

Biopolitics: A Feminist and Ecological Reader on Biotechnology. Edited by Vandana Shiva and Ingunn Moser. London and Atlantic Highlands, N.J.: Zed Books, 1995.

Bizot, Judith. "Vandana Shiva, Indian Scientist, Discusses Environmental Protection." [Interview] *UNESCO Courier* (March 1992): 8–11.

Close to Home: Women Reconnect Ecology, Health, and Development Worldwide. Edited by Vandana Shiva. London: Earthscan Publications, 1994.

Honebrink, Andrea. "Right Livelihood Award Winners: Alternative Nobel Prize Honors Five Courageous Winners (Five Women)." *Utne Reader* 62 (1994): 46.

Mies, Maria, and Vandana Shiva. *Ecofeminism.* London and Atlantic Highlands, N.J.: Zed Books, 1993.

Naramsihan, Shakuntala. "The Roots of a Movement: India Chipko Movement to Save Trees." *Connexions* 41 (Winter 1993): 22.

Shiva, Vandana. "Greening of India: Environmental Activism." *India Currents* (July 3, 1992). (Published online, Ethnic Newswatch Database.)

———. *Monocultures of the Mind: Perspectives on Biodiversity and Biotechnology.* London and Atlantic Highlands, N.J.: Zed Books, 1993.

———. *Staying Alive: Women, Ecology, and Development.* London: Zed Books, 1988.

———. *The Violence of the Green Revolution.* London and Atlantic Highlands, N.J.: Zed Books, 1991.

Shiva, Vandana, et al. *Biodiversity: Social and Ecological Perspectives.* London and Atlantic Highlands, N.J.: Zed Books, 1991.

Venkatram, Shree. "Neem Tree Is Closely Linked to the Life of Most Indians: Claim to Neem." *India Worldwide* (February 28, 1995): 38.

SAROJINI LOTLIKAR

JEAN'NE MARIE SHREEVE
(1933–)
Chemist

Birth	July 2, 1933
1953	B.A., Montana State University

Jean'ne Marie Shreeve. Photo courtesy of University of
Idaho Photo Services.

1956	M.S., analytical chemistry, University of Minnesota
1961	Ph.D., inorganic chemistry, University of Washington
1961–65	Assistant Professor of Chemistry, University of Idaho, Moscow
1965–67	Associate Professor of Chemistry, University of Idaho, Moscow
1967–73	Professor of Chemistry, University of Idaho, Moscow
1967–68	National Science Foundation Fellow, University of Cambridge; U.S. Ramsey Fellow
1970	Distinguished Alumni Award, University of Montana
1970–72	Alfred P. Sloan Fellow
1972	Honorable Alumna, University of Idaho; Garvan Medal, American Chemical Society
1973–87	Head, Dept. of Chemistry, and Professor, University of Idaho

1975	Outstanding Achievement Award, University of Minnesota
1975–81	Petroleum Research Fund Advisory Board
1978	Senior U.S. Scientist Award, Alexander Von Humboldt Foundation; Fluorine Award, American Chemical Society
1980	Excellence in Teaching Award, Chemical Manufacturers Association
1985–93	Board of Directors, American Chemical Society
1987–	Vice Provost of Research and Graduate Studies, Professor of Chemistry, University of Idaho
1991–	Board of Directors, American Association for the Advancement of Science
1992–	Board of Governors, Argonne (IL) National Laboratory; Harry and Carol Mosher Award, Santa Clara Valley Section, American Chemical Society

Jean'ne Marie Shreeve, internationally known and nationally recognized for her contributions to the understanding of synthetic fluorine chemistry, was born at Deer Lodge, Montana, in 1933 to Charles W. and Maryfrances (Briggeman) Shreeve. They named her after a popular tune of the time, "Jeannine, I Dream of Lilac Time." Her brother William Charles, who became a close pal, had arrived 22 months earlier. As elsewhere in the country during the Depression era, life was tough in rural Montana. Shreeve's father was employed three days a week by the Northern Pacific Railroad, but her mother, who was trained as an elementary schoolteacher, was unable to find a teaching position—primarily because of an idea prevalent in society that only one member of each family should be gainfully employed. In the late 1930s, however, her mother did resume a teaching career, which spanned 40 years in the public schools of western Montana. Her father continued working in supervisory roles for the Northern Pacific Railroad until he retired. Both parents encouraged Shreeve and her brother to enjoy school and its activities. They also taught them a maxim for successful living: "Anything worth doing is worth doing well."[1]

As a senior at Thompson Falls, Montana, High School, Shreeve, who was always an enthusiastic student, became interested in science—especially chemistry—when she "encountered" the science teacher Lewis Y. Leonard. Enrolling at Montana State University (now the University of Montana) as a freshman on a tuition scholarship, Shreeve juggled classes, a sorority, band, and intramural sports. She also worked in the university library with a documents librarian and scholar, Lucile Speer. Later she worked in the chemistry department stockroom. After earning her B.A. in 1953, Shreeve taught lower division mathematics for a year at Missoula County High School in Montana. Then she began work on an M.S. degree in analytical chemistry at the University of Minnesota and was granted that degree in 1956.

When Shreeve matriculated at the University of Washington as a graduate student, she met Professor George A. Cady, who introduced her to the beauty of fluorine chemistry. He became a strong influence in her professional life. Completing her Ph.D. in inorganic chemistry in 1961, she accepted a one-semester appointment as assistant professor of chemistry at the University of Idaho and a mutually beneficial lifelong relationship commenced. At the start of her appointment, the chemistry department was poorly equipped to support research. The state, however, had just designated the campus at Moscow as Idaho's research university and granted it permission to award doctoral degrees. A permanent appointment soon opened for Shreeve, and she began her career working with Dr. Malcolm M. Renfrew, head of the chemistry department. Renfrew had returned to the University of Idaho after a very distinguished career in the chemical industry, where he had developed Teflon as a DuPont company researcher. Shreeve found him to be a "superb role model." Soon her research was booming, her teaching was recognized by students, and her service activities were proliferating. By 1964 she had established an aggressive research group at Idaho, obtained funds and equipment, and was publishing research results from work undertaken and completed in her department.

In 1967, Shreeve earned a full professorship at the University of Idaho and spent a year at Cambridge University as a National Science Foundation Fellow and Ramsey Fellow. There she worked with the great scholar Professor H. J. Emeleus. Her year at Cambridge was replete with productive work and stimulating scientific visitors from around the world. There was still time for rugby, pubs, a fellow at Newnham College, spring break in Greece, and the formation of a lifelong fishing relationship with the professor.

By 1973 Shreeve had succeeded her mentor, Malcolm Renfrew, as head of the chemistry department at the University of Idaho. In 1977 she spent a sabbatical year at Bristol University in the United Kingdom and as a visiting professor and an Alexander von Humboldt Senior Scientist awardee at Göttingen, Germany. This experience allowed her to be associated with Dr. Oskar Glemser, the director of the Institute for Inorganic Chemistry at the University of Göttingen. In 1987 Shreeve was appointed vice-president for research and graduate studies at the University of Idaho. At the same time, she maintained an active research group in fluorine chemistry. In this position, her University responsibilities extended throughout that state as a result of her leadership in obtaining and directing a National Science Foundation Experimental

Program to Stimulate Competitive Research (EPSCoR) grant to fund expanded research activity at the University of Idaho.

The major emphasis of Shreeve's research in the area of synthetic fluorine chemistry has been the synthesis, characterization, and reactions of fluorine compounds containing nitrogen, sulfur, and phosphorus. Shreeve and her students made a significant "find" for chemistry when they discovered the compound perfluorourea, a long-sought oxidizer ingredient. In her laboratory, she also developed new synthetic routes to several important compounds, including chlorodifluoroamine and difluorodiazine. Preparation of both of these compounds, which are utilized in synthesizing rocket oxidizers, was hard to accomplish by previously known techniques. Shreeve also evinced that controlled hydrolysis of bis(perfluoroalkyl) sulfur difluorides results in totally fluorinated alkyl sulfoxides. Her research has been reported in more than 270 papers that have been published in at least 29 journals, advances, and reviews worldwide. Her work as a fluorine chemist earned her the 1972 American Chemical Society's Garvan Medal, one of only two national awards available at that time to recognize outstanding achievement in chemistry by an American woman chemist each year. The honor cited her contributions to the fundamental understanding of the behavior of inorganic fluorine compounds and to the synthesis of important new fluorochemicals. Other honors for her contributions to chemistry include an Alfred P. Sloan Foundation Fellowship in 1970; a doctor of science degree from the University of Montana in 1982; an Outstanding Achievement Award by the University of Minnesota in 1975; the American Chemical Society's Award for Creative Work in Fluorine Chemistry in 1978; and, recently, the Harry and Carol Mosher Award of the Santa Clara Valley Section of the American Chemical Society.

In addition to performing research, Shreeve has devoted her life to educating other chemists. By working to improve the curriculum at the departmental and university levels, she has drawn qualified undergraduates into a productive research program. She regards the most important contribution of her life to be the 120 to 130 M.S., Ph.D., and postdoctoral students whom she has guided and directed in their study and research and who now are co-workers in her academic family. All these former students, with one exception, practice chemistry or a closely related field in the United States or abroad. Some have successful chemical businesses; some are college professors; others are in the chemical industry. Following her example, many of her students participate in American Chemical Society activities and serve as councilors and as chairs of sections and divisions. Shreeve's effectiveness as a teacher of undergraduate and graduate students was recognized in 1979 when the Manufacturing Chemists Association selected her to receive the College Chemistry Teaching Award.

Shreeve's continued promotion of teaching and research in chemistry is paralleled only by her exceptional and distinguished service to the American Chemical Society as a member of editorial boards and local and national committees. Her activities have included service as Inorganic Program chairman for regional meetings; officer and/or councilor of the Washington-Idaho Border Section, the Fluorine Division, and the Inorganic Division; and director of Region VI. She also has served as a member of the Petroleum Research Fund Advisory Board and the Board of Directors of the American Association for the Advancement of Science. Currently she is a member of the University of Chicago Board of Governors for Argonne National Laboratory. Her service on editorial boards includes the *Journal of Fluorine Chemistry* since 1970; *Accounts of Chemical Research*, 1973–1975; *Inorganic Synthesis* since 1976; and *Heteroatom Chemistry* since 1988. This service reflects her deep and continuing interest in research, which she regards as "important and rewarding . . . to all who are faculty members at a research university."

Today, Shreeve enjoys puttering in the garden, fishing, skiing, and driving fast. Most of all, however, she delights in chemistry and her co-workers, both in the lab and in the research office. A true pioneer, she has left as her legacy to the University of Idaho a distinguished program in chemistry that has achieved international recognition. A co-worker stated that her achievements were possible because of her "own personal creativity, tenacity and love of chemistry."

Note

1. Most of the information for this essay came from typewritten notes faxed to Cassandra Gissendanner by Dr. Shreeve on December 18, 1995. Some additional information was provided by a colleague of Dr. Shreeve.

Bibliography

American Men and Women of Science, 1995–96. 19th ed. New Providence, N.J.: R.R. Bowker, 1994.

Burton, D.J., A.S. Nodak, R. Guneratne, Debao Su, Wenbiao Cen, Robert L. Kirchmeier, and Jean'ne M. Shreeve. "Synthesis of (Sulfodifluoromethyl) Phosphonic Acid." *Journal of the American Chemical Society* 111 (1989): 1723.

"Garvan Medal." *Chemical and Engineering News* 50 (April 10, 1972): 41.

Guo, Cai-Yun, Robert L. Kirchmeier, and Jean'ne M. Shreeve. "A Thirty-Two-Membered Fluorinated Multifunctional Heterocycle." *Journal of the American Chemical Society* 113 (1991): 9000.

Gupta, Om Dutt, Robert L. Kirchmeier, and Jean'ne M. Shreeve. "Reactions of Trifluoroamine Oxide: A Route to Acyclic and Cyclic Fluoroamines, and N-Nitrosoamines." *Journal of the American Chemical Society* 112 (1990): 2383.

Kitazume, Tomoya, and Jean'ne M. Shreeve. "Some Chemistry of Fluorinated Octahedral Sulfur Compounds." *Journal of the American Chemical Society* 100 (1978): 492.

Su, Debao, Wenbiao Cen, Robert L. Kirchmeier, and Jean'ne M. Shreeve. "Synthesis of Fluorinated Phosphonic, Sulfonic, and Mixed Phosphonic/Sulfonic Acids." *Canadian Journal of Chemistry* 67 (1989): 1795.

Who's Who of American Women, 1995–96. 19th ed. New Providence, N.J.: Marquis Who's Who, 1995.

"Women Chemists." *Chemtech: The Innovator's Magazine* 6 (December 1976): 738–743.

CASSANDRA S. GISSENDANNER

SOFIA SIMMONDS
(1917–)
Biochemist

Birth	July 31, 1917
1938	B.A., chemistry, Barnard College at Columbia University
1941–42	Assistant Biochemist, Medical College of Cornell University
1942	Ph.D., biochemistry, Cornell University
1942–45	Research Associate, Medical College of Cornell University
1945–46	Instructor in Physiological Chemistry, School of Medicine, Yale University
1946–49	Microbiologist, School of Medicine, Yale University
1949–50	Assistant Professor, School of Medicine, Yale University
1950–54	Assistant Professor of Biochemistry and Microbiology, School of Medicine, Yale University
1954–63	Associate Professor of Biochemistry and Microbiology, School of Medicine, Yale University
1962–69	Biochemist, School of Medicine, Yale University
1969	Garvan Medal, American Chemical Society
1969–75	Molecular Biophysicist and Biochemist, School of Medicine, Yale University
1973–85	Director of Undergraduate Studies, Dept. of Molecular Biophysics and Biochemistry, School of Medicine, Yale University
1975–88	Professor, School of Medicine, Yale University

1988	Associate Dean and Dean of Undergraduate Studies, School of Medicine, Yale University
1988–	Professor Emerita
1990–91	Lecturer and Dean of Undergraduate Studies, School of Medicine, Yale University

Scholar, chemist, researcher, author, professor, and administrator, Sofia Simmonds has enjoyed a productive career. She earned her B.A. in chemistry from Barnard College at Columbia University in 1938 and her Ph.D. in biochemistry from Cornell University in 1942. She began her professional life as an assistant biochemist at the Medical College of Cornell University. She remained in that position from 1941 to 1942, and then she served as a research associate until 1945. Since 1945, Simmonds has worked at Yale University, where she has steadily risen through the faculty ranks and has also held administrative positions.[1]

Simmonds's research work concentrated on the study of amino acid metabolism of bacteria. She determined the efficiency of specific peptides by mutant strains that require one of the component amino acids for growth. Evidence of Simmonds's research can be found on the pages of numerous scientific publications, including *Journal of Biological Chemistry*, *Science*, and *Biochemistry*. Simmonds has published over 40 articles during her lengthy career. She is a member of the American Society of Biological Chemists and the American Chemical Society. For five years she served on the editorial board of the *Journal of Biological Chemistry*.

Among her many accomplishments is *General Biochemistry*, the textbook she co-authored with her husband, Dr. Joseph S. Fruton. First published in 1954 and revised in 1958, this text is credited with helping educate a generation of biochemists who have been responsible for recent advances in understanding biology and biochemistry.

In 1969, Simmonds was named recipient of the American Chemical Society's Garvan Medal. The award, given yearly since 1937, recognizes distinguished service to the field of chemistry by women chemists who are citizens of the United States. The award (now called the Francis P. Garvan–John M. Olin Medal) consists of a prize of $5,000, an inscribed gold medal, a bronze replica of the medal, and a travel allowance to attend the meeting where the award is presented.

At the Medical School of Yale University, Simmonds is a recognized face and name. For over 40 years she has been a valuable member of the faculty. She taught one of the basic courses in biochemistry for several years. She helped build and shape the Department of Molecular Biophysics and Biochemistry. As a professor emerita, she stepped into the role of dean of undergraduate studies when the department needed an interim leader.

Simmonds is an inspiration to all who follow in her footsteps. Her scholarship and intelligence have benefited both those who know her and those who know her work. Even in retirement, she remains active at Yale.

Note

1. Information for this essay was taken from Dr. Simmonds's curriculum vitae, six pages, 1995.

Bibliography

Altman, Sidney. "Tribute to Sofia Simmonds." Citation read at Yale College Faculty Meeting, May 1988.
American Men and Women of Science. 19th ed. New York: R.R. Bowker, 1994.
Bailey, Martha P. "Sofia Simmonds: Biochemist," in *American Women in Science: A Biographical Dictionary*, pp. 357–358. Denver, Colo.: ABC-CLIO, 1994.
"Garvan Medal—Sofia Simmonds." *Chemical and Engineering News* 47, no. 8 (1969): 100.

SYLVIA NICHOLAS

CHARLOTTE EMMA MOORE SITTERLY
(1898–1990)
Astrophysicist

Birth	September 24, 1898
1920	A.B., mathematics, Swarthmore College
1920–25	Mathematics Computer, Princeton Observatory
1925–28	Mathematics Computer, Mt. Wilson Observatory
1928–29	Mathematics Computer, Princeton Observatory
1931	Ph.D., astronomy, University of California, Berkeley
1931–36	Research Assistant, Princeton Observatory
1936–45	Research Associate, Princeton Observatory
1937	Annie J. Cannon Prize; married Bancroft W. Sitterly on May 30
1945–68	Physicist, Atomic Physics Division, National Bureau of Standards

Charlotte Emma Moore Sitterly. Photo reprinted by permission of Friends Historical Library of Swarthmore College.

1951	Silver Medal, Department of Commerce
1960	Gold Medal, Department of Commerce
1961	Federal Woman's Award, U.S. Civil Service Commission
1962	D.Sc., Swarthmore College
1963	Annie Jump Cannon Centennial Medal, Wesley College
1966	Career Service Award, National Civil Service League
1968	Honorary Doctorate, Universität zu Kiel, Germany
1968–71	Consultant, Office of Standard Reference Data
1971	D.Sc., University of Michigan
1971–78	Consultant, U.S. Naval Research Laboratory
1972	William F. Meggers Award, Optical Society of America
Death	March 3, 1990

Astrophysicist Charlotte Emma Moore Sitterly was one of the foremost authorities in the world on the composition of the sun and atomic spectra. Her critical analyses and organization of knowledge of atomic spectra formed the definitive reference sources used for decades in such fields as astronomy, laser physics, and spectral chemistry.

Born in 1898 in Ercildoun, Pennsylvania, Charlotte was one of six children. Her father, George Winfield Moore, was the superintendent of schools in Chester County; her mother, Elizabeth Palmer (Walton) Moore, was a schoolteacher. She attended public schools and then Swarthmore College in Swarthmore, Pennsylvania. She excelled in her studies and was active in extracurricular activities such as the glee club, the hockey team, and student government. She held a working fellowship, tutored students, and assisted faculty members to help pay for her expenses.[1] She graduated with a bachelor of arts degree in mathematics in 1920 and was elected to Phi Beta Kappa.

Encouraged by her physics instructor at Swarthmore, Charlotte Moore accepted her first job as a mathematics computer at the Princeton University Observatory. This was one of the few avenues of employment open at the time to women interested in the sciences. She initially was assigned to the laborious measurement and calculation that formed the "women's work" of astronomy, but she also had the opportunity to work with the esteemed astronomer Henry Norris Russell (H.N.R.), who was involved in spectroscopic studies. She described her own experience:

In complete ignorance of the requirements needed for this job, I reported for duty and was cordially received, but without fanfare assigned my first task, the photographic determination of the position of the moon.

The photographs were taken at the Harvard College Observatory by E.S. King, but they were sent to Princeton to be reduced by the method devised by H.N.R. The measuring had to be done on the ancient machine that I believe he brought from Cambridge, England. This machine had two weights attached to fishing cords, which were wound around the threads of long screws that controlled the settings in the x- and y-coordinates. In the midst of measuring a plate, one became accustomed to having the plate holder take a sudden crash whenever a fishing cord broke. I had never encountered such a contraption. H.N.R. preferred ingenuity to good equipment.

I had a rude awakening, suddenly being confronted with the long and complicated method involved: least squares solutions of more equations than unknown quantities, endless use of logarithm tables, strange correction factors, etc. At that time there was not even a desk calculator available. When all went well, the best I could manage was one plate-reduction per day.[2]

Charlotte Moore later collaborated with Henry Norris Russell on the analysis of complex atomic spectra and the study of double stars. The experience of working with Russell was to inspire her throughout her entire career. Five years after the start of her career at Princeton, she went to the Mount Wilson Observatory in Pasadena, California, to work on the preparation and publication of a new analysis of the solar spectrum. Recent photographs of solar spectra had been made with the large equipment at Mount Wilson. In addition, on a worldwide basis, great progress was emerging in laboratory spectroscopy. These new laboratory spectral data were assembled and used to interpret the Mount Wilson images. The wavelengths were also redetermined on the new International Angstrom scale. The results were published in 1928 in a monograph entitled *Revision of Rowland's Preliminary Table of Solar Spectrum Wavelengths with an Extension to the Present Limit of the Infrared*. For nearly four decades, this book stood as the authoritative work on the solar spectrum.[3]

Moore returned to Princeton in 1928 and then resigned a year later to pursue an advanced degree in astronomy at the University of California at Berkeley, where she held the highly prized Lick Fellowship. She continued her analysis of solar spectra by making use of her extensive collection of data on the spectra of chemical elements and molecules. By examining the light from sunspots, she was able to separate atomic from molecular lines and to identify many previously unknown lines. This formed the basis for her dissertation.

In 1931, after completing her graduate studies, Moore returned again to Princeton to work with Russell as a research assistant. By 1936 she had been promoted to associate. From 1931 to 1945 she worked on the assembly and critical analysis of the rapidly increasing body of data on atomic spectra. Moore prepared new tables that would make this large mass of data easily available to astronomers and physicists. A preliminary edition was published in 1933, but the work continued even during the wartime years. The project was completed with the revised edition of *A Multiple Table of Astrophysical Interest*.

On May 30, 1937, Charlotte married a colleague—astronomer and physicist Dr. Bancroft Walker Sitterly, who was then a member of the faculty at Wesleyan University. That year she also received the prestigious Annie J. Cannon Prize, awarded at that time every three years for distinguished contributions to astronomy by a woman of any nationality. One of Charlotte Moore Sitterly's contributions was the identification of the element technetium in the spectrum of sunlight, the first indication that this element exists in nature. Previously, the existence of technetium had been theoretically predicted but only produced artificially in a particle accelerator.

Her work at Princeton caught the attention of William F. Meggers, who was considering assembling data on atomic energy levels. Russell urged the National Bureau of Standards to support the project and nominated Charlotte Moore Sitterly as the most qualified person to take on the preparation of the data. In 1945 she accepted employment at the National Bureau of Standards in the Spectroscopy Section of the Atomic Physics Division. To collect the necessary data, she not only relied on published sources but also

> implored, cajoled, and persisted until she got her hands on unpublished material. Then the work began. She carefully checked all data for internal consistency and for conformance with what she knew about complex spectra. She was anxious to make the tables as useful as possible and surveyed the scientific community with an extensive and well-thought-out questionnaire to determine the optimum format for the presentation of the data. She knew what data were needed to fill in gaps in our knowledge of these spectra, and she was most successful in convincing spectroscopists that they must work to fill in the blank spaces. She viewed her responsibility as going far beyond the assembly of available data. She was determined to produce compilations of critically reviewed energy-level data as complete as possible.[4]

The volumes of *Atomic Energy Levels* were published from 1949 through 1958. In 1960, Sitterly received the Department of Commerce Gold Medal for her outstanding contribution in organizing and publishing her volumes of data and for her dedication, integrity, judgment, and leadership. In 1961 she was one of the first six women to receive the Federal Woman's Award, and in 1963 she was awarded the Annie Jump Cannon Centennial Medal.

Sitterly was an active member in numerous professional societies and was the author or co-author of nearly 100 papers and monographs. She believed travel to be one of the most significant aspects of a scientist's work because it promotes international cooperation and collaboration among scientists. In 1949 she was the first woman elected foreign associate by the Royal Astronomical Society of London. In 1958 she traveled to Moscow, a significant event at the time, to attend the tenth general assembly of the International Astronomical Union on the Joint Commission on Spectroscopy. In addition to being awarded honorary degrees by her alma mater, Swarthmore College, and the University of Michigan, she again gained international recognition by receiving an honorary doctorate from the Universität zu Kiel in Germany.

Charlotte Moore Sitterly and her husband enjoyed gardening, music, and traveling together until his death in 1977. Although Dr. Bancroft Sitterly's career was somewhat overshadowed by that of his illustrious wife, he also had been an accomplished physicist and educator. For many years he had held the position of chair of the physics department at American University in Washington, D.C. Throughout her life, Charlotte maintained close ties to her family and to the Society of Friends. In 1982 she researched and published a genealogy of her mother's family.

Charlotte Moore Sitterly had a long and illustrious career as an astronomer and astrophysicist. When spectroscopic data began to become available from instruments carried above the earth's atmosphere, she was immediately sent prints and began work on ranges of the spectra that had previously not been analyzed. She continued to work well past her mandatory retirement at the age of 70 until her death on March 3, 1990. She died in her home in Washington, D.C., of heart failure at the age of 91.

Notes

1. See *Current Biography Yearbook, 1962*, ed. Charles Moritz (New York: H.W. Wilson, 1962).

2. Charlotte E. Moore, "Collaboration with Henry Norris Russell over the Years," in *In Memory of Henry Norris Russell*, eds. A.G. Davis Philip and David H. DeVorkin (Albany, N.Y.: Dudley Observatory, 1977), pp. 27–28.

3. R. Tousey, "The Solar Spectrum from Fraunhofer to Skylab—An Appreciation of the Contributions of Charlotte E. Moore Sitterly," *Journal of the Optical Society of America, Part B* 5, no. 10 (October 1988): 2231.

4. Karl G. Kessler, "Dr. Charlotte Moore Sitterly and the National Bureau of Standards," *Journal of the Optical Society of America, Part B* 5, no. 10 (October 1988): 2045.

Bibliography

American Men and Women of Science. 13th ed. New York: R.R. Bowker, 1976.

Current Biography Yearbook, 1962. Edited by Charles Moritz. New York: H.W. Wilson, 1962.

Edlen, B., and William C. Martin. "Atomic Spectroscopy in the Twentieth Century: A Tribute to Charlotte Moore Sitterly on the Occasion of her Ninetieth Birthday." *Journal of the Optical Society of America, Part B* 5, no. 10 (October 1988): 2043–2044.

Kessler, Karl G. "Dr. Charlotte Moore Sitterly and the National Bureau of Standards." *Journal of the Optical Society of America, Part B* 5, no. 10 (October 1988): 2045.

Moore, Charlotte E. *Atomic Energy Levels as Derived from the Analyses of Optical Spectra*. Washington, D.C.: National Bureau of Standards, 1945–1958.

———. "Collaboration with Henry Norris Russell over the Years," in *In Memory*

of Henry Norris Russell. Edited by A.G. Davis Philip and David H. De-
    Vorkin. Albany, N.Y.: Dudley Observatory, 1977.
Moore, Charlotte E., and Harold D. Babcock. *The Solar Spectrum*. Washington,
    D.C.: Carnegie Institution, 1947.
Moore, Charlotte E., and Henry Norris Russell. *The Masses of the Stars*. Chicago:
    University of Chicago Press, 1940.
Sullivan, Walter. "Charlotte Sitterly, 91, Physicist: Devoted Career to Sunlight
    Studies." [Obituary] *New York Times* (March 8, 1990): D25.
Tousey, R. "The Solar Spectrum from Fraunhofer to Skylab—An Appreciation
    of the Contributions of Charlotte E. Moore Sitterly." *Journal of the Optical
    Society of America, Part B* 5, no. 10 (October 1988): 2230–2236.
Who's Who of American Women. 8th ed. Chicago: Marquis Who's Who, 1974–1975.
                                                        NANCY SLIGHT-GIBNEY

MARY FAIRFAX SOMERVILLE
(1780–1872)
Astronomer

Birth	December 26, 1780
1790–91	Attended boarding school at Musselburgh
1804	Married Samuel Grieg (died 1807)
1812	Married William Somerville (died 1860)
1816	Somervilles moved to London
1826	Mary presented paper, "The Magnetic Properties of the Violet Rays of the Solar Spectrum," to the Royal Society
1831	Published *The Mechanism of the Heavens*
1834	Published *On the Connexion of the Physical Sciences*; became honorary member of the Royal Academy of Dublin and the Société de Physique et d'Histoire Naturelle (Geneva)
1835	With Caroline Herschel became first women honorary members of the Royal Astronomical Society; became honorary member of British Philosophical Institution
1838	Moved to Italy
1848	Published *Physical Geography*
1869	Published *Molecular and Microscopic Science*
Death	1872

Although the growing appreciation of science prevailing in the early nineteenth century made her career possible, it was her own deep and sincere interest that brought a remarkable Scotswoman, Mrs. Mary Fairfax Greig Somerville, to the study of science and mathematics and guided her through her long and distinguished career. Some have felt that she was not a creative scientist in the sense that men such as Wollaston, Herschel, or Faraday were. She herself recognized the limitations of her experimental work. In a draft of her autobiography, she wrote that "in the climax of my great success [and in] the approbation of some of the first scientific men of the age and of the public in general I was highly gratified, but much less elated than might have been expected, for although I had recorded in a clear point of view some of the most refined and difficult analytical processes and astronomical discoveries, I was conscious that I had never made a discovery myself, that I had no originality." She described her situation with a curious theology: "I have perseverance and intelligence but no genius, that spark from heaven is not granted to the sex, we are of the earth, earthy, whether higher powers may be allotted to us in another existence God knows, original genius in science at least is hopeless in this."[1]

Even though there were favorable attitudes toward science in general, Mary Somerville's thoughts were not unusual for the time. Women were allowed to study botany or describe the discoveries of male scientists, but they were considered incapable of conducting original experimentation. But Somerville knew that as she satisfied her own thirst for learning, she could make contributions to the studies that were important to her and bring knowledge of them to a wider public. Throughout her life she was especially supportive of efforts to improve the intellectual and social opportunities of women. In the dedication of her book *On the Connexion of the Physical Sciences*, she wrote that she had tried "to make the laws by which the material world is governed more familiar to my country women."[2] Always a supporter of women's education, it is fitting that a women's college at Oxford—Somerville—is named after her.

Mary Fairfax was born in Burntisland, Scotland, in 1780. Her mother's attitude, common for well-connected Scottish families of that time, was that women's education was not necessary beyond learning the Bible and reading newspapers and keeping household accounts. Mary's father returned from sea duty when she was 9 years old and sent her to a popular, expensive boarding school at Musselburgh when she was 10. The twelve months she spent there were her only experience with formal, full-time instruction. Although she wrote in her autobiography that she was "like a wild animal escaped out of a cage" after this educational experience, she did gain a taste for reading and arithmetic. Because of her brightness, curiosity, and determination, she later taught herself Latin and algebra. She became aware of algebra by glimpsing some strange symbols in a ladies' fashion magazine. When her parents found out about her interest in mathematics, her father forbade her study, expressing a common fear of the day "that the strain of abstract thought would injure the tender female frame."[3] This, however, did not discourage her interest in astronomy:

> We had two small globes, and my mother allowed me to learn the use of them from Mr. Reed, the village school master, who came to teach me for a few weeks in the winter evenings. . . .
>
> My bedroom had a window to the south, and small closet near had one to the north. At these I spent many hours, studying the stars by the aid of the celestial globe. . . .
>
> In Robertson's *Navigation* I flattered myself that I had got precisely what I wanted; but I soon found that I was mistaken. I perceived, however, that astronomy did not consist in star-gazing.[4]

In 1804 Mary married a distantly related cousin, Samuel Grieg, a young British naval officer. His death in 1807 left her with two sons, a modest inheritance, and a move from London back to her parents' home. Having financial support and a continued interest in learning, she began openly educating herself in trigonometry and astronomy. Relatives and acquaintances showed no support or interest and thought her foolish or eccentric. She did find support among some of her social circle, including the founders of the *Edinburgh Review*, a journal that urged the education of women. In 1811 she won a silver medal, the first of many awards and honors, for a solution to a prize problem.[5]

Mary then married her first cousin, Dr. William Somerville, in 1812. He had lived in Canada for many years. They discovered in becoming reacquainted as adults that they had much in common besides the death of spouses and sons to rear—an interest in intellectual matters and liberal, tolerant views. William always defended Mary's pursuit of mathematics and astronomy and actually encouraged her to broaden her studies to include Greek, botany, and mineralogy. William Somerville's role as a friend of science began in Edinburgh and continued as they lived in London and on the continent. His medical position with the Army and later with Chelsea Hospital gave him access to scientific circles and enabled Mary to further her scientific pursuits.[6]

Between 1815 and 1840, Mary was actively engaged in scientific study, experimentation, and writing in both London and Paris. Scientific men of Great Britain and Europe regarded her as a cherished colleague. It was with their encouragement that she wrote *On the Connexion of the Physical Sciences* (1834), which presented a comprehensive picture of the latest research in the physical sciences, calling to the attention of both specialists and general readers much scientific work that had been little noticed. The book helped to define the unsettled term "physical science." An earlier book, *Mechanism of the Heavens* (1831), was an important contribution to the modernization of English mathematics. Although dismissed by some twentieth-century historians as popular writing by a "mere woman," these two books were well received by specialists, informed readers, and many others who became interested in science because of them.

Because Mary Somerville held liberal views, she was attached to the reform elements in science, society, and politics. However, because of her sweet character and good sense, she maintained friendly and civil relations with conservative and even reactionary circles. As a strong, rational, and compassionate woman, she was never a violent activist but preferred to work within the system. Although occasionally criticized for her "unwomanly" pursuit of science, she was accepted at home and abroad as "the premier scientific lady of the ages." Her contemporaries also considered her fulfillment of her role as wife and mother to be exemplary. According to their accounts, she carefully saw to the education of her daughters and son and managed a household with economy and style. She seemed to preserve all the traditional female traits and graces and enjoyed the arts and social occasions.[7]

Although her life appears to have been much a case of being in the right places at the right times, Mary and her husband experienced poverty and tragedy. Two sons died in childhood, and Dr. Somerville was not always in a financially secure position. The Somervilles lived in Italy for many years because of William's bad health. The marriage lasted for almost 50 years until William's death in 1860.[8] After he died, Mary remained in Italy and continued to conduct experiments (e.g., one on the solar spectrum) and to write. In 1869, when she was 89 years old, she published her last book and also received the first gold medal awarded by the Italian Geographical Society and the Victoria gold medal of the Royal Geographical Society. When she died in 1872 she was called "the queen of science" by the London *Morning Post*.[9]

Notes

1. Elizabeth Chambers Patterson, *Mary Somerville and the Cultivation of Science, 1815–1840* (The Hague: Martinus Nijhoff Publishers, 1983), p. 89.

2. Patricia Phillips, *The Scientific Lady: A Social History of Women's Scientific Interests, 1520–1918* (New York: St. Martin's Press, 1990), p. 115.

3. Patterson, *Mary Somerville*, p. 2.

4. Janet Horowitz Murray, *Strong-Minded Women, and Other Lost Voices from Nineteenth-Century England* (New York: Pantheon Books, 1982), pp. 207–208.

5. Patterson, *Mary Somerville*, pp. 4–5.

6. Ibid., p. 6.

7. Ibid., p. xii.

8. Ibid., p. 6.
9. Ibid., p. 1.

Bibliography

Alic, Margaret. *Hypatia's Heritage: A History of Women in Science from Antiquity through the Nineteenth Century.* Boston: Beacon Press, 1986.

Anderson, Bonnie S., and Judith P. Zinsser. *A History of Her Own: Women in Europe from Prehistory to the Present.* New York: Harper & Row, 1988.

Mozans, H.J. *Women in Science.* New York: D. Appleton and Company, 1913. Reprint: Cambridge, Mass.: MIT Press, 1974.

Murray, Janet Horowitz. *Strong-Minded Women, and Other Lost Voices from Nineteenth-Century England.* New York: Pantheon Books, 1982.

Patterson, Elizabeth Chambers. *Mary Somerville and the Cultivation of Science, 1815–1840.* The Hague: Martinus Nijhoff Publishers, 1983.

Phillips, Patricia. *The Scientific Lady: A Social History of Women's Scientific Interests, 1520–1918.* New York: St. Martin's Press, 1990.

Schacher, Susan, coord. *Hypatia's Sisters: Biographies of Women Scientists, Past and Present.* Seattle, Wash.: Feminists Northwest, 1976.

CAROL B. NORRIS

BETTY SULLIVAN

(1902–　)

Biochemist

Birth	May 31, 1902
1922	B.Sc., chemistry, University of Minnesota; Assistant Chemist, Russell Miller Milling Co.
1927	Chief Chemist, Russell Miller Milling Co.
1935	Ph.D., biochemistry, University of Minnesota
1947	Vice President, Director of Research, Member of the Board of Directors, Russell Miller Milling Co.
1948	Thomas Burr Osborne Medal, American Association of Cereal Chemists
1953	Outstanding Achievement Award, University of Minnesota
1954	Garvan Medal, American Chemical Society

1958	Vice President, Director of Research, Milling Division, Peavey Co.
1967	Retired from Peavey Co.; Vice President, Experience, Inc.
1969	President, Experience, Inc.
1973	Chairman of Board, Experience, Inc.
1975	Member, Board of Directors, Experience, Inc.
1991	Retired

Betty Sullivan is an award-winning biochemist whose research led her to become an expert on wheat protein and baking technology.

Sullivan was born in Minneapolis, Minnesota, in 1902. She had a happy childhood, growing up within an extended family that included her parents, one brother, a grandmother, and two aunts. She was quite athletic as a child, apparently causing some concern to her mother, who worried about her daughter's preference for playing football and baseball with her brother and his friends. Her ability to excel in nontraditional areas continued throughout her educational and work life. She became interested in chemistry owing to her high school physics teacher. When she started at the University of Minnesota in 1918, classes were large because many who had served in the armed services were returning to school under the G.I. Bill. There were only about three other women studying with her at the time.[1]

While still an undergraduate at the University of Minnesota, she worked with George Frankforter, head of the chemistry department, on a project for the Russell Miller Milling Co. (RMMC). Dr. Frankforter was quite influential in the direction of her career. Sullivan was hired at RMMC as an assistant chemist following her graduation in 1922, the first woman to be employed in such a capacity at the company. Her undergraduate thesis was based on the chemical reaction in which smaller molecules combine to form larger ones in pinene, which is found in turpentine oils.

After working for two years, Sullivan traveled to Paris to pursue further education in biochemistry. She received an International Education Scholarship from the University of Paris and worked in the fermentation division of the Pasteur Institute with Auguste Fernbach. Sullivan was one of five or six women studying at the Medical School at the University of Paris at the time. Her course of study included the opportunity to hear **Mme. Marie Curie** lecture.

Soon after her return from France, Sullivan was named chief chemist at RMMC. She pursued graduate work at the University of Minnesota while working and completed her Ph.D. in biochemistry in 1935. Her dissertation was entitled "Lipids of the Wheat Embryo."

As a result of Sullivan's biochemical research, she became an expert in the development and marketing of products that used grains and oil-seeds as raw materials. Sullivan's research interests have included studies of the determination of moisture in wheat and flour, the inorganic constituents of wheat and flour and their relationship to gluten quality, lipids of the wheat embryo, and isolation of glutathione from wheat germ and its effect on the oxidation and reduction of flour. She became an authority on the chemistry of wheat proteins and the technology of baking. The author of more than 60 articles, Sullivan also holds the patent on a flour improver, accelerated moisture conditioning of grain, and the method of obtaining protein-rich flour (with George Dasher).

Sullivan was active throughout her professional life in many organizations. She was a member of the American Chemical Society, the American Association of Cereal Chemists, Sigma Xi, the American Association for the Advancement of Science, the American Society of Bakery Engineers, and La Société de Chimie Biologique. Additionally, Sullivan was awarded a national honorary membership in Iota Sigma Pi, the Society of Women Chemists. She chaired the Technical Advisory Committee of the Miller's National Federation and was treasurer of the Associates of the Food and Container Institute for the Armed Forces. She served as president of the American Association of Cereal Chemists in 1943–1944. In 1948, Sullivan was the first woman to receive the Thomas Burr Osborne Medal, the highest medal offered by that organization. This award was given by the association to those deserving of special recognition for their work in the field only six times in the previous twenty years.

Sullivan chaired the Minnesota section of the American Chemical Society in 1950–1951 and was a member of the advisory board of the *Journal of Agricultural and Food Chemistry*, the editorial board of *Cereal Chemistry*, and the board of trustees of *Biological Abstracts*. In 1953 she received the Outstanding Achievement Award from the University of Minnesota, given to graduates who have achieved distinguished success in their fields. In 1954 she received the Garvan Medal, established to honor American women for their service in the field of chemistry, from the American Chemical Society.

At the time she received this award, Sullivan was vice-president and director of research at RMMC. When Russell Miller Milling Co. merged with the Peavey Flour Mills in 1958, Betty Sullivan became vice-president and director of research at Peavey Co. In this position she was responsible for the development, market research, and process engineering of new foods and chemical derivatives for proteins, sugars, and starches.

Sullivan retired from Peavey in 1967 at the mandatory retirement age of 65 and became vice-president of a consulting firm that she helped found, called Experience, Inc. She worked with other former corporate executives, economists, and academicians. They consulted through the Agency for International Development and the World Bank with third world countries as well as for U.S. corporations in the field of agribusiness. She was president of Experience, Inc., from 1969 to 1973 and remained a member of the Board of Directors until 1990, when the firm merged with another consulting company.

According to Sullivan, the most significant challenge she faced was whether or not to get married. If she had married, she would not have been able to pursue a biochemical career in the same way. Her greatest satisfaction came from discovering "something no one knew before—in my own case, the finding of the presence and identification of glutathione in wheat germ and its effect on flour properties." She strongly encourages women to pursue careers in science.

An avid musician, Sullivan has enjoyed playing the violin and outdoor sports such as swimming and tennis. Throughout her professional and personal life she has worked hard, shown a genuine concern for others, and displayed a great sense of humor, as evidenced by a love of practical jokes. As a trail blazer for women in the field of biochemistry, she has served as a source of inspiration for many.

Note

1. This essay was completed with the assistance of Dr. Sullivan, who provided many helpful insights and answered several questions, December 1995.

Bibliography

American Men and Women of Science, 1995–96. 19th ed. New York: Bowker, 1995.
"Betty Sullivan." *Chemical and Engineering News* 26 (July 12, 1948): 2055.
"New Honors for Smith and Sullivan." *Chemical and Engineering News* 32 (March 22, 1954): 1138.
Sullivan, Betty. "The Mechanism of the Oxidation and Reduction of Flour." *Cereal Chemistry* 6, suppl. (1948): 16–31. [Thomas Burr Osborne Award address]
———. "Proteins in Flour." *Journal of Agriculture and Food Chemistry* 2 (1954): 1231–1234. [Garvan Medal address]
Sullivan, Betty, and Marjorie Howe. "The Isolation of Glutathione from Wheat Germ." *Journal of the American Chemical Society* 59 (1937): 2742.
Sullivan, Betty, et al. "On the Presence of Glutathione in Wheat Germ." *Cereal Chemistry* 13 (1936): 665–669.
———. "Relation of Particle Size to Certain Flour Characteristics." *Cereal Chemistry* 37 (1960): 436–455.

MARILYN MCKINLEY PARRISH

HENRIETTA HILL SWOPE
(1902–1980)
Astronomer

Birth	October 26, 1902
1925	Bachelor's degree, Barnard College
1928	Master's degree, Radcliffe College
1928–42	Staff, Harvard College Observatory
1943–47	Mathematician, U.S. Navy Hydrographic Office
1947–52	Associate in Astronomy, Barnard College
1952–77	Staff, Mount Wilson and Palomar Observatory
1968	Annie J. Cannon Prize, American Astronomical Society
1975	Distinguished Alumna Award, Barnard College; D.Sc., University of Basel
1980	Medal of Distinction, Barnard College
Death	November 24, 1980

Henrietta Swope won honors and awards for her precise measurements of variable stars. Perhaps more incredible, she discovered 2,000 stars herself.

Henrietta Hill Swope was born in St. Louis, Missouri, in 1902, one of five children of Gerard Swope and Mary Dayton (Hill) Swope. Gerard Swope was an electrical engineer and a former president of General Electric. After earning her bachelor's degree at Barnard in 1925 and a master's degree from Radcliffe in 1928, Henrietta spent the next 14 years at Harvard College Observatory. It has been noted that Henrietta "evidently . . . was not particularly inspired to continue in astronomy [after leaving Barnard] . . . for she entered the graduate school of Commerce and Administration at the University of Chicago."[1] She found, however, that this field was not to her liking, describing herself as a "wee mouse among many fierce cats."[2] Upon being encouraged by **Margaret Harwood**, then director of the Maria Mitchell Observatory, to "apply for a position at Harvard College Observatory . . . Swope became the most successful discoverer of variable stars since **Henrietta Leavitt**."[3]

Swope worked for Harvard College Observatory from 1928 until 1942. **Cecilia Payne-Gaposchkin** comments in her autobiography that Swope arrived there two years after her and at that time the work on variable stars was "placed in the capable hands of Henrietta Swope . . . [who] evinced an extraordinary flair for discovering variable stars."[4]

During World War II, Swope worked both as a member of the scientific staff of the radar laboratory at MIT (also referred to in some sources as the Radiation Laboratory) and as a mathematician in the U.S. Navy Hydrographic Office from 1943 to 1947. When the war ended, Swope attempted to return to Harvard. **Dorrit Hoffleit** has noted that funds at Harvard College Observatory were very scarce and that Swope refused to work for a token salary: "the millionairess did not follow my example by volunteering to work at a token salary, or even like our famous predecessor, **Annie J. Cannon**, who had the reputation of regularly receiving her just salary and then turning it back to the Department."[5] Swope was very disappointed not to be reappointed to Harvard. She then taught at Barnard as an associate in astronomy from 1947 to 1952.

In 1952, Swope was asked to go to Mount Wilson and Palomar Observatory in Pasadena, California, as assistant to "the ebullient and brilliant Walter Baade."[6] Ron Brashear has commented that Swope objected strongly to the title she would receive as Baade's assistant, which was that of "computer." Upon being informed that this was an archaic and never-updated term for her position, she accepted the job but asked that the title be changed, which it was.[7] She began as an assistant and then became a research fellow.[8]

It was Swope's prior work and publication in the *Harvard Annals* of her work on variable stars that caught Baade's eye, so when he needed a collaborator to study all the plates, Swope's work at Harvard made her a prime candidate.[9] Swope finished her career at the Mount Wilson and Palomar Observatory. Hoffleit noted that because of the larger and more modern equipment there, Swope's work at Mount Wilson and Palomar "ultimately led to greater recognition and honors than she could have acquired for her work at Harvard."[10]

Swope is listed in Hoffleit's "Women in the History of Variable Star Astronomy" as second only to Henrietta Leavitt in the number of variable stars discovered. She discovered 2,000 variable stars while working at Harvard College Observatory. While working with Harold Shapley, she helped to establish the "famous period-luminosity relation for Cepheid variable stars, thereby permitting determination of the Sun's position in our galaxy and of distances to other galaxies."[11] After World War II, because of the value of her work, Walter Baade hired Swope as a collaborator in his work at Mount Wilson on variable stars, first in the Andromeda nebula and then in several dwarf galaxies. In 1962, while using the plates taken with the 200–inch Hale Reflector by Walter Baade, Swope determined the distance to the Andromeda galaxy to be 2.2 million light years.[12] In 1968, Swope inferred that the dwarf galaxies show stars that lie between the Population I and II sequenced variable stars.

For the work she did at Harvard College Observatory and at Mount Wilson and Palomar, Swope was awarded the Annie Jump Cannon Prize in 1968 from the American Astronomical Society. Her alma mater, Barnard College, awarded her its Distinguished Alumna Award in 1975 and its Medal of Distinction "for her work in developing careful and precise methods for the measurement of faint stars" in 1980.[13] In 1975, Swope received an honorary doctor of science degree from the University of Basel.

Henrietta Swope gave a large gift to Las Campanas Observatory in Chile after her retirement, to enable the purchase of a 40–inch telescope for that observatory. This telescope was named in Swope's honor.[14] She died in 1980 in Pasadena, California.

Notes

1. Dorrit Hoffleit, *The Education of American Women Astronomers before 1960* (Cambridge, Mass.: American Association of Variable Star Observers, 1994), p. 31.

2. Barbara L. Welther, "Henrietta Hill Swope: Variable Stars in the Milky Way and Andromeda," *Mercury* (January–February 1992): 18.

3. Hoffleit, *The Education of American Women Astronomers*, p. 31.

4. Cecilia Payne-Gaposchkin, *Cecilia Payne-Gaposchkin: An Autobiography and Other Recollections*, ed. Katherine Haramundanis (New York: Cambridge University Press, 1984), p. 170.

5. Dorrit Hoffleit to Kimberly Laird, personal correspondence, December 17, 1995.

6. Ibid.

7. Ron Brashear, e-mail communication, December 13, 1995.

8. "Obituaries," *Physics Today* 34 (March 1981): 88.

9. Welther, "Henrietta Hill Swope," p. 18.

10. Dorrit Hoffleit to Kimberly Laird, personal correspondence, December 17, 1995.

11. "Obituaries," *Physics Today* 34 (March 1981): 88.

12. Ibid.

13. Welther, "Henrietta Hill Swope," p. 18; *CIW Newsletter* (April 1981): 1, 6.

14. *CIW Newsletter* (April 1981): 1, 6.

Bibliography

Hoffleit, Dorrit. *The Education of American Women Astronomers before 1960*. Cambridge, Mass.: American Association of Variable Star Observers, 1994.

"Obituaries," *Physics Today* 34 (March 1981): 88.

Payne-Gaposchkin, Cecilia. *Cecilia Payne-Gaposchkin: An Autobiography and Other Recollections*. Edited by Katherine Haramundanis. New York: Cambridge University Press, 1984.

KIMBERLY J. LAIRD and DORRIT HOFFLEIT

PAULA SZKODY

(1948–)

Astronomer

Birth	1948
1970	B.S., astrophysics, Michigan State University
1972	M.S., astronomy, University of Washington
1975	Ph.D., astronomy, University of Washington
1975–82	Research Associate, Lecturer, University of Washington
1976	Visiting Scientist, Kitt Peak National Observatory
1978	Annie J. Cannon Award; Visiting Assistant Professor, University of Hawaii
1978–80	Visiting Associate, Caltech
1980–81	Adjunct Assistant Professor, UCLA
1982–83	Senior Research Associate, University of Washington
1983–91	Research Associate Professor, University of Washington
1991–93	Research Professor, University of Washington
1993–	Professor and Research Professor, University of Washington, Seattle
1994	Fellow, American Academy of Arts and Sciences
1995–99	Member-at-Large, American Academy of Arts and Sciences

It wasn't until I got to high school that I had any significant science classes. I went to a private high school for girls, and the science teachers were excellent, had advanced degrees, and were very good at teaching. It was very encouraging. When you're at a girl's school, you do everything. There are no restrictive comparisons with boys in classes. But it's also socially limiting. I would not encourage my daughter to go to an all-girl school. She's in a small private school, so hopefully she's getting the in-depth experience that I had.

I see my daughter, who is 13 years old, going through this period of doubt: she started out sort of equal in all areas, and now she starts to say, "Oh, I can't do math." You have to go very actively against that mindset and say, "There's no reason to say that—you can do anything." I've been involved for a long time with a program called Expanding Your Horizons, which has workshops to help girls in their career choices.

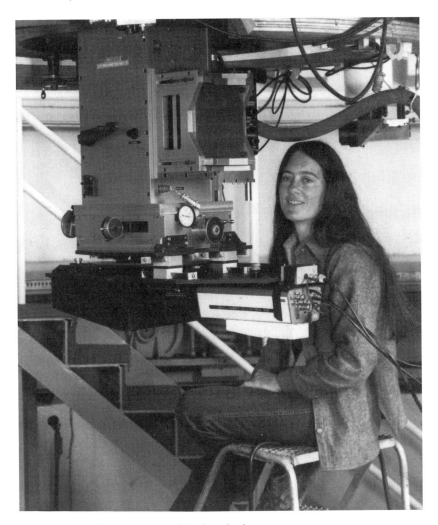

Paula Szkody. Photo courtesy of Paula Szkody.

When I went to university, I wasn't sure which area of science I was going to go into. I took an astronomy course, found it very interesting, and from that point on I was an astrophysics major. Then in my junior year I got to go to an observatory in Switzerland and help with research on solar astronomy, and that helped clinch my interest.

Astronomy at that time was a smaller field, and highly dependent on the bigger field of physics. To me physics was very impersonal, but astronomy—at that time, at least—was something an individual could do on their own and still make an important contribution, and I like that aspect of it.

Now, it's very difficult to work alone in the field. I'm on several team projects, with from five to twenty people. But on the other hand, my university has access to a telescope in New Mexico, and there are projects using it that I can do alone from here or with a graduate student. Especially with e-mail now, the scope of work is really international, and I work with Europeans on several projects.

Coming from a girl's high school, it was quite a switch for me to enter a field that is predominately men. I was the only woman astronomy major in my class, and I'm the only woman on the faculty in my astronomy department. It is getting better in graduate schools now; but when I finished my graduate work, my department had been in existence for ten years and I was the first woman to get a Ph.D. But now 25 percent of graduate students in our department are women.

I found a niche in astronomy that is still a fairly small subset. In the United States there are maybe 30 people who work in the field of cataclysmic variables [cataclysmic variables are involved with very close binary stars that transfer material from a normal, cool star to a hot, high-density companion], and internationally there are maybe 200 people at our meetings. It's a very personable field, the people are extremely nice, and as I started I was very much encouraged by the people who are active in it. Not all fields in astronomy are like that. But in the area of variable stars everybody works together, and that appeals to me a lot.

In scientific research you're constantly faced with reviews, you're constantly being evaluated, and that aspect of it I don't like very much. It's competitive, and that's what you have to do to survive. In my field, with only two or three other women, at major meetings we often go off with each other. But at some point you don't think of yourselves as women—you're basically all scientists.

When I was growing up, my family thought it was quite strange for me to be interested in this field. I was the only one in my extended family to get a Ph.D. When I told my father about wanting to go to college, he said "Why? You're just going to end up getting married." And that was the typical attitude in those days. But my parents supported me in whatever I wanted to do. My mother always wanted me to be a grade school teacher; to her, that was a safe career. But they never tried to stop me and always tried to help.

I'm married to another astronomer, which happens to a lot of women astronomers. I met my husband in graduate school, and we got married after I completed my Ph.D. I have two children, ages 13 and 10, so I have a family life as well as an astronomer's life. Astronomy is not a 9-to-5 job, and it's all I can do to pursue that and have a family.

What excites me in my work is the research aspects of it: that there are still new discoveries to be made. It's very exciting to find some new kind of physics that is going on in a star system, or a new kind of star that has never been seen before. I have graduate students, but the courses I teach are centered on the undergraduate astronomy majors, and I find that's very nice because the kids have a lot of energy when they're in their twenties, a lot of interest. All that is nice, but what keeps me going in astronomy is the research.

In terms of how my work has an impact on the field, everything that is found has ramifications. For example, one thing that we found in the last ten years is that some star systems have enormous magnetic fields—much greater than anything we encounter on the surface of the earth—and we're trying to understand how matter reacts in such strong fields. We're studying a new branch of physics. If we can really understand what happens to material under these conditions, it will have relevance to other areas of astronomy and physics.

Science has to be carefully planned. If you don't have the high school preparation, you can't do it in college; and without the college preparation, you can't do it in grad school. What I see now is a generation of kids starting to get the right preparation, so when they reach college they'll be able to make better decisions.

In your romantic relationships there's often a decision that has to be made: What's going to happen when one wants to go to grad school and the other doesn't? Who's going to be the person who contributes the income? Women in general tend to be the one who says, "Okay, I'll drop out now and maybe get back to school later." There has to be some encouragement that women have a right to their life, too. My husband and I spent several years working at different places just to get our careers going. But you reach a point, especially when you have kids, where you can't do that, and you have to reach a middle ground where you each have a piece of your career. So it's a juggling act.

At some level, I don't think girls are encouraged to work as hard toward something, so they tend to take the easy courses rather than the ones that require a lot of thought and homework. I think attitudes are changing, though—you do see more women in science classes, and not just the bookworm type or the ones who are thought to be strange in any way; it's a wide variety of women. Science is a calling, like anything else.

One advantage to being a woman scientist is the fact that when you go to a meeting of 200 or so people, with only five women, everyone knows who you are. You are much more easily remembered. Of course, that can serve as a disadvantage if you don't do a good job. Women may be criticized more, but they are also noticed more, so it's easier to get established. And I can interact with my students more fully as a woman. They ask me questions about what they are doing with their life, not just how they're doing in my course. I think men in general can be supportive but tend to be much more aloof. I think what women have to contribute is of benefit to the whole field.

Bibliography

Mateo, M., and Paula Szkody. "VW Hyi: The White Dwarf Revealed." *Astronomical Journal* 89 (1984): 863.

Szkody, Paula. "IUE and Einstein Observations of Cataclysmic Variables." *Bulletin of the American Astronomical Society* 12 (1980): 819.

———. "Observed Pulsations of Dwarf Novae at Maximum." *Astrophysical Journal* 207 (1976): 190.

———. "Stepnanian's Star: The Energy Distribution Reveals a Non-Typical Cataclysmic Variable." *Publications of the Astronomical Society of the Pacific* 93 (1981): 456.

———. "Systems with Peculiar Humps: Hot Spots? Magnetic Word Processing Disks?" *Annals of the Israel Physical Society* 10 (1993): 148.

Szkody, Paula, et al. "Outburst Spectra of Eleven Dwarf Novae." *Astrophysical Journal Supplement* 73 (1990): 441.

PAULA SZKODY, as told to DOUGLAS EBY

BEATRICE MURIEL HILL TINSLEY

(1941–1981)

Astronomer

Birth	January 27, 1941
1961	B.Sc., University of Canterbury, New Zealand; married Brian Tinsley
1963	M.Sc., University of Canterbury, New Zealand
1967	Ph.D., University of Texas, Austin
1969–74	Visiting Scientist, University of Texas, Dallas
1971	Received first National Science Foundation funding
1973–74	Assistant Professor, University of Texas, Austin
1974	Annie Jump Cannon Prize in Astronomy; Tinsleys divorced
1975–77	Sloan Foundation Research Fellowship
1975–78	Associate Professor, Yale University

| 1978–81 | Professor, Yale University; diagnosed with melanoma |
| Death | March 23, 1981 |

Beatrice Tinsley was the first person to make a realistic, computer-generated model of how the color and brightness of a galaxy change as the stars that make up the galaxy are born, grow old, and die. Before Tinsley's work, astronomers treated galaxies as static, unchanging objects; since her work, galaxies have been recognized as entities that develop and change. The study of the evolution of galaxies and stellar populations, which Tinsley played a central role in originating, is now an active field of astronomy. Because galaxies are the milestones that astronomers use to measure the universe as a whole, Tinsley's evolutionary models of galaxies had a profound impact on cosmology. Tinsley's short but extremely prolific and successful career exemplifies many of the most striking difficulties encountered by women scientists. Her life is coming to be one of the heroic and cautionary tales circulated among feminists in the science community.

Beatrice Hill was born in Chester, England, in 1941, during a time of air raids and rationing. Her father, Edward Hill, freshly ordained in the Church of England, soon moved his family from postwar privation to more comfortable living in New Zealand, where he rose from vicar of the rural town of Southbridge to mayor of the city of New Plymouth. Beatrice—intelligent, vivacious, and extraordinarily self-disciplined—enjoyed a childhood of girl's schools, ponies, and violin lessons, her days rigorously scheduled by self-imposed, minute-by-minute timetables. As a teenager she took math classes at New Plymouth Boys High School, because classes at the level for which she was ready were not offered at the Girls High School. At Canterbury University she studied physics. Although she became interested in astronomy and cosmology as an undergraduate, local research opportunities led her to write a master's thesis in chemical physics.

In 1961, Beatrice married fellow physicist Brian Tinsley and, after he completed his Ph.D., followed him to Dallas, Texas. Beatrice hoped the move would be temporary, because although Brian's position at the Southwest Center for Advanced Studies (now part of the University of Texas) enabled him to pursue nonmilitary research in atmospheric physics, Dallas offered few opportunities for her. After a year, Tinsley complained that she had reached a "dead end" trying to learn astrophysics and physical cosmology from teachers who treated general relativity as a "mathematical game. . . . I was getting depressed at my scientific stagnation, not having a baby either."[1] Enrolling in 1964 in the astronomy department at Austin, Tinsley faced a weekly 400–mile commute but was soon researching a Ph.D. dissertation on the evolution of galaxies.

Tinsley's motivation for studying the evolution of galaxies was cosmological. Astronomers use galaxies as markers to measure the most distant parts of the universe, seeking the answers to cosmological questions such as how rapidly the universe is expanding. Will it expand forever, or eventually collapse? As one looks out into the universe, however, one looks back in time; so to answer these questions, one needs to know how galaxies change over time. A few astronomers had modeled galaxies by population synthesis—assembling the population of stars of different ages, masses, and chemistry whose total light best matched the light of the galaxy. This could only be used to model a galaxy at a single moment. Tinsley's models were specifically evolutionary because she postulated reasonable initial conditions for the birth of stars, then let each star develop according to astronomers' best theories of the lives of stars. She added up the colors and luminosities of the evolving stars to find the total color and luminosity of the entire galaxy as it developed. She calculated from this a correction for astronomers to use when they employ galaxies in cosmological measurements.[2]

Tinsley finished her dissertation in late 1966. Her three-year, graduate-school immersion in astronomy must have been a sharp contrast to the next years of her life, for in 1966 the Tinsleys adopted a son, Alan, and in 1968, a daughter, Teresa. From Tinsley's family letters and friends' reports, it is evident that she found combining children and science oppressive. She knew that her graduate work had been creative and successful, and she hoped for a commensurately good career. However, with two small children and a husband whose research demanded frequent travel, she was tied to Dallas, where she was isolated from other astronomers. She could find no real astronomy job. For several years she gave up research entirely, occupying herself with political activism, mainly concerning family planning and population control.

In about 1970, Tinsley did something very few people ever accomplish in a profession as fast-moving as scientific research: she restarted her stalled career. As a visiting scientist at the newly forming University of Texas at Dallas, she applied for and received part-time National Science Foundation funding. Her rate of publication leaped in 1971, from 5 papers in the preceding seven years to 7–15 papers a year thereafter. She began to get attention from the wider astronomical community, although at Dallas and Austin (where she was given a part-time, temporary appointment) she remained, she wrote, "rejected and undervalued intellectually . . . a *gut* problem to me."[3] Conference travel and visiting appointments, at Mount Wilson and Palomar Observatories in 1972 and at the University of Maryland in 1973, were crucial to keeping her in communication with other researchers. She began to work with several astronomers who would become long-term collaborators and friends, including James Gunn, at Caltech, and Richard Larson, at Yale.

Tinsley found her situation in Dallas intolerable. The problem was partly professional. Although Tinsley received the American Astronomical Society's 1974 Cannon Prize, open to female astronomers, and offers of positions at Cambridge, Chicago, and Yale, the University of Texas made no move to regularize her appointment.

Yet the problem must also have been personal, because in late 1974 the Tinsleys divorced. On Christmas Day, leaving the children with Brian, Beatrice departed for a six-month visit to Santa Cruz, after which she became an assistant professor at Yale. Her position at Yale provided Tinsley for the first time with professional security and, she wrote, "a sense of hope and power over the future that has escaped me for years."[4]

At Yale, Tinsley extended her models of galaxies to take account of the effects of star formation, stellar evolution, gas inflow from the galaxy's environment, galaxies' interactions with their neighbors, and evolution of galaxy chemistry. At times she joined cosmological debates, arguing that the universe is open—its density is too low for gravity ever to cause it to fall back upon itself. By the late 1970s, however, she became increasingly convinced that galaxies evolve so much and in such complex ways that they cannot be used as static "standard candles" to measure cosmological distance. To use them to gain understanding of the structure of the universe, they must first be studied in their own right. Because Tinsley's models demonstrated how the results of work in many other areas of astronomy could be synthesized into models of the evolution of galaxies far more realistic than any previously constructed, they laid an effective foundation for future work. Rather than providing a lasting answer to cosmological questions, Tinsley was largely responsible for establishing the photometric evolution of galaxies as a field of study in astronomy.[5]

The fact of bringing together a wide body of work to bear on so large a question gave Tinsley an extensive professional and popular audience, the latter reached through the *Scientific American* magazine and radio appearances.[6] In 1977, Tinsley convened a watershed symposium on stellar populations in galaxies. The Yale symposium and its proceedings, edited by Tinsley and her Yale colleague Richard Larson, effectively clinched the status of the study of the evolution of galaxies as a field of astronomy.[7] Building the network that made her synthetic investigations possible and convening a conference that made this network not just her private resource but the core of a recognized field are among Tinsley's salient achievements as a mature scientist.

In 1978, Tinsley learned that a lesion on her leg had become a melanoma, a malignant skin cancer requiring immediate surgery. She survived for three years, continuing her research and teaching while undergoing extensive radiation and chemotherapy. For the last year of her life Tinsley brought her daughter, 11–year-old Terry, to live with her. It was the longest the two had been together since Tinsley left Dallas in 1974. Eventually Tinsley was hospitalized in the Yale infirmary, only blocks away from the astronomy department, where she completed her last paper days before her death on March 23, 1981, at the age of 40.

Tinsley is commemorated by a biennial prize awarded by the American Astronomical Society for exceptionally creative or innovative research, and by a visiting professorship of astronomy at the University of Texas at Austin.

Notes

1. Beatrice Tinsley to the Hills, 1964, quoted in Edward Hill, *My Daughter Beatrice: A Personal Memoir of Dr. Beatrice Tinsley, Astronomer* (New York: The American Physical Society, 1986), pp. 46–47.

2. B. M. Tinsley, "Evolution of the Stars and Gas in Galaxies," *Astrophysical Journal* 151 (1968): 547–565; B. M. Tinsley, "Possibility of a Large Evolutionary Correction, the Magnitude-Redshift Relation," *Astrophysics and Space Science* 6 (1970): 344–351.

3. Beatrice Tinsley to Edward Hill, 1974, quoted in Hill, *My Daughter Beatrice*, p. 76.

4. Beatrice Tinsley to Edward Hill, 1975, quoted in Hill, *My Daughter Beatrice*, p. 87.

5. B. M. Tinsley, "Evolution of the Stars and Gas in Galaxies," *Fundamentals of Cosmic Physics* 5 (1980): 287–388. This review article, written soon before Tinsley's death, lays out her final understanding of the field and provides an extensive bibliography, including her most significant publications.

6. J. Richard Got III et al., "Will the Universe Expand Forever?" *Scientific American* 234 (March 1976): 62–73.

7. *The Evolution of Galaxies and Stellar Populations*, eds. B. M. Tinsley and R. B. Larson (New Haven: Yale University Observatory, 1977).

Bibliography

Hill, Edward. *My Daughter Beatrice: A Personal Memoir of Dr. Beatrice Tinsley, Astronomer*. New York: American Physical Society, 1986. [This memoir contains many letters from Tinsley to her parents, an introduction by Sandra M. Faber, and a reprint of Larson and Stryker's obituary.]

Larson, Richard B., and Linda Stryker. "Beatrice Muriel Hill Tinsley." [Obituary] *Quarterly Journal of the Royal Astronomical Society* 23 (1982): 162–165.

Overbye, Dennis. *Lonely Hearts of the Cosmos: The Story of the Scientific Quest for the Secret of the Universe*. New York: HarperCollins, 1991. [See especially Chapter 10.]

Tinsley, B.M. "Evolution of the Stars and Gas in Galaxies." *Fundamentals of Cosmic Physics* 5 (1980): 287–388.

JOANN EISBERG

MARJORIE JEAN YOUNG VOLD

(1913–1991)

Colloid Chemist

Birth	October 25, 1913
1918	Moved to the United States
1921	Naturalized as a U.S. citizen
1934	B.S., University of California, Berkeley, Phi Beta Kappa, valedictorian, received the University Medal
1936	Ph.D., chemistry, University of California, Berkeley; married Robert Donald Vold on June 14; Lecturer at the University of Cincinnati
1937–41	Research Associate, Stanford University
1942–46	Industrial Chemist, Union Oil Co.
1947–57	Research Associate and Lecturer in Chemistry, University of Southern California
1953–54	Guggenheim Fellow, University of Utrecht, Netherlands
1955–57	Honorary reader in physical chemistry, Indian Institute of Science, Bangalore, India
1958	Diagnosed with multiple sclerosis
1958–74	Adjunct Professor, University of Southern California
1966	*Los Angeles Times* Woman of the Year
1967	Garvan Medal, American Chemical Society
Death	November 4, 1991

Marjorie Jean Young Vold, a prolific colloid chemist and distinguished professor, wrote or co-authored approximately 150 scientific papers, numerous book chapters, and two books during a career that lasted over five decades. Colloid science, the "science of large molecules, small particles, and surfaces," is defined in her text as "that branch of physical chemistry which determines and attempts to explain and predict the properties of substances in each of the two major divisions of the colloidal state. . . . Fibers and films are thus included as well as three dimensional solid particles, liquid droplets, and gas bubbles."[1]

Vold spent most of her career researching the phase behavior and kinetic properties of soaps, liquid crystals, and colloidal suspension. Her research foci ranged from the "less glamorous colloidal systems like greases to the more fashionable biopolymers" such as DNA.[2] She often collaborated with her husband, Dr. Robert D. Vold, who was also a chemist. Internationally recognized in the chemistry community for their contributions to colloid science, together the Volds established the renowned Center for Surface and Colloid Chemistry at the University of Southern California's chemistry department. Their book, *Colloid Chemistry: The Science of Large Molecules, Small Particles, and Surfaces* (1964), of which Marjorie was the first author, was considered such a useful and well-written text about colloid science that it was translated into Japanese.[3]

Marjorie Vold was born in 1913 in Ottawa, Ontario, to Reynold Kenneth and Whilhelmine (Aitken) Young. Members of her immediate family had been scientists or science teachers since 1895.[4] She moved to the United States in 1918 and was naturalized as an American citizen in 1921. In 1934 she earned her B.S. from the University of California at Berkeley, graduating Phi Beta Kappa and, as valedictorian of her class, received the University Medal. She earned the Ph.D. degree in chemistry from Berkeley two years later in 1936 at the remarkably young age of 23. That same year, she married Robert Donald Vold on June 14. A fellow graduate student, Robert had earned the Ph.D. in chemistry at Berkeley in 1935. They had three children: Mary Louise, Robert Lawrence, and Wylda Bryan.

In 1936, after finishing her Ph.D., Vold worked as a lecturer at the University of Cincinnati for a year. From 1937 to 1941 she held a position as a research associate at Stanford University. During World War II, from 1942 to 1946, she worked as an industrial chemist for Union Oil Company. She then took a position as a research associate and lecturer in chemistry at the University of Southern California from 1947 to 1957. From 1953 to 1954 she and her husband conducted research in the lab of fellow colloid chemist Professor Overbeek, at the University of Utrecht in the Netherlands. Marjorie received a Guggenheim Fellowship to pursue this work, and a Fulbright Fellowship supported her husband. From 1955 to 1957 she was an honorary reader in physical chemistry at the Indian Institute of Science in Bangalore, India, where, in 1957, she was the first woman to address the Institute.[5] From 1958 until her retirement in 1974 she held the position of adjunct professor of chemistry at the University of Southern California.

Vold was diagnosed with multiple sclerosis in 1958. From 1960 until her death, she was usually confined to a wheelchair. One of her former graduate students at the University of Southern California, K. L. Mittal (1974), remarked that despite "her poor health and this terrible handicap, she [had the] admirable stamina to deliver advanced colloid chemistry lectures continuously for two hours."[6] This was testimony to her extraordinary commitment to teaching science. Another less well known example of Vold's dedication to teaching young people about science was her decades of volunteer work for the Boy and Girl Scouts.[7]

Among her many distinctions, Vold chaired the American Chemical Society's Women's Service Committee from 1945 to 1952. The *Los Angeles Times* named her "Woman of the Year" in 1966.[8] In 1967 the National Lubricating Grease Institute awarded Vold the Author's Award for best paper presented at the organization's annual meeting. Also in 1967, the American Chemical Society awarded her the prestigious Garvan Medal for her outstanding contributions to the field of chemistry, including "computer simulation of kinetic and particulate phenomena; a theory of van der Waals' interaction between nonhomogeneous spheres; [and] elucidation of phase transitions in soap systems."[9] At that time, only 5 percent of the members of the American Chemical Society were women.

In a remark revealing both her optimism and realism, she was quoted in a 1967 newspaper article about herself and her husband, "Science: A Tie That Binds," as saying the following about the scientific community: "It is my conviction that in some far off Utopia there will be no distinction on the basis of nationality, sex or such irrelevant factors among scientists. Nevertheless, chemistry today is still a man's world."[10]

From April 2 to 5, 1974, members of the Division of Colloid and Surface Chemistry sponsored a symposium on colloid science at the 167th Meeting of the American Chemical Society in Los Angeles, California, "to pay tribute to these two outstanding workers in colloid science," Marjorie Vold and her husband Robert.[11] After working in the chemistry department at the University of Southern California for nearly three decades, in June 1974 they were both retiring from their positions.

Vold wrote her final scientific paper, "Micellization Process with Emphasis on Premicelles," when she was 78 years old and bedridden from multiple sclerosis at the Rancho Bernardo Convalescent Hospital in Poway, California.[12] She submitted this paper to the prestigious journal *Langmuir*, the American Chemical Society's journal on surfaces and colloids, a month before she died. *Langmuir* published it posthumously in 1992.

During a scientific career spanning more than half a century, Vold's extraordinary teaching and research endeavors, which at the end of her life included designing complex mathematical simulations on a computer while confined to her bed, are testimony to her profound, lifelong commitment to science. In her obituary her son, Dr. Robert Lawrence Vold, a chemical physicist, remarked: "Retirement, to Marjorie Vold, simply meant continuing her scientific endeavors in a different setting. . . . Her courage and persistence in the face of a 35–year battle with multiple sclerosis is truly remarkable and remains a permanent source of inspiration to all who knew her."[13]

Marjorie J. Vold died on November 4, 1991.[14]

Notes

1. M.J. Vold and R.D. Vold, *Colloid Chemistry: The Science of Large Molecules, Small Particles, and Surfaces* (New York: Reinhold Publishing Group, 1964), p. 1.

2. *Colloidal Dispersions and Micellar Behavior*, ed. K.L. Mittal (Washington, D.C.: American Chemical Society, 1975), p. x.

3. Ibid.

4. J. Maroney, "Science: A Tie That Binds," *Los Angeles Herald-Examiner* (November 19, 1967): C3, cols. 1–5.

5. Ibid.

6. *Colloidal Dispersions and Micellar Behavior*, p. xii.

7. R.D. Vold, "Obituary: Marjorie Jean Vold, 1913–1991," *Langmuir* 8 (1992b): 1234.

8. *Colloidal Dispersions and Micellar Behavior*, p. xii.

9. "Garvan Medal: Marjorie J. Vold," *Chemical and Engineering News*, 45, no. 15 (1967): 87.

10. Maroney, "Science: A Tie That Binds," p. C3.

11. *Colloidal Dispersions and Micellar Behavior*, p. ix.

12. See M.J. Vold, "Micellization Process with Emphasis on Premicelles," *Langmuir* 8 (1992a): 1082–1085.

13. Vold, "Obituary: Marjorie Jean Vold, 1913–1991," p. 1234.

14. The author acknowledges the library staff at Cornell University for their assistance.

Bibliography

American Men and Women of Science: Physical and Biological Sciences. 15th ed. New York: R.R. Bowker, 1982.

Bailey, M.J. *American Women in Science: A Biographical Dictionary*. Denver, Colo.: ABC-CLIO, 1994.

"Garvan Medal: Marjorie J. Vold." *Chemical and Engineering News* 45, no. 15 (1967): 87.

Maroney, J. "Science: A Tie That Binds." *Los Angeles Herald-Examiner* (November 19, 1967): C3.

Vold, R.D., and M.J. Vold. "A Third of a Century of Colloid Chemistry," in *Colloidal Dispersions and Micellar Behavior*. Edited by K.L. Mittal. Washington, D.C.: American Chemical Society, 1975.

Who's Who in Science. 1st ed. Edited by A.G. Debus. Chicago: Marquis Who's Who, 1968.

SHARON SUE KLEINMAN

EMMA T.R. WILLIAMS VYSSOTSKY

(1894–1975)

Astronomer

Birth	October 23, 1894
1916	B.A., mathematics, Swarthmore College
1916–17	Demonstrator in Mathematics, Swarthmore College
1917–19	Actuarial, Fidelity Mutual Life Insurance Co.
1919–21	Volunteer, Child Feeding Program, Germany
1921–24	Actuarial, Provident Mutual Life Insurance Co.
1925–27	Instructor in Mathematics, Swarthmore College
1926	Postgraduate, University of Chicago
1929	Married Alexander N. Vyssotsky
1929–44	Instructor in Astronomy, University of Virginia
1930	Ph.D., astronomy, Radcliffe College
1946	Annie J. Cannon Medal
1948	Publication of *An Investigation of Stellar Motions,* co-authored with husband, Alexander N. Vyssotsky
Death	May 12, 1975

In a career abbreviated by illness, Emma Williams was honored for her research on stellar spectra, but for much of her life she remained an unnamed resource to her husband's successful career in astronomy.

When Emma T.R. Williams, a native of Media, Pennsylvania, received her bachelor's degree in mathematics from Swarthmore College in 1916, she was unusual for that time and place. She was a woman passionately interested in mathematics and the physical sciences who intended to make their study her career. Employment choices for women with degrees in math were few and far between in the early twentieth century. Emma took her degree and went to work in the life insurance industry as an actuarial, calculating insurance and annuity premiums and dividends. With two years doing volunteer work in Germany in a children's postwar relief effort, Emma spent five years as an actuary, all the while yearning for better things.

In 1925, Emma signed on as an instructor in the mathematics department of her alma mater, Swarthmore College. She spent two years there

Emma T.R. Williams Vyssotsky. Photo reprinted by permission of Friends Historical Library of Swarthmore College.

before heading for Radcliffe College to take her Ph.D. in astronomy. On the Harvard campus she met a Russian astronomer, Alexander Vyssotsky. They were married in 1929, the same year that her Ph.D. requirements were finished. Alexander had been working at the University of Virginia, so Emma relocated there to begin her career as an astronomer.[1] Her Ph.D. was officially awarded in February 1930, when Emma was 35 years old.[2]

The Leander McCormick Observatory at the University of Virginia was founded in 1877. It was the first major scientific facility at the university, and the 26–inch refractor lens has been regarded as the most perfect large lens ever produced. The Observatory's second director, S. A. Mitchell, undertook an extensive photographic program with the 26–inch telescope beginning in 1914 that was to continue for quite some time. The

determination of stellar parallaxes using multiple photographic expo-
sures and trigonometry was the compelling professional program that
the Vyssotskys became involved in when they arrived in 1929. Alexander
became an assistant professor, and Emma was hired as an instructor.

Although a dozen or so other observatories became involved in the
study of stellar distances, the 2,300 parallaxes determined at McCormick
represent about 20 percent of the total known prior to the 1980s.[3] Many
of these parallaxes are for nearby dwarf stars discovered by the Vyssotskys using a special objective prism attached to the astrograph at the
observatory. Alexander was known for how quickly and accurately he
could classify spectra from the wide-field, multiple-exposure spectral
plates.[4]

In 1944, Emma left the university because of an increasingly debilitat-
ing illness. She and everyone else thought that her leaving would be
temporary, that the doctors would discover what was wrong with her
and effect a cure that would allow her to return to her work. Unfortunately, this did not happen for another 13 years. Alexander Vyssotsky
was promoted to professor of astronomy in 1945, a post he held until
his retirement in 1958. Emma never returned to the Observatory. She
was recognized for her achievements in stellar spectra and color indices
in 1946 with the Annie J. Cannon Medal given by the American Astronomical Society. In 1948 both her career and her husband's were capped
by their monograph *An Investigation of Stellar Motions*, considered a classic in its field.

Why did Emma Williams Vyssotsky never rise above the official level
of instructor? The Vyssotskys were a powerful team, producing accurate
determinations of stellar motions and of the constants of galactic structure from observations made at McCormick and other observatories.
State-supported positions were first provided for astronomy at the
University of Virginia in the 1920s, although no more than three positions were ever available before the 1960s.[5] One can speculate that it was
because she was female at a southern college, or that it was cheaper to
pay her as an instructor (even though her position was research-oriented
and she rarely had any teaching responsibilities), or that nepotism rules
prohibited two married professors in the same department—but the precise reasons may never be known.[6] She can only be remembered for what
she managed to accomplish in that time and place.

Emma T. R. Williams stood out in her short career for many reasons.
She was only the second woman to be awarded a Ph.D. in astronomy
from Radcliffe College and the third astronomy Ph.D. at either Harvard
or Radcliffe (the first was **Cecilia Payne-Gaposchkin** in 1925; the second
was Frank Hogg in 1929). Proud of her colonial English ancestry, she
refused to publish under her married name, instead using her birth name
for her entire career. She is remembered for her incisive and dis-

ciplined mind, her command of mathematics and statistics, and her life-long support of her husband's work. Their dinner-table shoptalk discussions of observational astronomy were very fruitful in providing a clear and rigorous framework for Alexander Vyssotsky's more intuitive outlining of research problems. This relationship continued through many papers in which Emma's name did not appear and throughout the long and difficult course of her illness. She was eventually diagnosed with a chronic case of brucellosis, a bacterial disease that can recur as frequently and virulently as malaria. Brucellosis, also known as undulant or Malta fever, is a formerly common disease of cattle that can be passed to humans via unpasteurized milk products. Emma was cured with a course of strong antibiotics, but the damage to her career and her health had been done. She eventually died in the retirement community of Winter Park, Florida, two years after her husband Alexander. She was 81 years old.[7]

Notes

1. That Emma met Vyssotsky at Harvard was confirmed by Dorrit Hoffleit at Yale, Victor Vyssotsky, and Bob McCutcheon of NASA.

2. *Who's Who* incorrectly listed Vyssotsky's Ph.D. as having been awarded in 1929. The date of 1930 was verified by Radcliffe College.

3. General information about the Leander McCormick Observatory was gathered through *Telescopes*, edited by Thornton Page and Lou Williams Page (New York: Macmillan, 1966), and the University of Virginia Astronomy Department home page on the World Wide Web.

4. Dorrit Hoffleit and Bob McCutcheon confirm Alexander's facility with spectral plates.

5. See University of Virginia Astronomy Department home page on the World Wide Web.

6. The Alumni Office confirms Emma's lack of teaching responsibilities.

7. Personal information on Emma from author's notes of conversation with Victor Vyssotsky.

Bibliography

Vyssotsky, A.N., and Emma T.R. Williams. "Color Indices and Integrated Magnitudes of 15 Bright Globular Clusters." *Astrophysical Journal* 77 (1933): 301.

———. "Galactic Structure and Kinetic Theory." *Astrophysical Journal* 98 (1943): 187.

———. "An Investigation of Stellar Motions; Additions and Corrections." *Astronomical Journal* 56 (1951): 68.

Who's Who of American Women, 1966–67. Chicago: Marquis Who's Who, 1967.

Williams, Emma T.R. "Evidence for Space Reddening from Bright B Stars." *Astrophysical Journal* 75 (1932): 386.

———. "A Spectrophotometric Study of Astars." Ph.D. dissertation, Radcliffe College, 1930.

———. "Systematic Errors in the Determination of the Contours of the Hydrogen Lines in A-Stars." *Astrophysical Journal* 72 (1930): 127.

Williams, Emma T.R., and A.N. Vyssotsky. "The Constants of Galactic Rotation and Precession." *Astronomical Journal* 53 (1948): 27–30.

REBECCA L. ROBERTS

ELIZABETH AMY KREISER WEISBURGER

(1924–   )

Chemist

Birth	April 9, 1924
1944	B.S., chemistry, *cum laude*, Phi Alpha Epsilon, Lebanon Valley College, Annville, PA
1947	Ph.D., organic chemistry, University of Cincinnati
1949–51	Postdoctoral Research Fellow, National Cancer Institute (NCI), National Institutes of Health (NIH), Bethesda, MD
1951–89	U.S. Public Health Services
1972–74	Consultant, National Academy of Sciences, National Research Council, Washington, DC
1973	Meritorious Service Medal, U.S. Public Health Service, NIH
1973–81	Chief, Laboratory of Carcinogen Metabolism, NCI, NIH, Bethesda, MD
1977–79	Chairperson, Carcinogenesis Working Group, NCI, NIH, Bethesda, MD
1981	D.Sc., University of Cincinnati; Garvan Medal, American Chemical Society; Hillebrand Prize, Chemical Society of Washington, DC
1985	Assistant Director of the Division of Cancer Etiology; first woman elected president of the Board of Trustees of Lebanon Valley College, Annville, PA

"Behind dark tinted glasses, with her coat collar pulled up, [Elizabeth] Weisburger could easily be mistaken for a private eye in an Agatha Christie novel. But in fact she is a real-life medical sleuth, world renowned for her research in chemical carcinogenesis."[1] From talking with the affable Weisburger, one senses the enthusiasm, energy, and commitment that she holds for her work. She spent many years studying chemical carcinogenesis, a field that explores the mechanisms by which environmental chemicals cause mutations and cancer. No wonder some refer to her as a pioneer in the field of chemical carcinogenesis. "Her studies have led to insight into chemical carcinogenesis on a molecular level, which is fundamental to understanding how to prevent and how to treat cancer," says **Joyce Kaufman** of Johns Hopkins University. In 1980, Kaufman submitted a letter to the American Chemical Society recommending Weisburger for the prestigious Garvan Medal, awarded to outstanding women in chemistry.[2] As a woman, Weisburger has climbed to the pinnacle of her profession, receiving many awards, significant appointments, and respect worldwide.

Elizabeth Amy Kreiser was born in 1924 in Finland, Bucks County, Pennsylvania. She is the daughter of Raymond Samuel Kreiser and Amy Elizabeth Snavely, both of German descent. Both parents were born in Lebanon County—Raymond on March 3, 1900, and Amy on July 8, 1901. They attended Millersville State College in Pennsylvania. After graduation Amy taught at Harrison, Webster, and Hershey rural schools before marrying Raymond on August 29, 1922. Raymond taught elementary school and sold Prudential life insurance in Bucks County. In 1924, a few weeks after the birth of Elizabeth Amy, the family returned to their original community in Lebanon County. Other Kreiser children were born as follows: Edith Alma (1926), Wesley Raymond (1928), Frederick Samuel (1930), Evelyn J. (1931), Eleanor Mary (1933), Elaine Shirley (1934), Thomas H. (1935), Alfred John (1939), and Ellen Pauline (1945).[3] Learning was encouraged because many of the Snavely siblings were also schoolteachers. Thus Elizabeth learned to read early and was taught at home by her mother until she was 8 years old. She later attended high school at Jonestown, where her aunt, Lottie Snavely, taught English and Latin. Despite Elizabeth's many chores, including helping with the other children, she did well in school. She also managed time to study old New York State Regents examinations provided by her Aunt Lottie. This provided insight into possible types of questions presented in scholarship examinations. Of course, Kreiser won not only a state scholarship but also one from Lebanon Valley College.[4]

Elizabeth Amy Kreiser pursued further education in 1940 at Lebanon Valley College in Annville, Pennsylvania, where she majored in chemistry. Although she also had an interest in biology, she was somewhat deterred by the perception of a biology professor who felt that in order to excel in biology, one must draw well. So Kreiser concentrated her studies on chemistry, mathematics, and physics. During summer vacation, she worked odd jobs to contribute toward her college expenses. Nevertheless she excelled in her studies, was inducted into a scholastic honor society, and graduated *cum laude* in 1944. In September 1944 she attended the University of Cincinnati on a graduate assistantship. Here she began work in cancer research in the laboratory of Dr. Francis Earl Ray. Kreiser earned her Ph.D. degree in organic chemistry in 1947 but continued to work in Ray's laboratory as a research associate. She synthesized analogs of the carcinogen 2–acetylaminofluorene for testing at the National Cancer Institute (NCI) in Bethesda, Maryland.[5]

After several years of graduate work, Kreiser combined work with family by marrying John Hans Weisburger, a fellow graduate student, on April 7, 1947. John was born on September 15, 1921, in Stuttgart, Germany. His family fled Germany during World War II to Belgium. He attended universities in Belgium, France, and Cuba before coming to the United States in 1943. As a result of his travels, he spoke English and several European languages. The Weisburgers had three children: William Raymond was born in August 1948; Diane Susan in March 1955; and Andrew John in July 1959.[6] Both John and Elizabeth accepted postdoctorate fellowships at the NCI in 1949. At the NCI, Elizabeth lived a very busy life: she became a commissioned officer in the U.S. Public Health Services in 1951; became a member of the Carcinogen Screening Section of the Experimental Pathology Branch of NCI in 1961; and gave lectures, attended meetings, and wrote and reviewed papers. In addition, she cared for the children with the help of various housekeepers and a nearby nursery. In 1962 her husband, oldest son, and she attended an international conference in Moscow. The Weisburgers took advantage of this opportunity to visit a number of nearby scientific facilities and to travel to other European cities.[7]

One of Elizabeth Weisburger's primary research efforts was directed at elucidation of carcinogenesis by aromatic amines and aminoazo dyes. These two chemical classes are members of one of two broad categories of carcinogens, those that require metabolic activation to achieve full oncogenic potency. She and her colleagues assessed the results of large-scale rodent testing of suspected carcinogens and found that the majority of chemicals that induced liver tumors in mice or rats also caused tumors in other tissues. They concluded that induction of liver tumors in rodents is an important indicator of potential tumorigenicity in humans.[8] Consequently, Weisburger investigated the relationship between mutagenesis and carcinogenesis and was one of the first scientists to point out that the potential danger of some of the principal drugs used in clinical cancer therapy were alkylating agents and mutagens.[9] Her most important contributions to the area of carcinogenesis and toxicology are: the application of chemistry to synthesize reference metabolites; and structure-activity relationships in the selection of compounds for testing as possible carcinogens.[10] For the chemical selection project she was a member of the government-established working group of scientists to review chemical compounds systematically and to suggest candidate compounds for toxicological testing.[11]

Although Weisburger led a very busy life and made many important contributions to the identification of hazardous chemicals, she managed to encourage others by sharing words of wisdom. For instance, when an experiment did not work, she always encouraged her team to take heart. A sign on her wall read: "Don't take life so seriously; you'll never get out of it alive."[12] This same philosophy was conveyed while encouraging young people to go into science careers. Weisburger said that "they should learn as much as they can, especially about research techniques. Understanding computer science is a necessary part of biomedical research these days . . . they shouldn't take it too seriously. By that I mean they shouldn't let it get them down when an experiment doesn't go well, even after trying it a dozen times."[13]

Weisburger also shared her wisdom on how progress is made in science:

> Progress in science is rarely made in spectacular ways. In fact it is usually by building with little bits of knowledge that scientists eventually arrive at the dramatic understanding the public hears about as "breakthroughs." The importance of those bits of knowledge is appreciated mainly by the scientist who has spent years working on a research problem. And those small contributions—made by individual scientists—are the links which culminate, ultimately, in a "breakthrough." It's a shared work.[14]

Weisburger admitted to Barbara Lynch that biomedical research is a competitive field, but she emphasized that this competition is constructive. She said, "If you keep an open attitude, you will find competition in the field inspiring instead of threatening, and we'll be the better for it."[15]

Elizabeth Weisburger is proof that women can excel in the field of science. She received many accolades for her work on chemical carcinogenesis; served on many important committees; was invited to give lectures and other presentations both nationally and internationally; and published over 255 papers in journals, chapters in books, and reviews. When she retired from the National Institutes of Health in 1989, she was NCI's assistant director for chemical carcinogenesis and held the rank of captain in the U.S. Public Health Services. In retirement she works as a consultant, reviews grants and proposals, and continues to write—since 1990 she has produced nine additional publications. For leisure she enjoys walking and hiking.

Notes

1. Francis X. Mahaney Jr., "Carcinogenesis Expert Elizabeth Weisburger Retires from NCI," *NIH Record* (February 21, 1989): 8.

2. Joyce J. Kaufman, "Recommendation Package" to the American Chemical Society, recommending Weisburger for Garvan Medal.

3. Elizabeth F. Washburn, *Snively—Snavely: The Swiss Ancestors and American Descendants of Johann Jacob Schnebele (1659–1743) and Other Snivelys and Snavelys of Southeastern Pennsylvania* (Baltimore: Gateway Press, 1986), p. 2.

4. See Ann A. Kaplan, "Elizabeth Amy Kreiser Weisburger (1924–)," in *Women in Chemistry and Physics: A Biobibliographic Sourcebook*, eds. Louise S. Grinstein et al. (Westport, Conn.: Greenwood Press, 1993).

5. See "Elizabeth Amy Kreiser Weisburger (1924–)," and Kaufman, "Recommendation Package."

6. See Washburn, *Snively—Snavely*.

7. See Kaplan, "Elizabeth Amy Kreiser Weisburger (1924–)."

8. Cheryl Marks, "Elizabeth Weisburger of N.I.H. Wins 1981 Hillebrand Award," *Capital Chemist* 31 (1981): 1, 10.

9. Kaufman, "Recommendation Package."

10. See Kaplan, "Elizabeth Amy Kreiser Weisburger (1924–)."

11. Barbara Lynch, "Elizabeth Weisburger: Making a Menu for Mice," *SciQuest* 52 (1979): 21–23.

12. Mahaney, "Carcinogenesis Expert Elizabeth Weisburger Retires from NCI," pp. 8–9.

13. Lynch, "Elizabeth Weisburger: Making a Menu for Mice," pp. 21–23.

14. Mahaney, "Carcinogenesis Expert Elizabeth Weisburger Retires from NCI," pp. 8–9.

15. Ibid.

Bibliography

Kaplan, Ann A. "Elizabeth Amy Kreiser Weisburger (1924–)," in *Women in Chemistry and Physics: A Biobibliographic Sourcebook*. Edited by Louise S. Grinstein et al. Westport, Conn.: Greenwood Press, 1993.

Kaufman, Joyce J. "Recommendation Package" to the American Chemical Society, recommending Weisburger for Garvan Medal.

Lynch, Barbara. "Elizabeth Weisburger: Making a Menu for Mice." *SciQuest* 52 (1979): 21–23.

Mahaney, Francis X., Jr. "Carcinogenesis Expert Elizabeth Weisburger Retires from NCI." *NIH Record* (February 21, 1989): 8–9.

Marks, Cheryl. "Elizabeth Weisburger of N.I.H. Wins 1981 Hillebrand Award." *Capital Chemist* 31 (1981): 1, 10.

Washburn, Elizabeth F. *Snively—Snavely: The Swiss Ancestors and American Descendants of Johann Jacob Schnebele (1659–1743) and Other Snivelys and Snavelys of Southeastern Pennsylvania*. Baltimore: Gateway Press, 1986.

RUTH A. HODGES

SARAH FRANCES WHITING

(1847–1927)

Astronomer, Physicist

Birth	1847
1865	B.A., Ingham University, Le Roy, NY
1865–76	Teacher of classics and mathematics, Brooklyn Heights Seminary
1876–1912	Professor of Physics, Wellesley College
1900–16	Director of Whitin Observatory, Wellesley College
1905	Honorary degree, Tufts College
Death	1927

Sarah Whiting was first and foremost a teacher, very interested and successful in promoting the education of women. Her wide-ranging interests, from electric lights to X-rays, the weather, and astronomy, led her to the singular distinction of having made the first X-ray pictures in America.

Sarah Frances Whiting was born in Wyoming, New York, the daughter of Elizabeth Comstock and Joel Whiting. Her mother's family had settled in Connecticut and Long Island in the 1600s, and the Whitings came from Vermont. Joel Whiting was a graduate of Hamilton College and served as teacher and principal in a number of academies in New York State. Sarah became interested in physics while helping her father prepare demonstrations for his classes. He taught Latin, Greek, mathematics, and physics and tutored his daughter himself.

Sarah was well advanced in these subjects when she entered Ingham University in Le Roy, New York. She graduated with a bachelor's degree in 1865 and continued on at Ingham, teaching classics and mathematics. Later, she moved to Brooklyn Heights Seminary for girls, where she taught until 1876.

While in Brooklyn, Sarah attended scientific lectures and visited laboratories and exhibitions of new equipment. "She became known as an enthusiastic and effective teacher who showed great ingenuity in improvising apparatus for her lectures and shared with her students her excitement over new discoveries."[1] In 1875 Henry Fowle Durant, founder of Wellesley College, sought out Sarah to teach physics at his new school

Sarah Frances Whiting. Photograph by Notman. Courtesy of Wellesley College Archives.

of higher education for women. She had gained a reputation as one of
the few women trained in physics at that time. Durant arranged for her
to audit laboratory physics courses at the Massachusetts Institute of
Technology in order to learn more about the subject matter and how to
use the instruments.

For two years Sarah attended the physics classes of Professor Edward
C. Pickering, who had started the first student physics laboratory in the
United States. During the next two years Sarah also traveled to other
New England colleges to observe their physics programs. She studied
Pickering's two-volume *Physical Manipulations* and made lists of appa-
ratus, much of which came from abroad. She purchased and installed
the necessary equipment and, in 1878, opened the second undergraduate
teaching laboratory of physics at Wellesley.

Whiting was always alert for new ideas, discoveries, and additions to her lab. She held demonstrations in optics and the use of the microphone, taught the fundamentals of the microscope, and attended demonstrations by Alexander Graham Bell. A Foucault pendulum and ozone machine were set up in her lab. After visiting Thomas Edison, she began demonstrating incandescent lamps. In 1895, when Roentgen's discovery was announced, Whiting assembled the necessary equipment and "produced the first X-ray photographs in the United States."[2]

Whiting was invited to join the New England Meteorological Society as its only woman member. Shortly thereafter, she initiated a course in meteorology. She purchased an anemometer, thermometers, rain gauges, and other equipment. For ten years there was no weather station in the area, so she became a voluntary observer and her students collected data from the instruments and submitted it to the U.S. Weather Bureau.

In 1879, Pickering invited Whiting to observe some of the new applications of physics to astronomy—in particular, the use of the spectroscope in the investigation of stellar spectra. While in Brooklyn, she had attended lectures by the British physicist John Tyndall on recent developments in optical instruments and had learned of the spectroscope. In 1880, Whiting decided to introduce an astronomy course at Wellesley. For 20 years she taught with little equipment, but eventually Mrs. John C. Whitin, a Wellesley trustee, made funds available to build an observatory, which was completed in 1900. Whiting drew the plans for the building, "which housed a 12–inch refracting telescope with spectroscope and photometer attachment, a transit instrument, and the usual accessories."[3] The observatory contained a spectroscopic laboratory, which enabled Whiting and her students to compare solar and stellar spectra with those of laboratory sources and contribute important research to the field.

Whiting had two sabbatical years abroad. During 1888–1889 she visited scientists and laboratories in Germany and England and studied at the University of Berlin. In England she attended lectures at the British Association. One of the lectures described Hertz's discoveries about the heat spectrum. She also met in Cambridge with Lord Kelvin, the British physicist who did pioneering work in thermodynamics and magnetism, as well as with other scientists. She wrote a reminiscence of Kelvin for the journal *Science*, noting in that article that "in contrast to many other English and German scientists, Sir William (Kelvin) seemed neither surprised nor alarmed that a woman should devote herself to mathematics and physics."[4]

During her 1896–1897 sabbatical, Whiting enrolled at Edinburgh University to study with a leading physicist, Peter Guthrie Tait. During this visit she often went to Tulse Hill Observatory, where Sir William Huggins had laid the foundations of astrophysics. She formed a strong friendship with Lady Huggins, the wife of the astronomer, who was called the "Herschel of the Spectroscope" after Sir William Herschel. The spectroscope had become one of Whiting's strongest interests. Lady Huggins assisted her husband in the same manner that **Caroline Herschel** had assisted her brother in his astronomical observations—as a tireless co-worker. Lady Huggins's name appeared as joint author of ten of William Huggins's scientific papers. She and Sarah Whiting shared an interest in "educational justice" for women.[5] After her husband's death, Huggins made several bequests to Wellesley College, including a number of astronomical instruments from the Tulse Hill Observatory.

Whiting joined the American Physical Society shortly after it was founded, one of very few women members. The Society at first would not even invite women to its banquets. In 1883, Whiting was elected a fellow of the American Association for the Advancement of Science, one of only five women.

Whiting was a very religious person, a devout Congregationalist who supported the Wellesley College Christian Association and its missionary programs. She was also a prohibitionist. She encouraged her students to be scholars but also women of influence in their communities.

It has been noted that Whiting "was not a research physicist but an inspiring teacher. She trained many women who later became influential in research areas."[6] Notable among these was **Annie Jump Cannon**, a distinguished astronomer who specialized in classification and spectroscopic analysis of stars. Whiting guided Cannon's study at Wellesley and interested her in spectroscopy. In 1893, Cannon assisted Whiting as a postgraduate. Whiting's main contribution, therefore, was to educate women in a field previously restricted to men. She focused most of her energies on this goal, her publications having dealt mainly with teaching methods. In 1905 she received an honorary degree from Tufts College for her contributions to teaching.

Whiting lived on the Wellesley campus, at first in college dormitories and later in Observatory House, built by her benefactor, Mrs. Whitin. She retired from the physics department in 1912 to concentrate on astronomy. In 1916 she retired as director of the Whitin Observatory and spent the last years of her life in Wilbraham, Massachusetts, where she died in 1927.

Notes

1. Gladys Anslow, "Sarah Frances Whiting," in *Notable American Women, 1607–1950*, ed. Edward T. James (Cambridge, Mass.: Belknap Press, 1971), p. 594.

2. Janet B. Guernsey, "The Lady Wanted to Purchase a Wheatstone Bridge: Sarah Frances Whiting and Her Successor," in *Making Contributions: An Historical Overview of Women's Role in Physics*, ed. Barbara Lotze (College Park, Md.: American Association of Physics Teachers, 1984), p. 76.

3. Anslow, "Sarah Frances Whiting," p. 594.

4. Sarah Whiting, "Reminiscences of Lord Kelvin," *Science* 60 (August 15, 1924): 150.

5. Sarah Whiting, "Lady Huggins," *Science* 41 (June 11, 1915): 854.

6. Marilyn Bailey Ogilvie, "Sarah Frances Whiting," in *Women in Science* (Cambridge, Mass.: MIT Press, 1986), p. 175.

Bibliography

Anslow, Gladys. "Sarah Frances Whiting," in *Notable American Women, 1607–1950*. Edited by Edward T. James, pp. 593–595. Cambridge, Mass.: Belknap Press, 1971.

Bailey, Martha. "Sarah Frances Whiting," in *American Women in Science: A Biographical Dictionary*, pp. 416–417. Denver: ABC-CLIO, 1994.

Cannon, Annie Jump. "Sarah Frances Whiting." *Popular Astronomy* 35 (December 1927): 539–545.

Guernsey, Janet B. "The Lady Wanted to Purchase a Wheatstone Bridge: Sarah Frances Whiting and Her Successor," in *Making Contributions: An Historical Overview of Women's Role in Physics*. Edited by Barbara Lotze, pp. 65–90. College Park, Md.: American Association of Physics Teachers, 1984.

Ogilvie, Marilyn Bailey. "Sarah Frances Whiting," in *Women in Science*, pp. 174–175. Cambridge, Mass.: MIT Press, 1986.

"Sarah Frances Whiting," in *Biographical Cyclopaedia of American Women*. Edited by Erma Conkling Lee and Henry C. Wiley. New York: Williams-Wiley, 1928.

Whiting, Sarah. "Daytime Work in Astronomy." *School Science and Mathematics* 11 (May/June 1911): 417–421, 494–499.

———. "Lady Huggins." *Science* 41 (June 11, 1915): 853–855.

———. "Pedagogical Suggestion for Teachers of Astronomy." *Popular Astronomy* 20 (March 1912): 156–160.

———. "Priceless Accessions to Whitin Observatory, Wellesley College." *Popular Astronomy* 22 (October 1914): 487–492.

———. "Reminiscences of Lord Kelvin." *Science* 60 (August 15, 1924): 149–150.

JOAN G. PACKER

LEE ANNE M. WILLSON

(1947–)

Astronomer

| Birth | March 14, 1947 |
| 1968 | A.B., physics, Harvard University |

Lee Anne M. Willson. Photo courtesy of News Service, University Relations, Iowa State University.

1969	Fulbright and American Scandinavian Foundation fellowships for study at the University of Stockholm
1970	M.S., astronomy, University of Michigan; Planetarium Demonstrator, University of Michigan
1973	Ph.D., astronomy, University of Michigan
1973–75	Instructor in Physics and Astronomy, Iowa State University
1975–79	Assistant Professor of Physics and Astronomy, Iowa State University
1979–88	Associate Professor of Physics and Astronomy, Iowa State University
1980–81	Annie Jump Cannon Award in Astronomy

1983–85	Coordinator, Astronomy and Astrophysics Program, Iowa State University
1987–90	Coordinator, Astronomy and Astrophysics Program, Iowa State University
1988–93	Professor of Physics and Astronomy, Iowa State University
1993–	University Professor, Dept. of Physics and Astronomy, Iowa State University

Dr. Lee Anne Willson has ideas that may change the way many people think about the universe, but investigating the stars while standing on planet Earth was not her first career wish. Willson learned about science early in her life. Her father was a scientist, and she spent much of her youth poring over science fiction books. When she was in junior high school, she thought it would be great to be an astronaut but figured there was little hope because she had bad vision and a crooked knee and was female. Instead, Willson shifted her interests to the next closest topic on her mind—astronomy. After an excellent education and hard work, she has become well known for her research on mass loss in stars. "I think what I am doing is actually much more fun than being an astronaut," she says now.[1]

In addition to an innate love of science, Willson learned early that perseverance can be nearly as important as knowledge. When she started college at Harvard, Willson wanted to take an advanced physics course for which she did not have the prerequisite classes, so she spoke with the professor. He told her that it was not a good course and that he didn't like girls. Willson's reply? "I'll see you in class on Monday." She had no problems with the class, either, except that she thought it wasn't very good.

Once she settled into graduate school at the University of Michigan, Willson was asked to do a seminar on the spectra of Mira variables. "Mira variables are red giant stars that vary tremendously in brightness with periods around one year," she explains. "The varying brightness is a result of stellar pulsation—the star gets bigger and smaller, cooler and hotter, sort of like a giant beating heart. This pulsation leads to mass loss, which strips away all but the small, dead core of the star—and when we see that emerge, we call it a white dwarf." Willson dove into the work fully and gave her presentation before the department and an eminent guest. She began to study fluorescent lines characteristic of these stars, and the topic evolved into part of her thesis. These stars continued to be the main part of Willson's research as she took a position at Iowa State University.

Willson came upon a barrier when her tenure-track position in the physics department at Iowa State was not fully accepted by all department members outside the astronomy group. "I had been given double the teaching load of other professors, since it was assumed that I had followed my husband to the university and was just teaching to keep busy," she said. Willson went through stressful times when she was not promoted according to schedule, but she continued to move forward in her own right and got beyond the troubles of these early years. "I remember noting to my mother that since I had not been able to join one of the stronger astronomy departments, my strategy would be to try to make our group into the kind of program I would have liked to join," said Willson. No matter what cards she was dealt, she was ready to find the best in the situation and build up from there.

Willson has collaborated with many researchers, demonstrating her ability to work with scientists in her own department and at other universities such as the University of Washington. She has found that coming up with new ideas in astronomy does not bring recognition overnight. When she and her colleague, Steven Hill, first decided that Mira stars pulsated in a certain way, their findings contradicted all others at the time. Ten years later, Willson and Hill's "radical suggestion" became a commonly accepted idea.

With her latest work, however, Willson has been putting many assumptions about the entire universe into question. She began investigating whether main-sequence stars like the sun pulsate and lose mass. She has noted that "mass-loss in these stars would have many consequences, including the possibility that the oldest stars may not be as old as people have thought. That would be good, because other ways of determining the age of the universe appear to be telling us that the universe is younger than some of the stars in it—and that can't be true!" Other consequences of this hypothesis raise the questions of why there are more stars at some masses and fewer at other masses than simple theories predict, and whether stars can create or only destroy the light elements such as lithium, beryllium, and boron. Willson and her colleagues continue to work on detailed calculations and compare their results with observations.

Aside from investigating ways to turn the universe on its ear, Willson has participated in many professional service activities, which she recognizes is important for increasing the visibility of women in science. These efforts show Willson's commitment to working with her peers to solve problems and increase understanding. Her peers have elected her to several influential positions, including councilor of the American Astronomical Society; council member for the American Association of Variable Star Observers; member of the Board of Directors of the Association of Universities for Research in Astronomy (AURA, the organization that operates the national observatories in Arizona, New Mexico, and Chile and the Space Telescope Science Institute); and chair of the Observatories

Council (the AURA group that has responsibility for running the national observatories).

As well as being elected to service positions, Willson was awarded the Annie Jump Cannon award in 1980. This award was established by **Annie Cannon**, a spectroscopic astronomer. Early recipients were eminent female astronomers and the award was a brooch. Around 1970, the American Association of Sciences and the American Association of University Women redesigned the award as a grant of research funds for young women astronomers.

Balancing professional life with personal development and family is also important to Willson. She credited saving room for "fun" classes during her studies with keeping her sane as well as giving her a well-rounded education. Her fun classes included French, music theory, comparative literature, and several design classes. Since taking her job in Iowa, Willson has enjoyed painting, sculpting ceramics, and wheel ceramics. "Throwing a pot on the pottery wheel is one of the most effective ways to get rid of stress," she says. In addition to her artistic interests outside of science, Willson has made time for her family.

Willson met her future husband, a mathematician, while she was studying at Harvard; they were able to attend graduate school together at the University of Michigan. When it came time to find two teaching jobs in the same place, Willson considered herself very lucky that Iowa State University had room for both of them. They have raised two children in Ames, with one now in college studying mathematics and the other pursuing graduate studies in Scandinavian languages and linguistics.

Since her children have become more independent, Willson has had more time to think about what it is that she does that benefits humanity most. "Of all my activities, I believe it is the teaching of large numbers of students who have little other experience with science that has the largest potential impact on our future," said Willson. "So many of the issues that must be solved in the next generation need the ability to think clearly and logically about consequences of various actions." Willson continues to work on creative ways to help students learn how to think, with her commitment to excellence never ceasing.

Note

1. All quotations are taken from a personal interview with Lee Anne Willson by Joy Schaber, 1995.

Bibliography

Guzik, J., L.A. Willson, and W. Brunish. "Effects of Main Sequence Mass Loss on Solar Models." *Astrophysical Journal* 319 (1987): 957–965.

Willson, L.A. "Ancient Monuments and Pre-Historic Astronomy." *Encounters* 5, no. 3 (1982): 7.

———. "The Dark Sky Paradox and the Origin of the Universe." *Astronomy* 6 (September 1978): 52–57.

———. "Main Sequence Mass Loss and the Age of the Universe," in *Fourteenth Texas Symposium on Relativistic Astrophysics.* New York Academy of Sciences, 1989.

Willson, L.A., and G.H. Bowen. "Effects of Pulsation, Mass Loss, and Stellar Evolution." *Nature* 312 (November 29, 1984): 429–431.

Willson, L.A., and S.M. Richardson. "Cosmic Contamination: Elemental Clues to the Origin of the Solar System." *Astronomy* 8 (June 1980): 6–22.

JOY SCHABER

CHIEN-SHIUNG WU

(1912–1997)

Physicist

Birth	1912
1934	Graduated from the National Central University with degree in physics
1936	Left China for the United States
1940	Ph.D., physics, University of California, Berkeley
1941	Lecture tour of United States
1942	Married Luke Yuan
1944	Began working in war-related research, Columbia University
1952	Became member of faculty, Columbia University
1954	Wu and husband became U.S. citizens
1958	D.Sc., Princeton University
1959	D.Sc., Smith College
1960	D.Sc., Goucher College
1961	D.Sc., Rutgers University
1964	Comstock Award, National Academy of Sciences
1971	D.Sc., Yale University
1972–81	Endowed professorship, Columbia University

| 1981 | Retired from Columbia; traveled as lecturer |
| Death | February 16, 1997 |

Chien-Shiung Wu is one of the foremost experimental physicists of the modern era. Her 1956 experiment on β-decay on cobalt-60 nuclei confirmed the nonconservation of parity in weak interactions. β-decay is a common process in which a neutron becomes a proton through the emission of an electron. This type of radiation was historically referred to as "β-rays." The conservation of parity assumes the symmetry of real physical events with their mirror images. Wu's finding that this was not always the case stunned the scientific world. An inspired and dedicated experimentalist, Wu has demonstrated throughout her academic and scientific career a love of her field and a persistence in the face of difficulties, which is characteristic of great scientists.

Born near Shanghai, China, in 1912, Chien-Shiung Wu was encouraged from childhood by her parents, particularly her father, to solve interesting problems and to pursue her education as far as she could. When she finished the highest grade offered at the local elementary school, she continued her education at a boarding school not far from Shanghai, the Suzhou Girls' School, where she enrolled in the teacher training program. Dissatisfied with the amount of science that students in the teacher training program were learning, Wu taught herself physics, chemistry, and mathematics from the books and notes of her friends who were in the academic program at Suzhou. It was during this time that she began to realize that she had a special affinity for physics.

Wu excelled in all her high school subjects and was also very active in student political organizations. After graduating with the highest grades in her class, she was selected to attend the National Central University in Nanjing. When asked by her parents what she would like to study, Wu replied "physics" but added that since her background was not very good she would probably study education instead.[1] Wu's father, however, felt that she should do all of which she was capable. He brought home math and physics texts. Wu studied that summer; and when the term started, she began studies in mathematics. She transferred to the physics program soon thereafter. After graduating from the National Central University in 1934, Wu worked for two years, first teaching physics at the University level, then doing research in X-ray crystallography. Her advisor encouraged her to pursue graduate studies in the United States, where she herself had earned a Ph.D. at the University of Michigan. In 1936 Wu left for the United States, hoping to finish a Ph.D. quickly at Michigan and go back to China. She would not return, however, until 1973.

Once in the United States, Wu decided instead to attend Berkeley. At that time Berkeley's physics department boasted such faculty as Ernest Lawrence and Robert Oppenheimer. There Wu began to work in the field of nuclear physics, joining Emilio Segrè's research group.[2] In her classes and in the lab Wu was determined to do all she could, working long hours to achieve her goal. Her thesis examined two areas: the electromagnetic energy given off by a charged particle as it passes through matter and slows down, and the inert gases emitted in the fission of a uranium nucleus. She received her Ph.D. in 1940 and continued to work at Berkeley in the area of nuclear fission. She was soon recognized as an authority on the subject and was consulted by top scientists such as Enrico Fermi on aspects of fission.

Wu was not hired by Berkeley despite her growing reputation. Certainly her situation was not aided by her gender and her nationality, at a time when top research universities had no women physics faculty at all and discrimination against Asians was strong. Wu decided not to continue as a research assistant at Berkeley. Instead, she and fellow Chinese physicist Luke Yuan, whom she had known since her first days in the United States, were married, and the two moved to the East Coast. Yuan was to work on radar devices at RCA laboratories, and Wu was offered a teaching position at Smith College. Although Wu enjoyed teaching, she missed research. After seeing Ernest Lawrence at a conference in the East, she admitted that she felt "sort of out of the way" not doing any experiments.[3] Lawrence recommended her to a number of universities. With the shortage of physicists owing to the war, Wu received offers even from universities such as MIT, Princeton, and Columbia, none of which accepted female students at the time. Wu decided to go to Princeton, where she became Princeton's first woman instructor.[4]

After several months at Princeton, Wu was interviewed and offered a job by the Division of War Research at Columbia, where she worked from 1944 until the end of the war. After the war she became a senior researcher at Columbia. In 1947 her son, Vincent Wei-chen Yuan, was born. Soon afterwards, both she and Yuan were offered positions at the National Central University in China. On the advice of her father (with whom Wu had been able to correspond since the end of the war with Japan), and to avoid raising Vincent in a communist country, Wu and Yuan decided not to return. After the communist victory in 1949, return was not possible. Yuan and Wu became American citizens in 1954. When Wu was finally able to visit China in 1973, her parents and brothers had already died.

At Columbia in the postwar era, Wu settled into research. She chose to concentrate on β-decay and developed a reputation for careful, reliable experiments. Wu confirmed one of Fermi's theories of β-decay, that most of the electrons ejected from the nucleus in β-decay would have very high speeds. Previous experiments had not been able to confirm this, because the physicists had used radioactive films of uneven thickness and electrons lost energy colliding with atoms in the thicker areas. These experiments helped to establish Wu's reputation as a brilliant experimental physicist.

Wu demanded reliability and hard work of herself and of her graduate students. Possessing a great enthusiasm for her work, she spent long hours in the lab and was completely committed to carrying out careful experiments in order to produce reliable, definitive results that would be a meaningful contribution to the understanding of her field. She expected a similar commitment from her graduate students and soon acquired a reputation for working her students very hard. She also acquired a nickname, "The Dragon Lady," although it was used more by those who were not her students. Her students thought of her more as a strong mother figure. She is remembered by some as "the most human" of the professors at Columbia at the time.[5]

Since the end of the war, Wu had worked as a researcher at Columbia. Until she was given a part-time teaching assignment arranged by William Havens, director of Columbia's Nuclear Physics Laboratories, Wu was repeatedly overlooked for promotion to a faculty position. Finally in 1952 she became a faculty member.

From 1952 to 1956, Wu explored areas of research other than β-decay. But Wu's work in β-decay was not over. In the early 1950s, while Wu was working on other problems, experiments on the decay of elementary particles known as K-mesons had given rise to a dilemma for physicists. The K-mesons were found to be identical in every measurable quantity, but they differed in the way they decayed: some K-mesons decayed into two π-mesons, others into three π-mesons. Since the π-mesons had odd parity, decay into two π-mesons implied that the K-mesons had even parity, whereas decay into three π-mesons meant a K-meson with odd parity. Thus either the two otherwise identical particles had to be different, or parity was not conserved in the decay.

Physicists were reluctant to accept the second alternative, as that would mean the overthrow of a fundamental principle of physics: that the mirror image of a physical process should also be a plausible physical process. An example of such symmetry is a round ball spinning clockwise. If we were to see the ball in a mirror, it would seem to be spinning counterclockwise; however, unless we knew beforehand which was the real ball and which was the mirror image, we could not know from the images alone which was real and which was not—we could just as well be seeing a real ball spinning in the opposite direction and its mirror image. A ball spinning counterclockwise and one spinning clockwise are equally plausible in nature; neither is "aphysical." This basic symmetry between real events and their mirror images, which leads to the law of conservation of parity, was demonstrated in other atomic and nuclear processes, and physicists assumed that β-decay and other weak interactions would also conserve parity. However, while trying to solve the dilemma posed by the K-meson decays, theoretical physicists Tsung Dao Lee and Chen Ning Yang realized that no experiment had ever confirmed the conservation of parity in weak interactions. They published a paper suggesting several experiments that might settle the question.

When Lee approached Wu about the experiment, Wu recognized that this was an opportunity to use her expertise in β-decay to confirm or disprove a fundamental natural law. Wu decided to investigate the emission of electrons by radioactive cobalt-60 (Co-60) in order to determine if there was a preferred direction of emission. In order to make the measurements, the Co-60 had to be cooled to very low temperatures to fix the positions of the nuclei and the nuclei then had to be aligned with a strong magnetic field. If parity were conserved, the electrons would be ejected in the direction of the magnetic field and opposite the magnetic field in equal amounts. This would be analogous to a spinning ball ejecting particles from both ends; the mirror process would be indistinguishable from the real process. If parity were not conserved, more electrons would be ejected from one end than the other. Like a spinning ball ejecting particles only from one end, the mirror image of the process and the real process would then be distinguishable. This would mean that for nuclear processes like β-decay, there would be an asymmetry between the physical universe and its mirror image.

It was a difficult experiment. An electron detector needed to be put inside a cryostat, which acts like a thermos to isolate the interior from outside heat, to function as a β-spectrometer at very low temperatures; and the Co-60 β-source had to be in a very thin surface layer and remain polarized long enough for measurements to be taken. Neither of these difficult techniques had ever been used before. There were very few places in the United States equipped to do low-temperature nuclear orientation experiments like this one. Wu called Ernest Ambler at the National Bureau of Standards in Washington, D.C., who immediately accepted the idea of a collaboration to carry out the experiment.

For the rest of the year, the research team resolved problem after problem in order to reach a definitive answer to the question of parity conservation in β-decay. Finally, in January 1957 the result was announced: more electrons were emitted opposite the direction of the magnetic field than along it. Parity had fallen.

The result astonished most of the physics community. Many physicists had been convinced only six months earlier that such an experiment would yield the opposite result. Wu herself had given parity nonconservation a "one in a million chance." Other experiments followed that showed the nonconservation of parity in other weak interactions. Theorists began to investigate the possibility of other conservation laws, such as CP (the product of parity and charge parity) invariance and vector current conservation in β-decay. Wu's experiment was a milestone in nuclear physics. For the theoretical work that had prompted the experiment, Yang and Lee were awarded the Nobel Prize.

After the parity experiment, Wu moved on to other areas. She returned occasionally to β-decay, for example, to confirm Feynman and Gell-Mann's theory of conservation of vector current in β-decay. This theory, which was developed in the burst of research that followed Wu's discovery of parity violation, is expressed in a formula that describes observed phenomena of weak interactions that are charged. Wu was recognized on numerous occasions for her outstanding work, including many awards, acceptance into the National Academy of Sciences, a full professorship (and later an endowed professorship) at Columbia, and many honorary degrees. After her retirement in 1981 she lectured and taught in many places, addressing issues of policy as well as science. Concerning the lack of women in science, she maintained that "unimpeachable tradition," rather than women's intellectual capacity or socioeconomic factors, was the reason that there were so few women in science. It is characteristic of Chien-Shiung Wu to formulate the problem with such clarity. As in her scientific endeavors, Wu sought to identify the real question and seek a definitive answer.

With a keen ability to find fundamental questions and with no fear of the difficult, Chien-Shung Wu carried out many difficult and important experiments that shaped modern physics, particularly in the area of nuclear physics. Like her idols, **Marie Curie** and **Lise Meitner**, Wu combined her love of physics and an incredible determination to change the way we understand nature. Although she could be completely absorbed by her work, she never lost her sense of wonder at the interesting world she sought to understand. As Wu later said of the parity conservation experiment, "These were moments of exhilaration and ecstasy! A glimpse of this wonder can be the reward of a lifetime. Could it be that excitement and ennobling feelings like these have kept us scientists marching forward forever!"[6]

Notes

1. Sharon Bertsch McGrayne, *Nobel Prize Women in Science: Their Lives, Struggles and Momentous Discoveries* (New York: Carol Publishing Group, A Birch Lane Press Book, 1993), p. 260.

2. Ibid., p. 262.

3. Ibid., p. 265.

4. Ibid., p. 266.

5. Ibid., p. 271.

6. Chien-Shung Wu, "The Discovery of Nonconservation of Parity in Beta De-

cay," in *Thirty Years since Parity Nonconservation: A Symposium for T. D. Lee*, ed. Robert Novick (Boston and Basel: Birkhäuser, 1988), p. 35.

Bibliography

Boorse, Henry A., et al. *The Atomic Scientists: A Biographical History.* New York: John Wiley and Sons, 1989.

Gardner, Martin. *The Ambidextrous Universe: Left, Right, and the Fall of Parity.* New York: New American Library, 1969.

McGrayne, Sharon Bertsch. *Nobel Prize Women in Science: Their Lives, Struggles and Momentous Discoveries.* New York: Birch Lane Press, 1993.

Segrè, Emilio. *From X-Rays to Quarks: Modern Physicists and Their Discoveries.* New York: W.H. Freeman, 1980.

Wehr, M. Russell, James A. Richards Jr., and Thomas W. Adair III. *Physics of the Atom.* Reading, Mass.: Addison-Wesley, 1984.

Wu, Chien-Shung. "The Discovery of Nonconservation of Parity in Beta Decay," in *Thirty Years since Parity Nonconservation: A Symposium for T.D. Lee.* Edited by Robert Novick. Boston and Basel: Birkhäuser, 1988.

URSULA ALLEN

ROSEMARY WYSE

(1957–   )

Astrophysicist

Birth	January 26, 1957
1977	B.Sc. (first class honours), physics with astrophysics, University of London, Queen Mary College, Physics Dept.
1978	Passed with Distinction Part Three of the Mathematics Tripos, University of Cambridge, Dept. of Applied Mathematics
1978–81	Bachelor Scholarship, Emmanuel College, Cambridge
1978–82	University of Cambridge, Institute of Astronomy
1981–82	Amelia Earhart Fellowship, ZONTA International
1982–83	Lindemann Fellow of the English Speaking Union of the Commonwealth
1983	Ph.D., astrophysics, Cambridge University
1983–85	Parisot Postdoctoral Fellow, University of California, Berkeley

Rosemary Wyse. Photo courtesy of Emil Venere, Office of News and Information, Johns Hopkins University.

1985	University of California President's Fellow, University of California, Berkeley
1986	Annie Jump Cannon Award, American Association of University Women; Postdoctoral Research Fellow, Academic Affairs, Space Telescope Science Institute
1986–88	University of California President's Fellow, University of California, Berkeley
1988–90	Assistant Professor, Johns Hopkins University
1990–92	Fellowship, Alfred P. Sloan Foundation
1990–93	Associate Professor, Johns Hopkins University
1993–	Professor, Johns Hopkins University

Rosemary Wyse is a Scottish-born, Cambridge-educated astrophysicist whose investigation of the formation of stars, the structure of galaxies, and the composition of the universe has explored some of the most puzzling and enduring mysteries of astronomy.

Wyse received her bachelor of science degree in physics from the University of London, Queen Mary College, in 1977 and began her graduate education in Cambridge's Department of Applied Mathematics. There, she earned part three of the Mathematics tripos, which is similar to an American master's degree, focusing her studies on theoretical physics. She continued at Cambridge in its Institute of Astronomy, deciding to specialize in astrophysics rather than particle physics because of her preference for greater individual involvement with the scholarly product. She earned the Ph.D. in astrophysics in 1983. Wyse was awarded the Annie Jump Cannon prize from the Association of American University Women in 1986 in recognition of her potential to contribute to the field of astronomy.

With her supervisor at Cambridge, Bernard T. Jones, Wyse began her scholarly career exploring the structure and formation of galaxies, particularly elliptical galaxies. The two examined the physical parameters of galaxies (including surface brightness, velocity, and metal content), noting relationships between these parameters in elliptical galaxies. To explain these relationships, they theorized that the varying qualities develop in response to the amount of energy dissipated when clouds of gas collapse to form the galaxies.

After her work at Cambridge, Wyse held a series of postdoctoral and fellowship positions, the majority of these at the University of California at Berkeley. She maintained a scholarly connection to Cambridge, however, collaborating with Gerard Gilmore on a number of research projects. Throughout her career Wyse has continued to pursue explanations for star formation and galaxy evolution, but more recently she has concentrated on the motion and chemical composition of stars and their rates of formation as well as the "missing matter" mystery, which has plagued and intrigued modern-day astronomers.

Wyse's work with Gilmore on the distribution of metals (elements that are heavier than hydrogen and helium) in the Milky Way has been an important contribution to the study of star and galaxy formation; it is revealing for what it reflects about the age and history of the stars as well as that of the galaxy as a whole. Together, Wyse and Gilmore measured the vertical distribution of metallicity in the galaxy, determining that there are three distinct layers: an inner layer at the disk's central plane, called the thin disk; a thick disk that surrounds it and generally has a somewhat lower metal content; and the extreme spheroid, which marks the vertical visible limits of the galaxy and has the lowest metallicity levels.

The proposition of a thick disk went against classical models and as such was somewhat controversial. However, Wyse and Gilmore were able to extend and confirm the work with further studies, both in the metallicity and the motion of the stars. They incorporated these data into a proposition that served as a relatively simple solution to the so-called "G-dwarf problem"—the conflict between the metallicity predicted and that actually found in the oldest of the sun's neighboring stars, which are, like the sun, members of the thin disk. In their model the thick disk has provided the extra metallicity, losing gases to its inner layer in time to enrich the stars during the formation process.

The presumption of a thick disk helped resolve yet another conflict between expected and measured metallicity in the galaxy. In classical models, stars that do not lie in the thin disk, called "Population II" stars, necessarily belong to the extreme spheroid, which is dominated by metal-poor subdwarfs with relatively random motion. The chemical composition and motion data for extreme spheroid subdwarfs in the Milky Way conflicted with observations of the dominant spheroid in other galaxies. With the new model of galactic structure, most of the stars assigned by previous models to the extreme spheroid could be identified as members of the thick disk, with higher metal content and more organized motion.

Wyse has also, often in collaboration with Berkeley's Joseph Silk, investigated star formation rates (SFRs), which are intimately entwined with the chemical makeup and structure of galaxies. This work has explored both the factors that influence these rates as well as the result, in terms of galactic chemical composition and structure, of rate variations. Wyse accepts the classical Schmidt model of star formation, which holds that the rate depends heavily on gas density; but she has extended it to consider other possible compounding factors, especially the frequency with which gas clouds collapse. The SFR has profound implications for the ultimate shape and composition of a galaxy. The process of star formation depletes gases in a galaxy. So galaxies where star formation has proceeded rapidly and inefficiently have few remaining gases. Furthermore, stars with greater masses, which tend to form more rapidly, are less stable; they rapidly burn their energy, becoming star remnants such as neutron stars and black holes. Finally, there is a general observation that rapid star formation appears to correlate with an elliptical, rather than spiral, galactic shape.

Since 1988, Wyse has been on the faculty at Johns Hopkins University, where her interest in star and galaxy formation has continued. This interest has led naturally to a problem that has preoccupied many modern-day astronomers: the probability that a large portion of the mass of the universe, as determined both theoretically and by gravitational measurements, has not been visually discovered. This missing mass is called "dark matter," and although there is general agreement that the universe must contain much more matter than is represented in the visible stars—

perhaps even ten times as much—there is hot debate about what form that matter might take. The matter we are familiar with is "baryonic" in form; that is, it is ordinary matter, made up of protons and neutrons. One of the key issues of the dark matter debate is whether dark matter is baryonic yet for some reason hard to find, or whether it is nonbaryonic, an exotic kind of matter unlike anything scientists have yet discovered. Wyse accepts the possibility that some portion of the matter may be nonbaryonic; however, she also recognizes that there are many possible and somewhat more ordinary hiding places for the missing matter. She has explored these in pursuit of the solution to this puzzle. In recent years, her interest in galaxy formation has been focused more on the formation of dwarf and other low-visibility galaxies, which are particularly notable because they may be homes for some of the dark matter. She has also examined other possible locations, including white dwarfs (stars that have burned out their thermonuclear energy), brown dwarfs (stars that, due to their low mass, cannot generate thermonuclear energy), planets, and other nonluminous objects. Because the evolution of galaxies determines the existence of these as well as other possible hiding places, such as intergalactic gas, the dark matter debate is both a continuation and a natural outgrowth of her previous work.

Bibliography

Gilmore, Gerard, and Rosemary F.G. Wyse. "The Abundance Distribution in the Inner Spheroid." *Astronomical Journal* 90, no. 10 (October 1985): 2015–2026.

———. "The Chemical Evolution of the Galaxy." *Nature* 322, no. 6082 (August 28, 1986): 806–807.

Jones, Bernard J.T., and Rosemary F.G. Wyse. "The Formation of Disc Galaxies." *Astronomy and Astrophysics* 120, no. 2, pt. 2 (April 1983): 165–180.

———. "The Ionisation of the Primeval Plasma at the Time of Recombination." *Astronomy and Astrophysics* 149, no. 1, pt. 1 (August 1985): 144–150.

Wyse, Rosemary F.G. "Dark Matter in the Galaxy." *International Journal of Modern Physics D* 3, suppl. (1994): 53–61.

———. "On the Epoch of Elliptical Galaxy Formation." *Astrophysical Journal* 299, no. 2, pt. 1 (December 15, 1985): 593–615.

Wyse, Rosemary F.G., and Gerard Gilmore. "Kinematics of the Galaxy from a Magnitude-Limited Proper-Motion Sample." *Astronomical Journal* 91, no. 4 (April 1986): 855–869.

———. "The Structure of the Galaxy." *Annals of the New York Academy of Science* 596 (June 1990): 8–19.

ELIZABETH VILLEPONTEAUX

XIDE XIE

(1921–)

Physicist

Birth	March 19, 1921
1946	B.S., physics, Xiamen University, Xiamen, Fujian, China
1949	M.A., physics, Smith College
1951	Ph.D., physics, Massachusetts Institute of Technology
1978–83	Vice President of Fudan University, Shanghai, China
1980	Member of the Chinese Academy of Sciences (Academician)
1981	D.Sc., Smith College; CCNY of the City University of New York
1983–88	President of Fudan University, Shanghai, China
1985	D.Sc., University of Leeds, Leeds, United Kingdom
1985–	Director of the Center for American Studies, Fudan University, Shanghai, China
1986	D.Sc., Mt. Holyoke College; Fellow, American Physical Society
1987	D.Sc., Kansei University, Osaka, Japan; Beloit College; State University of New York, Albany
1988	Fellow, The Third World Academy of Sciences, Trieste, Italy
1988–	Adviser of Fudan University, Shanghai, China
1991	Foreign Honorary Member of the American Academy of Arts and Sciences
1993	D.Sc., Suffolk University; McMaster University, Hamilton, Ontario

In China, Xide Xie needs no introduction. In the international physics community, she is well known. An outstanding Chinese woman physicist, her passion and courage to challenge accepted limits has become an inspiration for generations of Chinese scientists. Xide Xie (in the order of Chinese, it is Xie Xide, pronounced approximately as She-eh Shee De) has received numerous honors from the Chinese government as well as awards and honorary degrees from Smith College, Leeds University, the

Xide Xie. Photo reprinted by permission of the Sophia Smith Collection, Smith College.

State University of New York, and Beloit College. A distinguished educator, she has cultivated many students into elite scientists, some of whom are now members of China's Academy of Science and some of whom are now making important contributions in the international physics community. Her dedication and commitment to physics and education have set an example for many scientists to follow.

Xide was born to an intellectual family in Fujian province in 1921. When she was 4 years old, her mother passed away. She stayed with her grandmother for two years until her father, a distinguished physicist who received his Ph.D. in physics at the University of Chicago, returned from the United States to teach at Yenjing University in Beijing. There, Xide completed most of her early education and discovered great interest in physics, math, and medical science. During World War II the family fled Beijing and moved south. They stayed in Wuchang for half a year and then went to Changsha and Guiyang.

Despite her great interest in science, Xide had to put her college dreams aside in 1938. Serious hip joint problems—she actually had tuberculosis in the joint—confined her to bed for four years, and the family had to move further south to Guiyang to escape the war. During that bedridden time Xide taught herself English, calculus, and physics. She enrolled in Xiamen University in 1942 and graduated four years later with a degree in math and physics.

A year later, Xide left for advanced study in the United States. She received her master's degree from Smith College in 1949 and her Ph.D. in physics from MIT in 1951. Xide went to England in 1952 to marry her high school sweetheart, Cao Tianqin, who had just received a Ph.D. in biochemistry from the University of Cambridge. Despite strong opposition from Xide's father and many bureaucratic difficulties, they returned to China in the same year to serve their country.

Soon Xide started work in Shanghai as a lecturer at Fudan University, one of the most prestigious universities in China. It did not take long for her to show her talent as a teacher and a scholar. She developed various classes in optics, mechanics, solid state physics, and the like. Four years later, Xide was promoted to associate professor and director of the solid state physics lab. Her devotion to teaching science was highly recognized on campus. She was well known for spending endless time coaching students and writing course material to keep students informed of the latest trends and progress in physics. She was also well respected for providing young colleagues with opportunities for advancement by letting them teach courses she had established while she explored difficult new ones. In 1958 she became the adjunct vice-president of the Shanghai Institute of Technical Physics. That year she also co-authored the book *Physics of Semiconductors*, the first book published on this subject in China. In the early 1960s she started organizing a research group at Fudan University to study electronic states of solids, both experimentally and theoretically.

Like many other scientists, Xide was an unfortunate victim of the political turmoil during the Cultural Revolution (1966–1976). In addition to detention and public humiliation, Xide had to face the pain caused by breast cancer. After she was released from prison she was forced to perform janitorial and factory labor until 1972, when she was allowed to return to academic life. Looking back, one cannot help marveling at her great strength and her beliefs, which pulled her through the darkest period of her life. S. V. Meschel wrote the following in his article "Teacher Keng's Heritage":

During the conversation with Xie Xide one could not but be impressed with her gentle character, the lucidity of her ideas and penetrating insight of the true scholar. It was difficult to visualize this petite lady of infinite gentleness as a survivor of years of extreme adversity, who against tremendous odds became one of her country's leading scientists, administrator and policy maker.[1]

A year after the end of the Cultural Revolution, Xide discovered tremendous interest in surface physics from reading Western journals. She soon convinced the Ministry of Education and the State Commission of Science and Technology to fund the establishment of the Modern Physics Research Institute and eight new labs at Fudan. In 1978, Xide was named vice-president of Fudan University, president of the Modern Physics Research Institute, and vice-chairman of the Chinese Physical Society. Her prolific research results and outstanding administrative leadership soon brought her greater career advancement. In 1980 she was elected to be a member of the Chinese Academy of Sciences, and in 1983 she became president of Fudan University.

As a physicist, Xide's main research areas are in solid state physics and semiconductor surface physics. She has published more than 60 scientific research papers with her colleagues, four of which won the Chinese National Award. She has also co-authored four volumes in solid state physics: *Group Theory and Its Applications In Physics; Solid State Physics*, Volumes 1 and 2; and *Semiconductor Physics*.

In appreciation of Xide's achievements, the Chinese government has awarded her several policy-making posts. In the 1980s she was the chairperson of the Chinese People's Political Consultative Conference—Shanghai Committee and a member of the Central Committee of the Chinese Communist Party. In 1993 she was elected to the Standing Committee of the Chinese People's Political Consultative Conference.

Despite her busy schedule, Xide never failed to spend time with her family. While coping with her cancer, Xide demonstrated extraordinary love and courage in tending her husband, Cao, who had been ill and bedridden from 1987 until he passed away in early 1995. Xide insisted on feeding and reading to him in person every day when she was in town. Her courage and optimism has become a well-publicized legend by the media. Xide has a son, a daughter-in-law, and a granddaughter, whom she visits as often as she can.

Today, at age 74, Xide is more assertive and vigorous than ever. She continues to be very active in the international physics community. She has attended conferences and given speeches in many countries in Asia, North America, South America, and Europe to discuss her thoughts on science and social issues. She has been invited to the March meeting of the American Physics Society every year since 1983. She played a key role in the 21st International Conference on Semiconductor Physics in Beijing in 1991, one of the most important international conferences in physics. She continues to provide full support to her numerous students and colleagues, both in China and abroad, for their scientific achievements and their career advancement. As a woman scientist, she is also very concerned about the status of the education of women. Xide's achievements in scientific research, education, and public service highlight her remarkable career. Her passion and the strength she summons to meet all challenges have made her an outstanding role model for women in this new era.

Note

1. S.V. Meschel, "Teacher Keng's Heritage," *Journal of Chemical Education* 69, no. 9 (1992): 725.

Bibliography

Meschel, S.V. "Teacher Keng's Heritage." *Journal of Chemical Education* 69, no. 9 (1992): 723–730.
Wang, Zhen-Fan. "Woman Physicist—Xie Xide." *China Historical Materials of Science and Technology* 14, no. 3 (1993).
                    YIMIN WANG ZIMMERER and SHANG-FEN REN

JUDITH SHARN YOUNG
(1952–   )
Astronomer

Birth	September 15, 1952
1974	B.A. (with honors), astronomy, Harvard University
1975–90	Married Michael Young
1977	M.S., physics, University of Minnesota
1979	Ph.D., physics, University of Minnesota
1979–82	Postdoctoral Research Associate, University of Massachusetts, Amherst
1982–83	Annie J. Cannon Prize, American Astronomical Society

Judith Sharn Young. Photo courtesy of Judith S. Young.

1982–84	Visiting Assistant Professor, University of Massachusetts, Amherst
1984–87	Assistant Professor, University of Massachusetts, Amherst
1986–88	Maria Goeppert-Mayer Award, American Physical Society
1986–89	Sloan Research Fellowship
1987–93	Associate Professor, University of Massachusetts, Amherst
1991	Visiting JCMT Fellow, University of Hawaii
1993–	Professor, University of Massachusetts, Amherst

"I did not grow up thinking that I wanted to be an astronomer," Judith Young admits, "although I loved the sky and I remember even at age 11 just sitting outside and looking at the sky and being awed by the beauty of the stars."[1] Some might think that fate was sealed for Judith Sharn Rubin when she was born on September 15, 1952, in Washington, D.C.—that she was destined to pursue science. Young's parents are Robert J. Rubin, a physical chemist, and Vera Cooper Rubin, an astronomer. **Vera Cooper Rubin** is known for proving the existence of dark matter, a ground-breaking contribution that led to many other equally important discoveries in astronomy.

It just so happens that Young and each of her three brothers have chosen careers in the sciences, but Young did not feel compelled to develop her nascent interest in astronomy until her senior year in high school. Until then, she preferred chemistry and biochemistry. But she experienced a defining moment during an astronomy class: "the moment that I decided I wanted to be an astronomer was when I learned about black holes. They were so neat to me that I needed to be an astronomer. I was lucky to be taught the course by really an outstanding teacher and role model, and she is also my mother."

Young did not set out to emulate her mother, however. She was determined to contribute in her own way and without the potential benefit of the Rubin name. So, when she married in 1975, she took her husband's name to prevent instant recognition as Vera Rubin's daughter. She chose to keep the name "Young" even when she divorced after nearly 14 years of marriage.

Young's path to her current position at the University of Massachusetts at Amherst was a trying one, but she persevered. She chose to study astronomy as an undergraduate and was accepted at Harvard. She received her B.A. there with honors in 1974 but was advised by one of her professors in her junior year to drop out and get married. At best, Young was told, she would get a job in a junior college—and even those jobs were difficult to get. She was disappointed but not swayed by the advice.

Young was not readily admitted into graduate school because her physics background was considered to be weak. With encouragement from her parents, she persisted and was admitted to the University of Minnesota graduate program in astronomy in 1974.

In her second year there, even after she had completed the Ph.D. qualifying exam, she was told by the faculty in the department that she should stop with the master's degree. The sole impetus for their decision was that she was to be married to Michael Young, a graduate geology student also at the University of Minnesota. Young was told that she should follow her husband in his career and not pursue a career of her own. She felt compelled not only to ignore that advice but to switch from the astronomy program to the physics program. "There was one faculty member there who bridged the gap between physics and astronomy," Young recalls, "and she happened to be the only woman in the [physics] department at the time, Phyllis Freier. . . . I ended up pursuing for my Ph.D. cosmic ray physics, the study of the different isotopes in the cosmic rays that are showering our solar system, planet, and galaxy."

Young completed her dissertation, "The Isotopic Composition of Cosmic Rays," in 1979. That same year, she applied for and got a postdoctoral position at the Five College Radio Astronomy Observatory (FCRAO) of the University of Massachusetts at Amherst. The postdoctoral position led shortly thereafter to her appointment as the first woman astronomy faculty member at the university.

Studying galaxies was Young's goal. She explains that "in 1980 here at UMass, the receivers improved significantly and it was because of that that I was able to begin the galaxy work. . . . I think that if that hadn't happened, I might not have even stayed in astronomy because I remember thinking that maybe I would switch to another science like geology if the galaxy work did not become possible."

Early on at the FCRAO, Young was mentored by Nick Z. Scoville, a visiting radio astronomer with whom Young collaborated often. Much of their work involved studying the carbon monoxide and cold gas content of galaxies, because this is the material from which stars form. Among several other discoveries, they determined that the distribution of gas is proportional to the distribution of light in the galaxies. That is, as the amount of gas that is present in the galaxies increases, the more stars are formed. This work led to Young's receipt of the Annie J. Cannon prize, an award given in recognition of a young woman astronomer for her achievement and potential for research.

Young has been prolific throughout her 15–year professional career, publishing over 100 papers and giving nearly as many talks and seminars. In addition to the aforementioned galaxy studies, her interests lie in galaxy formation and evolution, star formation, and interstellar matter (the matter that lies between the stars in galaxies). She has received several professional awards and honors, including, in 1986, the Maria Goeppert-Mayer Award of the American Physical Society. Young was the first recipient of this award, which is given to a woman in the early stages of her career who has made a significant contribution to physics.

The Goeppert-Mayer Award carried with it the stipulation that she give a number of talks to encourage women in science. When she arrived to give the talks, she invariably encountered audiences of virtually all men. Young took the opportunity, therefore, to remind the men in the audiences that "encouraging women in science has at least as much to do with getting encouragement from the men in our lives as the women."

Although Young very much enjoys conducting research, teaching is just as important and rewarding to her. "I enjoy teaching the undergraduate courses a lot. I teach the graduate courses, and the students there are more interested; but there is a part of me that really likes bringing . . . just a very basic understanding of astronomy and the universe to people who are not used to thinking about science."

To bring the world of science to a new generation, in addition to her postsecondary-level teaching, Young has begun to write a children's book on the phases of the moon. Also, for the past six years she has been teaching astronomy at the elementary school that her daughter, Laura Rose Young (who is now 11 years old), attends.

From the time she was a girl, Young has been intrigued by and had an aptitude for not only astronomy and physics but chemistry and biochemistry as well. Since she attained full professor status in 1993, she has had the freedom to pursue more than astronomy. Recently she has forayed into biomedical research, with her sights set on a gentle cure for cancer or research that helps other investigators to that end.

Like George Gamov, for instance, who was a twentieth-century scientific polymath in astronomy, physics, and biochemistry, Young will contribute whatever she can, using all her talents to better the world. Young sees the parallels between research in astronomy and biomolecular research in that, she says, "galaxies are the cells of [a] large-scale structure, and the cells that I'm looking at that are cancers are [of a] small-scale structure. . . . It's very inspiring to me to be able to do both at the same time—and rewarding."

As for the future, Young plans to continue her astronomy research and teaching along with her biomedical research. She is a member of the American Astronomical Society and its Committee for the Status of Women, the International Astronomical Union, the American Physical Society, and the Association for Women in Science. She will also continue to teach in her daughter's school every year and to encourage Laura to pursue as a career whatever she most enjoys. This is something that Vera Rubin impressed on her when Judith was a child. "I remember my mother telling me when I was young that I could do anything I wanted to in my life," Young remarks, "if I kept my mind to it."

Note

1. All quotes are taken from audiotape answers by Judith S. Young to a questionnaire prepared by Sue Ann Lewandowski, Amherst, Massachusetts, June 1995.

Bibliography

Devereaux, N.A., J.D.P. Kenney, and J.S. Young. "Molecular Clouds in the Nuclear Region of NGC 3351." *Astronomy Journal* 103 (1992): 784.

Tacconi, L., and J.S. Young. "The Distribution of the ISM in the Scd Galaxy NGC 6946. II. The CO Data." *Astrophysics Journal Supplement* 71 (1989) 455.

Young, J.S., and N.Z. Scoville. "Extragalactic CO: Gas Distributions Which Follow the Light in IC 342 and NGC 6946." *Astrophysics Journal* 258 (1982): 467.

Young, J.S., F.P. Schloerb, J. Kenney, and S. Lord. "CO Observations of Infrared Bright Galaxies." *Astrophysics Journal* 304 (1986): 443.

Young, J.S., M. Claussen, and N.Z. Scoville. "Molecular Clouds in the Nuclear Region of NGC 3079." *Astrophysics Journal* 324 (1988): 115.

Young, J.S., N.Z. Scoville, and E. Brady. "The Dependence of CO Content on Morphological Type and Luminosity for Spiral Galaxies in the Virgo Cluster." *Astrophysics Journal* 288 (1985): 487.

Young, J.S., et al. "The FCRAO Extragalactic CO Survey: Global Properties of Galaxies," in *The Interstellar Medium in External Galaxies*. D.J. Hollenbach and Harley A. Thronson, Jr., eds. Washington, D.C.: National Aeronautics and Space Administration, Office of Management, Scientific and Technical Information Division, 1990.

Xie, S., J.S. Young, and F.P. Schloerb. "A ^{12}CO, ^{13}CO and CS Study of NGC 2146 and IC 342." *Astrophysics Journal* 421 (1994): 434.

SUE ANN LEWANDOWSKI

Appendix I: Scientists by Profession

Alchemists

Miriam the Alchemist

Analytical Chemists

Jeanette Grasselli (Brown)

Astronomers

Ida Barney

Jocelyn Bell Burnell

Annie Jump Cannon

Allie Vibert Douglas

Sandra Moore Faber

Williamina Paton Fleming

Wendy Laurel Freedman

Catharine D. Garmany

Margaret Joan Geller

Heidi Hammel

Margaret Harwood

Caroline Herschel

E. Dorrit Hoffleit

Helen Sawyer Hogg

Henrietta Swan Leavitt

Antonia Maury

Maria Mitchell

Cecilia Payne-Gaposchkin

Helen W. Dodson Prince

Dorothea Klumpke Roberts

Elizabeth Roemer

Nancy Grace Roman

Vera Cooper Rubin

Mary Fairfax Somerville

Henrietta Hill Swope

Paula Szkody

Beatrice Muriel Hill Tinsley

Emma T.R. Williams Vyssotsky

Sarah Frances Whiting

Lee Anne M. Willson

Judith Sharn Young

Astrophysicists

E. Margaret Burbidge

Allie Vibert Douglas

Charlotte Emma Moore Sitterly

Rosemary Wyse

Bacteriologists

Mary Engle Pennington

Biochemists

Mildred Cohn

Gerty Cori

Helen M. Dyer

Gertrude Belle Elion

Arda Alden Green

Icie Gertrude Macy Hoobler

Ines Mandl

Gertrude Perlmann

Mary Locke Petermann

Sarah Ratner

Florence B. Seibert

Sofia Simmonds

Betty Sullivan

Biophysicists

Mildred Cohn

Cancer Researchers

Jenny Pickworth Glusker

Kamal J. Ranadive

Chemists

Ruth Mary Roan Benerito

Leonora Neuffer Bilger

Mary Letitia Caldwell

Emma Perry Carr

Marjorie Caserio

Marie Sklodowska Curie

Catherine Clarke Fenselau

Mary Fieser

Helen M. Free

Ellen Gleditsch

Mary Lowe Good

Jeanette Grasselli (Brown)

Dorothy Anna Hahn

Anna Jane Harrison

Allene Rosalind Jeanes

Madeleine M. Joullié

Joyce Jacobson Kaufman

Pauline Beery Mack

Grace Medes

Agnes Fay Morgan

Dorothy Virginia Nightingale

Mary Engle Pennington

Marguerite Perey

Lucy Weston Pickett

Ellen Swallow Richards

Mary Lura Sherrill

Jean'ne Marie Shreeve

Marjorie Jean Young Vold

Elizabeth Amy Kreiser Weisburger

Colloid Chemists

Marjorie Jean Young Vold

Crystallographers

Jenny Pickworth Glusker

Dorothy Crowfoot Hodgkin

Isabella L. Karle

Jane Richardson

Ecologists

Ellen Swallow Richards

Environmentalists

Vandana Shiva

Home Economists

Ellen Swallow Richards

Molecular Spectroscopists

Leona Woods Marshall Libby

Nuclear Chemists

Ellen Gleditsch

Nuclear Physicists

Maria Goeppert-Mayer

Nutritional Chemists
Mary Letitia Caldwell

Physical Chemists
Ruth Mary Roan Benerito

Physicists
Laura Bassi
Katharine Burr Blodgett
Marie Sklodowska Curie
Ingrid Daubechies
Helen T. Edwards

Maria Goeppert-Mayer
Gertrude Scharff Goldhaber
Irène Joliot-Curie
Vera E. Kistiakowsky
Leona Woods Marshall Libby
Margaret Eliza Maltby
Lise Meitner
Vandana Shiva
Sarah Frances Whiting
Chien-Shiung Wu
Xide Xie

Appendix II: Scientists by Honors and Awards Received

Achievement and Service Award for Teaching and Research (Goucher College)

Helen M. Dyer

Achievement Award (American Association of University Women)

Katharine Burr Blodgett

Achievement Award (Society of Women Engineers)

Isabella L. Karle

Achievement Award for Biomedical Research in the Field of Cancer (George Washington University)

Helen M. Dyer

Actonian Prize (Royal Institution of Great Britain)

Marie Sklodowska Curie

Albert A. Michelson Prize (Franklin Institute)

Jocelyn Bell Burnell

Albert Einstein World Award of Science Medal

E. Margaret Burbidge

Alumnae Recognition Award (Radcliffe College)

E. Dorrit Hoffleit

American Cyanamid Faculty Award (University of Pennsylvania)

Madeleine M. Joullié

Anne Molson Gold Medal (McGill University)

Allie Vibert Douglas

Annenberg Foundation Award (American Astronomical Society)

E. Dorrit Hoffleit

Annie J. Cannon Prize

Ida Barney

Catharine D. Garmany

Margaret Harwood

Helen Sawyer Hogg

Antonia Maury

Cecilia Payne-Gaposchkin

Helen W. Dodson Prince

Charlotte Emma Moore Sitterly

Henrietta Hill Swope

Paula Szkody

Beatrice Muriel Hill Tinsley

Emma T.R. Williams Vyssotsky

Lee Anne M. Willson

Rosemary Wyse

Judith Sharn Young

Annie Jump Cannon Centennial Medal (Wesley College)

Charlotte Emma Moore Sitterly

Astronauts Silver Snoopy Award

Pauline Beery Mack

Austrian Honor Cross, First Class

Ines Mandl

Award for Cancer Research (Medical Council of India)

Kamal J. Ranadive

Award in Chemical Education (American Chemical Society)

Anna Jane Harrison

Award in College Chemistry (Manufacturing Chemists Association)

Anna Jane Harrison

Beatrice Tinsley Prize (American Astronomical Society)

Jocelyn Bell Burnell

Benjamin Apthorp Gould Prize (National Academy of Sciences)

Elizabeth Roemer

Bernard Gold Medal (Columbia University)

Irène Joliot-Curie

Berthelot Medal (French Academy of Sciences)

Marie Sklodowska Curie

Bijroet Medal (University of Utrecht, The Netherlands)

Isabella L. Karle

Biochemists Award (International Organization of Women)

Mildred Cohn

Borden Award (American Home Economics Association)

Icie Gertrude Macy Hoobler

Borden Award (American Institute of Nutrition)

Agnes Fay Morgan

Borden Award (Association of Medical Colleges)

Gerty Cori

Bower Award (Franklin Institute)

Isabella L. Karle

Cain Award (American Association of Cancer Research)

Gertrude Belle Elion

Cameron Prize (University of Edinburgh)

Marie Sklodowska Curie

Career Service Award (National Civil Service League)

Charlotte Emma Moore Sitterly

Carl Neuberg Medal (American Society of European Chemists and Pharmacists)

Ines Mandl

Sarah Ratner

Carolyn Wilby Prize (Radcliffe College)

E. Dorrit Hoffleit

Catherine Wolfe Bruce Medal (Astronomical Society of the Pacific)

E. Margaret Burbidge

Centennial Alumnae Award (Mount Holyoke College)

Vera E. Kistiakowsky

Centennial Medal (Canada)

Helen Sawyer Hogg

Certificate of Merit, Group 1 (American Medical Association)

Icie Gertrude Macy Hoobler

Charles Lathrop Parsons Award (American Chemical Society)

Mary Lowe Good

Chevalier Légion d'Honneur

Irène Joliot-Curie

Dorothea Klumpke Roberts

Commandeur, Ordre National du Mérite

Marguerite Perey

Commemorative Medal, 125th Anniversary of the Confederation of Canada

Helen Sawyer Hogg

Companion, Order of Canada

Helen Sawyer Hogg

Comstock Award (National Academy of Sciences)

Chien-Shiung Wu

Copley Medal (The Royal Society)

Dorothy Crowfoot Hodgkin

Cresson Medal (Franklin Institute)

Mildred Cohn

Davy Medal (Royal Society of London)

Marie Sklodowska Curie

Distinguished Achievement in Science Award (Secretary of the Navy)

Isabella L. Karle

Distinguished Alumna Award (Barnard College)

Henrietta Hill Swope

Distinguished Alumnae Citation (University of Arkansas)

Mary Lowe Good

Distinguished Alumni Award (College of Wooster)

Helen M. Free

Distinguished Alumni Award (University of Montana)

Jean'ne Marie Shreeve

Distinguished Alumnus Award (Polytechnic Institute of Brooklyn)

Ines Mandl

Distinguished Award (College of Physicians)

Mildred Cohn

Distinguished Daughters of Pennsylvania Medal

Pauline Beery Mack

Florence B. Seibert

Distinguished North Carolina Chemist Award (American Chemical Society)

Gertrude Belle Elion

Distinguished Service Award (American Academy of Achievement)

Mary Locke Petermann

Distinguished Service Award (Michigan Public Health Association)

Icie Gertrude Macy Hoobler

Distinguished Service Award (New Orleans Federal Executives Association)

Ruth Mary Roan Benerito

Distinguished Service Award (Society for Applied Spectroscopy)

Jeanette Grasselli (Brown)

Distinguished Service Award (U.S. Department of Agriculture)

Ruth Mary Roan Benerito

Allene Rosalind Jeanes

Distinguished Service Award (University Council of Queen's University)

Allie Vibert Douglas

Distinguished Service Citation (Bryn Mawr College)

Grace Medes

Distinguished Service Citation (University of Kansas)

Grace Medes

Distinguished University Scholar (University of Maryland)

Catherine Clarke Fenselau

Distinguished Woman Award (Benaras Hindu University)

Kamal J. Ranadive

Dorothea Klumpke Roberts Prize (University of California, Berkeley)

Elizabeth Roemer

Dr. Basantidevi Amirchand Award (Indian Council of Medical Research)

Kamal J. Ranadive

Draper Gold Medal (National Academy of Sciences)

Annie Jump Cannon

Earth Day International Award

Vandana Shiva

Ellen Richards Research Prize (Association to Aid Scientific Research by Women)

Annie Jump Cannon

Lise Meitner

Elliott Gresson Gold Medal (Franklin Institute)

Marie Sklodowska Curie

Engineering and Science Hall of Fame

Gertrude Belle Elion

Enrico Fermi Prize (U.S. Atomic Energy Commission)

Lise Meitner

**Ernest O. Lawrence Award
(U.S. Department of Energy)**

Helen T. Edwards

**Excellence in Teaching Award
(Chemical Manufacturers
Association)**

Jean'ne Marie Shreeve

**Faculty Distinguished
Achievement Award
(University of Michigan)**

Helen W. Dodson Prince

**Fankuchen Award (American
Crystallographic Association)**

Jenny Pickworth Glusker

**Federal Woman's Award (U.S.
Civil Service Commission)**

Ruth Mary Roan Benerito

Allene Rosalind Jeanes

Isabella L. Karle

Charlotte Emma Moore Sitterly

**First Achievement Award
(American Association of
University Women)**

Florence B. Seibert

**Fisher Award in Analytical
Chemistry (American
Chemical Society)**

Jeanette Grasselli (Brown)

**Fluorine Award (American
Chemical Society)**

Jean'ne Marie Shreeve

French Medal of Recognition

Irène Joliot-Curie

**Garvan Medal (American
Chemical Society)**

Ruth Mary Roan Benerito

Leonora Neuffer Bilger

Katharine Burr Blodgett

Mary Letitia Caldwell

Emma Perry Carr

Marjorie Caserio

Mildred Cohn

Gerty Cori

Helen M. Dyer

Gertrude Belle Elion

Catherine Clarke Fenselau

Mary Fieser

Helen M. Free

Jenny Pickworth Glusker

Mary Lowe Good

Jeanette Grasselli (Brown)

Arda Alden Green

Icie Gertrude Macy Hoobler

Allene Rosalind Jeanes

Madeleine M. Joullié

Isabella L. Karle

Joyce Jacobson Kaufman

Pauline Beery Mack

Ines Mandl

Grace Medes

Agnes Fay Morgan

Dorothy Virginia Nightingale

Mary Engle Pennington

Gertrude Perlmann

Mary Locke Petermann

Lucy Weston Pickett

Sarah Ratner

Florence B. Seibert

Mary Lura Sherrill

Jean'ne Marie Shreeve

Sofia Simmonds

Betty Sullivan

Marjorie Jean Young Vold

Elizabeth Amy Kreiser
Weisburger

George van Biesbroeck Award (University of Arizona)

E. Dorrit Hoffleit

Gimbel Philadelphia Award

Florence B. Seibert

Global 500 Award (U.N. Environmental Program)

Vandana Shiva

Gold Medal (Royal Astronomical Society)

Caroline Herschel

Gold Medal (U.S. Department of Commerce)

Charlotte Emma Moore Sitterly

Gold Medal for Science (awarded by the King of Prussia)

Caroline Herschel

Golden Honor Emblem of the City of Vienna

Ines Mandl

Graduate Achievement Medal (Radcliffe College)

Helen Sawyer Hogg

Graduate Society Medal (Radcliffe College)

E. Dorrit Hoffleit

Grand Prix de la Ville de Paris

Marguerite Perey

Gregori Aminoff Prize (Royal Swedish Academy of Science)

Isabella L. Karle

Guadalupe Almendaro Medal (Astronomical Society of Mexico)

Williamina Paton Fleming

Hall of Fame (Poultry Historical Society)

Mary Engle Pennington

Harry and Carol Mosher Award (Santa Clara Valley Section, American Chemical Society)

Jean'ne Marie Shreeve

Heineman Prize (American Astronomical Society)

Sandra Moore Faber

Helen M. Free Public Outreach Award

Helen M. Free

Helene-Paul Helbronner Prize (French Academy of Sciences)

Dorothea Klumpke Roberts

Henri Chretien Award (American Astronomical Society)

Wendy Laurel Freedman

Henri Wilde Prize (France)

Irène Joliot-Curie

Henry Hill Award (American Chemical Society)

Madeleine M. Joullié

Henry Norris Russell Prize (American Astronomical Society)

Cecilia Payne-Gaposchkin

Herschel Medal (Royal Astronomical Society)

Jocelyn Bell Burnell

Hillebrand Award (American Chemical Society)

Isabella L. Karle

Hillebrand Prize (Chemical Society of Washington, D.C.)

> Elizabeth Amy Kreiser Weisburger

John Elliott Award

> Florence B. Seibert

John Scott Award

> Florence B. Seibert

Judd Award (Sloan-Kettering Institute)

> Gertrude Belle Elion

Klumpke-Roberts Award (Astronomical Society of the Pacific)

> Heidi Hammel

> Helen Sawyer Hogg

L. and B. Freedman Foundation Award (New York Academy of Sciences)

> Sarah Ratner

Lavoisier Prize (Académie des Sciences, France)

> Marguerite Perey

Leibnitz Medal (Berlin Academy of Sciences)

> Lise Meitner

Lieban Prize (Vienna Academy of Sciences)

> Lise Meitner

Lindback Award for Distinguished Teaching (University of Pennsylvania)

> Madeleine M. Joullié

***Los Angeles Times* Woman of the Year**

> Marjorie Jean Young Vold

Louis Empain Prize for Physics

> Ingrid Daubechies

MacArthur Foundation Fellowship

> Ingrid Daubechies

> Helen T. Edwards

> Margaret Joan Geller

> Jane Richardson

***Mademoiselle* Merit Award**

> Elizabeth Roemer

Maria Goeppert-Mayer Award (American Physical Society)

> Judith Sharn Young

Marquet Prize (Academy of Sciences, Paris, France)

> Irène Joliot-Curie

Martin Company Gold Medal for Outstanding Scientific Accomplishments

> Joyce Jacobson Kaufman

Maryland Chemist Award (American Chemical Society)

> Catherine Clarke Fenselau

Maryland Section Outstanding Chemist (American Chemical Society)

> Joyce Jacobson Kaufman

Mateucci Medal (Italian Society for Sciences)

> Irène Joliot-Curie

Medal for Distinguished Achievement (Radcliffe College)

> Margaret Harwood

Medal of Distinction (Barnard College)

> Henrietta Hill Swope

Medal of Honor (American Cancer Society)

Gertrude Belle Elion

Medal of Service, Order of Canada

Helen Sawyer Hogg

Member of the Order of the British Empire

Allie Vibert Douglas

Merit Award (National Institutes of Health)

Catherine Clarke Fenselau

Meritorious Service Medal (U.S. Public Health Service)

Elizabeth Amy Kreiser Weisburger

Michigan Women's Hall of Fame

Isabella L. Karle

Midwest Award (American Chemical Society)

Gerty Cori

Modern Medicine Award for Distinguished Achievement (Children's Hospital of Michigan)

Icie Gertrude Macy Hoobler

NASA Group Achievement Award for Voyager Science Investigation

Heidi Hammel

NASA Group Achievement Award, Hubble Space Telescope Wide-Field Planetary Camera Team

Sandra Moore Faber

NASA Special Award

Elizabeth Roemer

National Achievement Award

Florence B. Seibert

National Inventors Hall of Fame

Gertrude Belle Elion

National Medal of Science (U.S.)

E. Margaret Burbidge

Mildred Cohn

Gertrude Belle Elion

Vera Cooper Rubin

National Women's Hall of Fame

Gertrude Belle Elion

Florence B. Seibert

Navy Superior Civilian Service Award

Isabella L. Karle

Newcomb-Cleveland Prize (American Academy of Arts and Sciences)

Margaret Joan Geller

Nobel Prize

Gerty Cori

Marie Sklodowska Curie

Gertrude Belle Elion

Maria Goeppert-Mayer

Dorothy Crowfoot Hodgkin

Irène Joliot-Curie

Norlin Achievement Award (University of Colorado)

Icie Gertrude Macy Hoobler

Notable Service Medal (awarded by President Herbert Hoover)

Mary Engle Pennington

Nova Medal (American Association of Variable Star Observers)

Annie Jump Cannon

Officer of the Order of Canada

Allie Vibert Douglas

Officier of the Légion d'Honneur

Marguerite Perey

Ohio Sciences and Technology Hall of Fame

Jeanette Grasselli (Brown)

Ohio Women's Hall of Fame

Jeanette Grasselli (Brown)

Order of Merit (England)

Dorothy Crowfoot Hodgkin

Ordre pour le Mérite, Civilian Class (West Germany)

Lise Meitner

Osborne and Mendel Award (American Institute of Nutrition)

Icie Gertrude Macy Hoobler

Otto Hahn Prize (West Germany)

Lise Meitner

Outstanding Achievement Award (University of Minnesota)

Jean'ne Marie Shreeve

Betty Sullivan

Outstanding Alumna Award (Baylor University)

Allene Rosalind Jeanes

Outstanding Professional Award (Organization of Professional Employees, USDA)

Ruth Mary Roan Benerito

"Padmabhushan" Award (Government of India)

Kamal J. Ranadive

Philadelphia Organic Chemists Club Award

Madeleine M. Joullié

Philadelphia Section Award (American Chemical Society)

Jenny Pickworth Glusker

Madeleine M. Joullié

Philadelphia Section Award (Association of Women in Science)

Madeleine M. Joullié

Pioneer Award (American Institute of Chemists LHD, Georgetown University)

Isabella L. Karle

Pittsburgh Spectroscopy Award

Catherine Clarke Fenselau

Planck Medal (German Physical Society)

Lise Meitner

Prix des Dames (Astronomical Society of France)

Dorothea Klumpke Roberts

Prix Prize (French Academy of Sciences)

Marie Sklodowska Curie

Prize in Sciences (City of Vienna)

Lise Meitner

Professional Achievement Award (American Society for Medical Technology)

Helen M. Free

Progress Medal (Photographic Society of America)

Katharine Burr Blodgett

Rear Admiral William S. Parsons Award (Navy League of United States)

Isabella L. Karle

Remsen Award (American Chemical Society)

Mildred Cohn

Research Career Development Award (National Institutes of Health)

Catherine Clarke Fenselau

Ricketts Prize

Florence B. Seibert

Rittenhouse Medal (Rittenhouse Astronomical Society)

Helen Sawyer Hogg

Robert Dexter Conrad Award

Isabella L. Karle

Robertson Award (University of Toronto)

Wendy Laurel Freedman

Royal Medal (The Royal Society)

Dorothy Crowfoot Hodgkin

Sandford Fleming Medal (Royal Canadian Institute)

Helen Sawyer Hogg

Sandoz Award and Gold Medal

Kamal J. Ranadive

Schlozer Medal (University of Göttingen)

Lise Meitner

Scroll Award (American Institute of Chemists)

Madeleine M. Joullié

Senior U.S. Scientist Award (Alexander Von Humboldt Foundation)

Jean'ne Marie Shreeve

Service Award Medal (Royal Astronomical Society of Canada)

Helen Sawyer Hogg

Silver Medal (Société Chimique de France)

Marguerite Perey

Silver Medal (U.S. Department of Commerce)

Charlotte Emma Moore Sitterly

Sloan Award in Cancer Research

Mary Locke Petermann

Southern Chemist Award (American Chemical Society)

Ruth Mary Roan Benerito

Southwest Regional Award (American Chemical Society)

Ruth Mary Roan Benerito

Squibb Award in Endocrinology

Gerty Cori

Steele Prize (American Mathematical Society)

Ingrid Daubechies

Sugar Research Prize

Gerty Cori

Superior Service Award to Biopolymer Research Team (U.S. Department of Agriculture)

Allene Rosalind Jeanes

Tata Memorial Hospital Jubilee Award

Kamal J. Ranadive

Thomas Burr Osborne Medal (American Association of Cereal Chemists)

Betty Sullivan

Trudeau Medal (National Tuberculosis Association)

Florence B. Seibert

Une Dame Chevalier (Chapitre Centre National de la Recherche Scientifique)

Joyce Jacobson Kaufman

Vincent du Vigneaud Award

Isabella L. Karle

Vladimir Karapetoff Award (MIT)

Heidi Hammel

Watumull Award

Kamal J. Ranadive

William F. Meggers Award (Optical Society of America)

Charlotte Emma Moore Sitterly

Woman of Achievement (1984 World's Fair)

Ruth Mary Roan Benerito

Woman of the Century (National Council of Jewish Women)

Allie Vibert Douglas

Women Pioneers in Nuclear Science (50th Anniversary Honoree)

Leona Woods Marshall Libby

Index

Page numbers in **bold** refer to main entries.

About the Editors and Contributors

Editors

BENJAMIN F. SHEARER is Vice-President for Student Life at Neumann College in Aston, Pennsylvania. He and his wife, Barbara, are the editors of *Notable Women in the Life Sciences: A Biographical Dictionary* (Greenwood, 1996) and the authors of *State Names, Seals, Flags, and Symbols* (Greenwood, 1987, rev. ed. 1994), as well as several other books published by Greenwood Press.

BARBARA S. SHEARER is Director of Public Services and External Relations at the Scott Memorial Library, Thomas Jefferson University in Philadelphia. She is a co-editor, with her husband, Benjamin, of *Notable Women in the Life Sciences: A Biographical Dictionary* (Greenwood, 1996) and co-author of *State Names, Seals, Flags, and Symbols* (Greenwood, 1987, rev. ed. 1994), as well as several other books published by Greenwood Press. She is co-author (with Geneva Bush) of *Finding the Source of Medical Information: A Thesaurus-Index to the Reference Collections* (Greenwood, 1985). She has also published several articles on medical database searching and on bibliometrics.

Contributors

NANCY ALLEE is Head of the Public Health Library at the University of Michigan. She holds a B.A. in English, an M.L.S., and an M.P.H. She is a member of the Academy of Health Information Professionals and is a reviewer for *Video Rating Guide for Libraries*.

URSULA ALLEN is a graduate student in physics at the University of Colorado. She is involved in research in solar physics.

FRANCIE BAUER is Chemistry Librarian at the University of Wyoming Science Library.

K. C. BENEDICT is Librarian at the College of Eastern Utah San Juan Campus Library. She has contributed articles to regional and national publications, including the *Utah Historical Encyclopedia*.

TERESA BERRY is Reference Librarian at the University of Tennessee, Knoxville. She holds a B.S. in chemistry and an M.S.L.S. She has published reviews in *Library Journal* and is editor of the East Tennessee Library Association newsletter.

CYNTHIA A. BILY teaches writing at Adrian College and writes extensively for reference works. She has most recently contributed to a chronology of women's history, a catalogue of multicultural children's books, and an encyclopedia of women's issues.

STEFANIE BUCK is Reference Librarian at the Thomas Cooper Library, University of South Carolina, Columbia.

JUDY F. BURNHAM, M.L.S., is UMC Site Coordinator for the Biomedical Library at the University of South Alabama in Mobile. Her research interests lie in information management and bibliometrics, on which she has authored and co-authored several articles.

CLARA A. CALLAHAN, M.D., is Clinical Associate Professor of Pediatrics and Associate Dean for Student Affairs at Jefferson Medical College in Philadelphia, Pennsylvania.

FAYE A. CHADWELL is Head of Collection Development at the University of Oregon in Eugene. Her most recent publications include a bibliography on the feminist transformation of the physical science classroom in Sue Rosser's *Teaching the Majority: Breaking the Gender Barrier in Science, Mathematics, and Engineering*; and "Thriving in the Information Environment: Your Campus Library," co-authored with Marilee Birchfield at the University of South Carolina, for *Your College Experience* (2nd edition).

GARY L. CHEATHAM is Assistant Professor of Library Services at Northeastern State University, Tahlequah, Oklahoma. Since 1988 he has published more than 20 reviews and articles in library science and history journals, including articles on the Civil War in Kansas in *Kansas History: A Journal of the Central Plains*.

MARIA CHIARA earned her doctoral degree in anthropology. She lives in Evanston, Illinois, where she is building a research and consulting practice devoted to understanding the career motivations and work experiences of professional women.

E. TINA CROSSFIELD was born and educated in Quebec, where she studied health sciences and worked for several years as a research assistant in molecular biology. For her B.A. she studied the relationships between science, technology, and society with a special focus on women's experiences in science. Currently she is an interdisciplinary graduate student and is working on the biography of Elsie Cassels (1863–1936), a Scottish immigrant and naturalist in western Canada.

DOUGLAS EBY lives in Beverly Hills, California, and has contributed multiple filmed entertainment production articles to magazines including *Hollywood Reporter, Post, Film & Video, Quake*, and *Cinefantastique*.

JOANN EISBERG is Visiting Researcher in the Department of History at the University of California at Santa Barbara. Her specialties are the history of science, especially the history of astrophysics and the history of women, gender, and science. She has taught history of science at the University of Wisconsin—Madison and has run the Women in Science program of the University of Wisconsin system. She is currently writing a feminist, scientific biography of astronomer Beatrice Tinsley, who is profiled in this volume.

DOLORES FIDISHUN is Head Librarian at Penn State—Great Valley in Malvern, Pennsylvania.

CASSANDRA S. GISSENDANNER, M.S.L.S., is Catalog Management Librarian at Thomas Cooper Library, University of South Carolina.

KATALIN HARKÁNYI is the Chemistry and Engineering Bibliographer and a Science Reference Librarian at the Malcolm A. Love Library at San Diego State University in San Diego, California. She has authored numerous chemistry and engineering bibliographies, as well as the monograph *The Natural Sciences and American Scientists in the Revolutionary Era*, and is co-author of an expert system, ChemRas.

KELLY HENSLEY is Media/Microcomputers Librarian and Assistant Professor at the Quillen College of Medicine Medical Library of East Tennessee State University in Johnson City, Tennessee.

RUTH A. HODGES is Chief Librarian at the District of Columbia Medical Library in Washington, D.C. Her recent publications include "Thurgood Marshall: Power of His Legacy" in the *Howard Law Journal* and articles on Helen Appo Cook and Myrtle Foster Cook in *Notable Black American Women*.

DORRIT HOFFLEIT is Senior Research Astronomer, retired, Yale University.

JILL HOLMAN holds a B.A. in history and computer science and an MILS. She is currently working on a World Wide Web page for the American Library Association's Feminist Task Force. Previously she was a Humanities and Social Sciences Reference Librarian at the University of South Carolina. Women in science has long been an interest ever since she was a girl in Minnesota feeling discouraged in her interest in science.

MAY MONTASER JAFARI is Reference Librarian at Indiana University Purdue University, Indianapolis (IUPUI) University Library. She has been the liaison librarian for both the Women's Studies Program and the School of Engineering and Technology at the IUPUI campus.

KATHLEEN PALOMBO KING is an Instructor at the Pennsylvania Institute of Technology, Computers, Engineering and Science and a Computer and AutoCAD Consultant. She is also technical editor of the *Journal of Afro-Latin American Studies & Literatures*.

TIMOTHY WILLIAM KLASSEN is Electronic Services Librarian at the University of Oregon Science Library in Eugene.

SHARON SUE KLEINMAN is a doctoral student in the Department of Communication at Cornell University. Her research interests focus on gender and communication, especially with regard to science and emerging communication technologies. Her publications include biographical essays on women in science, articles on communication theory, interviews with literary figures, and music reviews.

WALTER S. KOSKI is the Bernard N. Baker Professor of Chemistry at Johns Hopkins University in Baltimore, Maryland. He has published approximately 200 papers in technical journals concerning physics and chemistry.

ANNE-MARIE WEIDLER KUBANEK holds a Ph.D. in chemistry, Diploma in Collegial Teaching. She has been teaching chemistry at the college level in the Montreal area since 1975 and joined the chemistry department of John Abbott College in 1985. She is the author or coauthor of 15 scientific papers, holds two U.S. patents, and has done research and written on the history of women in science. Kubanek is an associate member of the Simone de Beauvoir Institute of Concordia University and of the McGill Centre for Research and Teaching on Women.

NATALIE KUPFERBERG is Coordinator of the Health Science Library at Ferris State University. She has written the Science section for the last three editions of *Magazines for Libraries* and has published an article, "Alternative Medicine Goes Mainstream," in the July 1994 issue of *Library Journal*. She has reviewed many books in the health sciences for *Choice* and *Library Journal*.

KIMBERLY J. LAIRD is Technical Services Librarian, East Tennessee State University, Quillen College of Medicine Library.

KRISTINE M. LARSEN, Ph.D., is Assistant Professor of Physics and Astronomy at the University of Connecticut in Storrs. She lectures regularly on astronomical topics at libraries, schools, and correctional facilities and is a frequent contributor to professional publications.

SUE ANN LEWANDOWSKI is the Science Cataloger at the University of Nebraska—Lincoln libraries. She holds a B.S. in geosciences and an M.L.S. She has recently served as editor of the *Bayou Chronicle*, the monthly newsletter of the Louisiana Universities Marine Consortium, and has published bibliographies on the effects of offshore oil and gas development.

JIE LI, M.L.S., is Information Services Librarian at the University of South Alabama Biomedical Library in Mobile.

DANIEL LIESTMAN is Associate Professor and the Bibliographic Specialist for the Social Sciences at Seattle Pacific University. He has written several articles on immigrants in America and was a contributor to the ninth and tenth supplements of the *Dictionary of American Biography*.

JENNIFER LIGHT holds an M.Phil. in history and philosophy of science. She is currently a Resident Tutor in History of Science at Eliot House, Harvard University, where she is a Ph.D. candidate in the Department of the History of Science. Her most recent publication, "The Digital

Landscape: New Space for Women?" appears in the September 1995 issue of *Gender, Place & Culture*.

ELEANOR L. LOMAX is Reference Librarian at Florida Atlantic University in Boca Raton, Florida.

MICHAEL LORENZEN, B.A., M.L.S., is Reference Librarian for Ohio University at Zanesville. His work has appeared in *Library Journal, Information Technology and Libraries, American Libraries*, and *Ohioana Quarterly*.

SAROJINI LOTLIKAR is Catalog and Reference Librarian at Millersville University in Millersville, Pennsylvania.

ELLEN F. MAPPEN holds a Ph.D. in history and has written about social feminism in England at the turn of the century, including *Helping Women at Work: The Women's Industrial Council, 1889–1914* (1985) and "Strategists for Change: Social Feminist Approaches to the Problems of Women's Work," in *Unequal Opportunities: Women's Employment in England 1800–1918*, edited by Angela V. John (1986). She is currently Director of the Douglass Project for Rutgers Women in Math, Science, and Engineering at Douglass College.

HEATHER MARTIN is Reference and Outreach Librarian at the Paul Laurence Dunbar Library, Wright State University, Dayton, Ohio. She holds both an M.L.I.S and an M.A. in English and has written biographical sketches on Evelyn Ashford and Margaret Avery for *African American Women: A Biographical Dictionary* (1993).

MARY ANN MCFARLAND is Assistant Director of Access Services and Associate Professor at the St. Louis University Health Sciences Center Library in St. Louis, Missouri. She has published articles and reviews in several journals.

USHA MEHTA is Library Assistant at the Business and Government Information Center, Getchell Library, University of Nevada, Reno. She holds an M.A. in psychology and an M.L.S.

SYLVIA NICHOLAS, M.L.S., is Reference Librarian at the Galter Health Sciences Library of Northwestern University in Chicago.

CONNIE H. NOBLES, Ph.D., is Assistant Professor at Southeastern Louisiana University and teaches both graduate and undergraduate courses in the Department of Teacher Education. She is a science/social studies educator whose research interests include women in science, archaeology (of gender), and graphic organizers. She has presented and published in all of these areas both nationally and internationally.

CAROL B. NORRIS is an Associate Professor and Online Searching Librarian at the Sherrod Library, East Tennessee State University, in Johnson City. She has published annotated bibliographies and articles in professional journals, as well as poetry.

JANE OPALKO is a freelance writer based in Houston. She holds degrees in physics and astronomy and frequently writes about women scientists for a variety of publications.

JANET OWENS is Reference Librarian at Montana State University—Bozeman. She is the author of other biographies of notable women.

JOAN G. PACKER is Head of the Reference Department at Central Connecticut State University Library in New Britain. She has published bibliographies on Margaret Drabble (1988) and Rebecca West (1991).

MARILYN MCKINLEY PARRISH has been a librarian for the past 15 years, working in a variety of public and academic libraries. Through her consulting firm, Information Transfer, she provides broker services and grant writing assistance to nonprofit organizations.

ANNE MARIE PERRAULT is Reference Librarian at the University of Buffalo.

SHANG-FEN REN received her undergraduate education in physics in China and her Ph.D in physics in the United States. She is now Assistant Professor of Physics at Illinois State University and a lifetime member of the American Physics Society. She is on the Committee of International Relations of the Association for Women in Science (AWIS) and also a co-president of the Heart of Illinois Chapter of AWIS.

REBECCA L. ROBERTS is a Staff Member at the Pennsylvania State University Medical School in Hershey, Pennsylvania. She is a three-time winner of the Doctors Kienle Literature Prize through Penn State, is a breeder of national champion Morgan horses, and is also attending law school.

NANCY G. ROMAN, Ph.D., is Head of NASA's Astronomical Data Center.

NINA MATHENY ROSCHER, Ph.D., is Professor and Chair of the Department of Chemistry at the American University in Washington, D.C.

MIRIAM ROSSI, Ph.D., is Associate Professor and Chair of Chemistry at Vassar College, Poughkeepsie, New York. Her research interests are the structure determination of small molecules of biological interest, particularly anti-tumor agents. She is also interested in chemical education and the history of chemistry.

DARIN C. SAVAGE, M.L.I.S., is Science and Engineering Reference Librarian at Utah State University's Science and Technology Library.

JOY SCHABER did her undergraduate work in biology and English and holds an M.S. in ecology. She has written articles on scientific issues for the University of California Natural Reserve System, the local Davis newspaper, and the UC—Davis News Service. Currently she is working on the ecology of Giant Sequoia groves for the Sierra Nevada Ecosystem Project.

CAROLE B. SHMURAK is Associate Professor of Education at Central Connecticut State University. She holds a B.A. in chemistry, an M.A. in biochemistry, and a Ph.D. in science education. Her recent publications include: "Emma Perry Carr: The Spectrum of a Life," *AmbEx* 41 (July 1994) and " 'Castle of Science': Mount Holyoke College and the Preparation of Women in Chemistry, 1837–1941," *History of Education Quarterly* 32, no. 3 (Fall 1992).

FLORA SHRODE, M.L.I.S., is Reference Coordinator, Science and Technology, at the Hodges Library, University of Tennessee—Knoxville.

NANCY SLIGHT-GIBNEY is Head of the Acquisition Department for the University of Oregon Library System in Eugene. She holds a B.A. and an M.A. in anthropology and an M.I.L.S. She is a contributor to professional publications.

TAMI I. SPECTOR is Associate Professor of Chemistry at the University of San Francisco. Both her A.B. degree and her Ph.D. are in chemistry. Her research interests are organic photochemistry, spectroscopic analysis of organic ions, and the history of chemistry and chemical education.

EILEEN HORN STANLEY is Director, Library Services, of the medical library at the Earl K. Long Medical Center in Baton Rouge, Louisiana. Ms. Stanley, who holds a B.S. in chemistry from Rockhurst College, has worked as a chemist, chemistry librarian, and medical librarian.

NANCY F. STIMSON is Assistant Librarian at the Health Sciences Library at the State University of New York at Buffalo. She holds a B.A. in biology and an M.L.S. degree. Ms. Stimson and Wendy Y. Nobunaga recently published an article entitled "The Life and Times of John H. Hickcox: Government Publications History Revisited" in the *Journal of Government Information* (September 1995).

ELLA N. STRATTIS is Assistant Branch Librarian for the Camden Campus of Rowan College of New Jersey.

VALERIE L. THOMAS holds a B.S. in physics, an M.A. in engineering administration, and an honorary Ph.D., awarded by George Washington University. A retired mathematician from NASA's Goddard Space Flight Center, she is currently Director for Strategic Planning at Dimensional Media Associates.

MARY HAWORTH THOMPSON is co-author of *The Scientist within You: Experiments and Biographies of Distinguished Women in Science*, Vol. 1, and *The Scientist within You: Women Scientists from Seven Continents*, Vol. 2. She is a teacher, journalist, and publisher. Her affiliations include the American Association of University Women and the National Federation of Press Women.

ELIZABETH VILLEPONTEAUX is Assistant Director of the U.S. Environmental Protection Agency Library, Research Triangle Park, North Carolina. She holds a B.A. in history and an M.S. in library science.

REBECCA LOWE WARREN is co-author of *The Scientist within You* series, vols. 1 and 2 (1994 and 1995), and adjunct faculty member at Marylhurst College. Her science class, "Beyond Marie Curie," spotlights the contributions of women in science and mathematics. Her affiliations include the American Association of University Women.

IRMGARD WOLFE, M.L.S., is Cataloger and Preservation Librarian at Cook Memorial Library, University of Southern Mississippi, in Hattiesburg.

YIMIN WANG ZIMMERER spent her first 20 years in Shanghai, China. In 1989 she came to the United States, where she received a B.A. in psychology and an M.S. in telecommunications. Since graduation, she has been working at MCI in Colorado Springs. She is also the treasurer of the Colorado Springs Asian Professional Network and teaches at the Colorado Springs Chinese School.